T0342564

Fundamentals of Numerical Mathematics for Physicists and Engineers

Fundamentals of Numerical Mathematics for Physicists and Engineers

Alvaro Meseguer
Department of Physics
Universitat Politècnica de Catalunya – UPC BarcelonaTech

This edition first published 2020

© 2020 John Wiley & Sons, Inc. All rights reserved.

The right of Alvaro Meseguer to be identified as the author of this work has been asserted in accordance with law.

Registered Office
John Wiley & Sons, Inc., 111 River Street, Hoboken, NJ 07030, USA

Editorial Office
111 River Street, Hoboken, NJ 07030, USA

For details of our global editorial offices, customer services, and more information about Wiley products visit us at www.wiley.com.

Wiley also publishes its books in a variety of electronic formats and by print-on-demand. Some content that appears in standard print versions of this book may not be available in other formats.

Library of Congress Cataloging-in-Publication Data

Names: Meseguer, Alvaro (Alvaro Meseguer), author.
Title: Fundamentals of numerical mathematics for physicists and engineers / Alvaro Meseguer.
Description: Hoboken, NJ : Wiley, 2020. | Includes bibliographical references and index.
Identifiers: LCCN 2019057703 (print) | LCCN 2019057704 (ebook) | ISBN 9781119425670 (hardback) | ISBN 9781119425717 (adobe pdf) | ISBN 9781119425755 (epub)
Subjects: LCSH: Numerical analysis. | Mathematical physics. | Engineering mathematics.
Classification: LCC QA297 .M457 2020 (print) | LCC QA297 (ebook) | DDC 518–dc23
LC record available at https://lccn.loc.gov/2019057703
LC ebook record available at https://lccn.loc.gov/2019057704

Cover Design: Wiley
Cover Image: Courtesy of Alvaro Meseguer, (background) © HNK/Shutterstock

Set in 9.5/12.5pt CMR10 by SPi Global, Chennai, India

Printed in the United States of America

V10018489_051420

Contents

About the Author

Alvaro Meseguer, PhD, is Associate Professor at the Department of Physics at Polytechnic University of Catalonia (UPC BarcelonaTech), Barcelona, Spain, where he teaches Numerical Methods, Fluid Dynamics and Mathematical Physics to advanced undergraduates in Engineering Physics and Mathematics. He has published more than 30 articles in peer-reviewed journals within the fields of computational fluid dynamics, and nonlinear physics.

Preface

Much of the material in this book is derived from lecture notes for two courses on numerical methods taught over many years to undergraduate students in *Engineering Physics* at the Universitat Politècnica de Catalunya (UPC) BarcelonaTech. Its volume is scaled to a one-year course, that is, a two-semester course. Accordingly, the book has two parts. Part I is addressed to first or second year undergraduate students who have a solid foundation in differential and integral calculus in one real variable (including Taylor series, $O(h)$ notation, and improper integrals), along with elementary linear algebra (including polynomials and systems of linear equations). Part II is addressed to slightly more advanced undergraduate or first-year graduate students with a broader mathematical background, including multivariate calculus, ordinary differential equations, functions of a complex variable, and Fourier series. In both cases, it is assumed that the students are familiar with basic Matlab commands and functions.

The book has been written thinking not only of the student but also of the instructor (or instructors) that is supposed to teach the material following an academic calendar. Each chapter contains mathematical topics to be addressed in the lectures, along with Matlab codes and computer hands-on practicals. These practicals are problem-solving tutorials where the students, always supervised and guided by an instructor, use Matlab on a local computer to solve a given exercise that is focused on the topic previously seen in the lectures. From my point of view, teaching numerical methods should encompass not only theoretical lectures, addressing the underlying mathematics on a blackboard, but also practical computations, where the student learns the actual implementation of those mathematical concepts. There are certain aspects of numerical mathematics, such as *conditioning* or *order of convergence*, that can only be properly illustrated by experimentation on a computer. These hands-on practicals may also help the instructor to efficiently assess the performance of a student. This can be easily carried out by using Matlab's `publish` function, for example. The end of each chapter also includes a short list of problems and exercises of theoretical (labeled with an A) and/or computational (labeled with an N) nature. The solutions to many of the exercises (and practicals) can be found at the end of the book. Finally, each chapter includes a Complementary Reading section, where the student may find suitable bibliography to broaden his or her knowledge on different aspects of numerical mathematics. Complementary lists of exercises can also be found in many of these recommended references.

This book is mainly written for mathematically inclined scientists and engineers, although applied mathematicians may also find many of the topics addressed in this book interesting. My intention is not simply to give a set of recipes for solving problems, but rather to present the underlying mathematical concepts involved in every numerical method. Throughout the eight chapters, I have tried to write a readable book, always looking for an equilibrium between practicality and mathematical rigor. Clarity in presenting major points often requires the supression of minor ones. A trained mathematician may find certain statements incomplete. In those passages where I think this may be the case, I always refer the rigorous reader to suitable bibliography where the key theorem and its corresponding proof can be found.

Whenever it has been possible, I have tried to illustrate how to apply certain numerical methodologies to solve problems arising in the physical sciences or in engineering. For example, Part I includes some practicals involving very basic Newtonian mechanics. Part II includes practicals and examples that illustrate how to solve problems in electrical networks (Kirchhof's laws), classical thermodynamics (van der Waals equation of state), or quantum mechanics (Schrödinger equation for univariate potentials). In all the previous examples, the mathematical equations have already been derived, so that those readers who are not necessarily familiar with any of those areas of physics should be able to address the problem without any difficulty.

Many of the topics covered throughout the eight chapters are fairly standard and can easily be found in many other textbooks, although probably in a different order. For example, Chapter 1 introduces topics such as nonlinear scalar equations, root-finding method, convergence, or conditioning. This chapter also shows how to measure in practice the order of convergence of a root-finding method, and how ill-conditioning may affect that order. Chapter 2 is devoted to one of the most important methods to approximate functions: *interpolation*. I have addressed three different interpolatory formulas, namely, monomial, Lagrange, and barycentric, the last one being the most computationally efficient. I devote a few pages to introduce the concept of *Lebesgue constant* or condition number of a set of interpolatory nodes. This chapter clearly illustrates that global interpolation, performed on a suitable set of nodes, provides unbeatable accuracy. Chapter 3 is devoted to numerical differentiation, introducing the concept of *differentiation matrix*, which is often omitted in other textbooks. From my point of view, working with differentiation matrices has two major advantages. On the one hand, it is a simple and systematic way to obtain and understand the origin of classical finite difference formulas. On the other hand, differentiation matrices, understood as discretizations of differential operators, will be very important in Part II, when solving boundary value problems numerically. Chapter 4 is devoted to numerical integration or *quadratures*. This chapter addresses the classical *Newton–Cotes* quadrature formulas, along with *Clenshaw–Curtis* and *Fejér* rules, whose accuracy is known to be comparable to that of *Gaussian* formulas, but much simpler and easier to implement. This chapter is also devoted to the numerical approximation of integrals with periodic integrands, emphasizing the outstanding accuracy provided by the trapezoidal rule, which will be exploited in Part II, in the numerical approximation of Fourier series. Finally, Chapter 4 briefly addresses the numerical approximation of improper integrals.

Part II starts with Chapter 5, which is an introduction to *numerical linear algebra*, henceforth referred to as NLA. Some readers may find it unusual not to find NLA in Part I. Certain topics such as matrix norms or condition number of a matrix implicitly involve multivariate calculus, and are therefore unsuitable for Part I, addressed to first or second year undergraduates. Chapter 5 exclusively focuses on just one topic: solving linear systems of equations. The chapter first addresses direct solvers by introducing LU and QR factorizations. A very brief introduction to iterative matrix-free *Krylov* solvers can be found at the end of the chapter. I think that the concept of *matrix-free* iterative solver is of paramount importance for scientists and engineers. I have tried to introduce the concept of Krylov subspace in a simple, but also unconventional, way. Owing to the limited scale of the book, it has been impossible to address many other important topics in NLA such as eigenvalue computation, the singular value decomposition, or preconditioning. Chapter 6 is devoted to the solution of multidimensional nonlinear systems of equations, also including parameters. This chapter introduces the concept of *continuation*, also known as *homotopy*, a very powerful technique that has been shown to be very useful in different areas of nonlinear science. Chapter 7 is a very brief introduction to numerical Fourier analysis, where I introduce the *discrete Fourier transform* (DFT), and the phenomenon of *aliasing*. In this chapter, I also introduce *Fourier differentiation matrices*, to be used later in the numerical solution of boundary value problems. Finally, Chapter 8 is devoted to numerical discretization techniques for *ordinary differential equations* (ODE). The chapter first addresses how to solve *boundary value problems* (BVP) within bounded, periodic, and unbounded domains. In this first part of the chapter, I exploit the concept of global differentiation matrix seen previously in the book. Local differentiation formulas are used in the second part of the chapter, exclusively devoted to the solution of *initial value problems* (IVP). The limited scale of the book has only allowed including a few families of time integrators. The chapter ends with a brief introduction to the concept of *stability* of a time-stepper, where I have certainly oversimplified concepts such as *consistency* or *0-stability*, along with *Dahlquist's equivalence theorem*. However, I have emphasized other, in my point of view, also important but more practical aspects such as *stiffness* and *A-stability*.

Barcelona, September 2019   *A. Meseguer*

Acknowledgments

Many people read early drafts at various stages of the evolution of this book and contributed many corrections and constructive comments. I am particularly grateful to Daniel Loghin, Mark Embree, and Francisco (Paco) Marqués for their careful reading of some parts of the manuscript. Other people who helped with proofreading include Erefila Sousa, Juan M. López, and Julia Amorós. I would also like to thank Toni Castillo for efficiently adapting Wiley LaTeX template to different Debian and Ubuntu's Linux environments I have been working with during the writing. Finally, I wish to thank Kathleen Santoloci, Mindy Okura-Marszycki, Elisha Benjamin, and Devi Ignasi of John Wiley & Sons, for their splendid editorial work.

A. Meseguer

Part I

1

Solution Methods for Scalar Nonlinear Equations

1.1 Nonlinear Equations in Physics

Quite frequently, solving problems in physics or engineering implies finding the solution of complicated mathematical equations. Only occasionally, these equations can be solved *analytically*, i.e. using algebraic methods. Such is the case of the classical quadratic equation $ax^2 + bx + c = 0$, whose exact *zeros* or *roots* are well known:

$$x = \frac{1}{2a}(-b \pm \sqrt{b^2 - 4ac}).$$ (1.1)

An obvious question is whether there is an expression similar to Eq. (1.1) providing the roots or zeros of the cubic equation $ax^3 + bx^2 + cx + d = 0$. The answer is yes, and such expression is usually termed as *Cardano's formula*.[1] We will not detail here the explicit expression of Cardano's formula but, as an example, if we apply such formulas to solve the equation

$$x^3 - x - 4 = 0,$$ (1.2)

we can obtain one of its roots:

$$\alpha = \sqrt[3]{2 + \frac{1}{9}\sqrt{321}} + \sqrt[3]{2 - \frac{1}{9}\sqrt{321}}.$$ (1.3)

There are similar formulas to solve arbitrary quartic equations in terms of radicals, but not for quintic or higher degree polynomials, as proposed by the Italian mathematician Paolo Ruffini in 1799 but eventually proved by the Norwegian mathematician Niels Henrik Abel around 1824.

---

1 First obtained in the mid-sixteenth century by the Italian mathematician Scipione del Ferro, inspired by the work of Lodovico Ferrari, a disciple of Gerolamo Cardano.

Fundamentals of Numerical Mathematics for Physicists and Engineers, First Edition. Alvaro Meseguer.
© 2020 John Wiley & Sons, Inc. Published 2020 by John Wiley & Sons, Inc.

What we have described is just a symptom of a more general problem of mathematics: the impossibility of solving arbitrary equations analytically. This is a problem that has serious implications in the development of science and technology since physicists and engineers frequently need to solve complicated equations. In general, equations that cannot be solved by means of algebraic methods are called *transcendental* or *nonlinear* equations, i.e. equations involving combinations of rational, trigonometric, hyperbolic, or even special functions. An example of a nonlinear equation arising in the field of classical *celestial mechanics* is Kepler's equation[2]

$$a + x - b \sin x = 0, \tag{1.4}$$

where a and b are known constants. Another popular example can be found in *quantum physics* when solving Schrödinger's equation for a particle of mass m in a *square well potential* of finite depth V_0 and width a. In this problem, the admissible energy levels $\mathcal{E}_n$ corresponding to the bounded states are the solutions of any of the two transcendental equations[3]:

$$\mathcal{E}_n \tan \mathcal{E}_n = \sqrt{\frac{\beta}{4} - \mathcal{E}_n^2} \quad \text{or} \quad \mathcal{E}_n \cot \mathcal{E}_n = -\sqrt{\frac{\beta}{4} - \mathcal{E}_n^2}, \tag{1.5}$$

where $\beta = 2ma^2 V_0/\hbar^2$ and $\hbar$ is the reduced Planck constant. In this chapter, we will study different methods to obtain *approximate* solutions of algebraic and transcendental or nonlinear equations such as (1.2), (1.4), or (1.5). That is, while Cardano's formula (1.3) provides the *exact* value x_1 of one of the roots of (1.2), the methods we are going to study here will provide just a numerical *approximation* of that root. If you have a rigorous mathematical mind you may feel a bit disappointed since it seems always preferable to have an exact analytical expression rather than an approximation. However, we should first clarify the actual meaning of *exact solution* within the context of physics.

It is obvious that if the coefficients appearing in the quadratic equation $ax^2 + bx + c = 0$ are known to infinite precision, then the solutions appearing in (1.1) are exact. The same can be said for Cardano's solution (1.3) of cubic equation (1.2). However, equations arising in physics or engineering such as (1.4) or (1.5) frequently involve universal constants (such as Planck constant h, the gravitational constant G, or the elementary electric charge e). All universal constants are known with limited precision. For example, the currently accepted value of the *Newtonian constant of gravitation* is, according to NIST,[4]

$$G = (6.674\ 30 \pm 0.000\ 15) \times 10^{-11} \ \text{m}^3 \ \text{kg}^{-1} \ \text{s}^{-2}.$$

---

2 First derived by the German astronomer Johannes Kepler in 1609.
3 See Appendix H of Eisberg and Resnick (1985).
4 National Institute of Standards and Technology.

As of 2019, the most accurately measured universal constant is

$$R_\infty = (10\ 973\ 731.568\ 160 \pm 0.000\ 021)\ \mathrm{m}^{-1},$$

known as *Rydberg constant*. In other words, the most accurate physical constant is known with 12 digits of precision. Other equations may also contain empirical parameters (such as the thermal conductivity κ or the magnetic permeability μ of a certain material), which are also known with limited (and usually much less) precision. Therefore, the solutions obtained from equations arising in empirical sciences or technology (even if they have been obtained by analytical methods) are, intrinsically, inaccurate.

Current standard double precision floating point operations are nearly 10 000 times more accurate than the most precise universal constant known in nature. In this book, we will study how to take advantage of this precision in order to implement computational methods capable of satisfying the required accuracy constraints, even in the most demanding situations.

1.2 Approximate Roots: Tolerance

Suppose we want to locate the zeros or roots of a given function $f(x)$ that is continuous within the interval $[a, b]$. Mathematically, the main goal of *exact* root-finding of $f(x)$ in $[a, b]$ is as follows:

Root-finding (Exact): Find $\alpha \in [a, b]$ such that $f(\alpha) = 0$.

However, computers work with finite precision and the condition $f(\alpha) = 0$ cannot be exactly satisfied, in general. Therefore, we need to reformulate our problem:

Root-finding (Approximate): For a given $\varepsilon > 0$, find $c \in (a, b)$ such that $f(x) = 0$ for some $x \in [c - \varepsilon, c + \varepsilon]$.

This reformulation introduces a new component in the problem: the positive constant ε, usually termed as *tolerance*, whose meaning is outlined in the plot on the right. Since the root condition $f(x) = 0$ cannot be satisfied exactly, we must provide an interval containing the root x. In the figure on the right, the root x lies within the interval $[c - \varepsilon, c + \varepsilon]$.

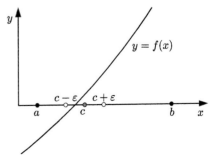

1.2.1 The Bisection Method

In order to clarify the concept of tolerance we introduce here the *bisection method* (also called the *interval halving method*). Let us assume that $f(x)$ is a continuous function within the interval $[a, b]$ within which the function has a *single* root α. Since the function is continuous, $f(a)f(b) \leqslant 0$. First, we define $a_0 = a$, $b_0 = b$, and $I_0 = [a, b]$ as the starting interval whose midpoint is $x_0 = (a_0 + b_0)/2$. The main goal of the bisection method consists in identifying that half of I_0 in which the change of sign of $f(x)$ actually takes place. Use the simple rules:

Bisection Rules:

$$\text{If } f(a_0)f(x_0) \leqslant 0 \quad \text{then set} \quad a_1 = a_0 \quad \text{and} \quad b_1 = x_0,$$
$$\text{If } f(x_0)f(b_0) \leqslant 0 \quad \text{then set} \quad a_1 = x_0 \quad \text{and} \quad b_1 = b_0.$$

Finally set $x_1 = \dfrac{1}{2}(a_1 + b_1)$ and $I_1 = [a_1, b_1]$.

Figure 1.1 shows the result of applying the bisection rule to find roots of the cubic polynomial $f(x) = x^3 - x - 4$ already studied in Eq. (1.2). As shown

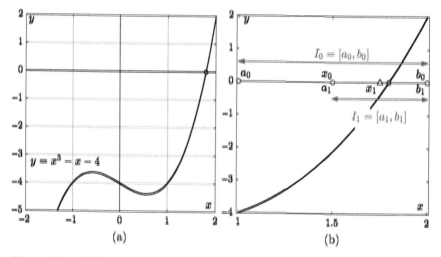

Figure 1.1 (a) A simple exploration of $f(x) = x^3 - x - 4$ reveals a change of sign within the interval $[1, 2]$ (the root has been depicted with a gray bullet). (b) We start the bisection considering the initial interval $I_0 = [1, 2]$. The bisection rule provides the new halved interval $I_1 = [a_1, b_1]$.

In Figure 1.1a, this polynomial takes values of opposite sign, $f(1) = -4$ and $f(2) = 2$, at the endpoints of the interval $I_0 = [a_0, b_0] = [1, 2]$ whose midpoint is $x_0 = 1.5$. Since $f(x_0) = -2.125$, the new (halved) interval provided by the bisection rules is $I_1 = [a_1, b_1] = [1.5, 2]$. The midpoint $x_1 = (a_1 + b_1)/2 = 1.75$ of I_1 (white triangle in Figure 1.1b) provides a new estimation of the root of the polynomial. We could resume the process by checking whether $f(a_1)f(x_1) < 0$ or $f(x_1)f(b_1) < 0$ in order to obtain the new interval I_2 containing the root and so on. After k bisections we have determined the interval $I_k = [a_k, b_k]$ with midpoint $x_k = (a_k + b_k)/2$. In general, the interval I_{k+1} will be obtained by applying the same rules to the previous interval I_k:

Bisection Method: Given $I_k = [a_k, b_k]$ such that $f(a_k)f(b_k) < 0$, compute $x_k = \dfrac{1}{2}(a_k + b_k)$ and set

$$I_{k+1} = [a_{k+1}, b_{k+1}] = \begin{cases} [a_k, x_k], & \text{if } f(a_k)f(x_k) < 0, \\ [x_k, b_k], & \text{otherwise.} \end{cases}$$

This general rule that provides I_{k+1} from I_k is an example of what is generally termed as *algorithm*,[5] i.e. a set of mathematical calculations that sometimes involves decision-making. Since this algorithm must be repeated or *iterated*, the bisection method described above constitutes an example of what is also termed as an *iterative algorithm*.

Now we revisit the concept of tolerance seen from the point of view of the bisection process. For $k = 0$, the estimation of the root was $x_0 = 1.5$, which is the midpoint of the interval $I_0 = [1, 2]$, whereas for $k = 1$ we obtain a narrower region $I_1 = [1.5, 2]$ of existence of such root, as well as its corresponding improved estimation $x_1 = 1.75$. In other words, for $k = 0$ the root lies within $[1.5 - \varepsilon, 1.5 + \varepsilon]$, with a tolerance $\varepsilon = 0.5$, whereas for $k = 1$ the interval containing the root is $[1.75 - \varepsilon, 1.75 + \varepsilon]$, with $\varepsilon = 0.25$. Overall, after k bisections, the tolerance is $\varepsilon = |a_k - b_k|/2$, becoming halved in the next iteration.

If we evaluate the radicals that appear in (1.3) with a scientific calculator, we can check that Cardano's analytical solution is approximately $\alpha \approx 1.796\ 321\ 899$. Expression (1.3) is mathematically elegant, but not very

---

5 The etymology of the word is the name of the Persian mathematician and astronomer Al-Khwārizmī- from the eighth century.

practical if one needs an estimation of its numerical value. For $k = 1$ we already have that estimation nearly within a 1% or relative error. The reader may keep iterating further to provide better approximations x_k of α, overall obtaining the sequence $\{x_2, x_3, \ldots, x_7, x_8, \ldots\} = \{1.875, 1.8125, \ldots, 1.792\,968\,7,$ $1.794\,921\,875, \ldots\}$. A natural question is whether this sequence has a limit. This leads to the mathematical concept of *convergence* of a sequence:

Convergence (Exact): The sequence $\{x_k\}$ is said to be *convergent* to α if $\lim\limits_{k \to \infty} x_k = \alpha$ or, equivalently, if $\lim\limits_{k \to \infty} e_k = 0$, where $e_k = x_k - \alpha$.

The difference $e_k = x_k - \alpha$ appearing in the previous definition is called the *error* associated with x_k. However, since e_k may be positive or negative and we are just interested in the absolute discrepancy between x_k and the root α, it is common practice to define the *absolute error* $\varepsilon_k = |x_k - \alpha| = |e_k|$ of the kth iterate. Checking convergence numerically using the previous definition has two main drawbacks. The first one is that we do not know the value of α, which is precisely the goal of root-finding. The second is that in practice we cannot perform an infinite number of iterations in order to compute the limit as $k \to \infty$, and therefore we must substitute the convergence condition by a numerically feasible one. For example, looking at the sequence resulting from the bisection method, it is clear that the absolute difference of two consecutive elements decreases when increasing k (for example, $|x_3 - x_2| = 0.0625$ and $|x_8 - x_7| = 0.002$). Since, by construction, the bisection sequence satisfies $|x_k - x_{k-1}| = |a_k - b_k|/2 = \varepsilon$, it is common practice to consider a sequence as *converged* when this difference becomes smaller than a prescribed threshold or tolerance ε:

Convergence (Practical): A sequence $\{x_k\}$ is said to be *converged to the desired tolerance* $\varepsilon > 0$ after K iterations if $|x_k - x_{k-1}| < \varepsilon, \forall k \geq K$.

The criterion above only refers to the convergence of a sequence, and not necessarily to the convergence to a root. After an iterative method has apparently converged to some stagnated value x_K, it is always advisable to check the magnitude of $|f(x_K)|$ in order to confirm if the method has succeeded. The code below is a simple implementation of the bisection method using the tolerance criterion previously described:

```
% Code 1: Bisection method for solving f(x) = 0 in [a,b]
% Input:   1. [a,b]: interval (it assumes that f(a)f(b) < 0)
%          2. tol: tolerance so that abs(x_k+1 - x_k) < tol
%          3. itmax: maximum number of iterations allowed
%          4. fun: function's name
% Output: 1. xk: resulting sequence
%          2. res: resulting residuals
%          3. it: number of required iterations

function [xk,res,it] = bisection(a,b,tol,itmax,fun)
  ak = a; bk = b; xk = []; res = [];
  it = 0; tolk = abs(bk-ak)/2 ;
  while it < itmax & tolk > tol
    ck = (ak + bk)/2; xk = [xk ck];
    if it > 0; tolk = abs(xk(end)-xk(end-1)); end
    fa = feval(fun,ak); fc = feval(fun,ck); res = [res abs(fc)];
    if fc*fa < 0; bk = ck; else ak = ck;  end
    it = it + 1 ;
  end
```

In the previous code, the iterations stop either when the number of iterations reaches itmax or when $|x_x - x_{k-1}| < \varepsilon =$ tol, i.e. the condition for practical convergence. This condition therefore provides what is usually known as a *stopping criterion* for root-finding methods such as the bisection and other algorithms. However, two words of caution deserve to be mentioned at this point. First, we are implicitly assuming that since $\{x_k\}$ seems to be a *Cauchy Sequence*[6] (that is, $|x_x - x_{k-1}|$ decreases with increasing k), its convergence is guaranteed. While a convergent sequence must necessarily be of Cauchy type, the reverse statement is, in general, not true (although here it will be assumed to be). Second, the quantity α has disappeared from the convergence criteria. In other words, once $|x_K - x_{K-1}| < \varepsilon$ for some K and beyond, we only know that the sequence has converged (in the practical sense) to some value x_K, that henceforth will play the role of the numerical root within the prescribed tolerance.

Finally, the bisection code also provides the vector res containing the sequence of absolute quantities $\{|f(x_k)|\}$, usually called the *residuals*. Since this

---

6 $\{x_k\}$ is said to be of Cauchy-type if $\forall \varepsilon > 0$, $\exists N \in \mathbb{N}$ such that $|x_m - x_n| < \varepsilon$, $\forall m, n > N$.

quantity should approach zero when $\{x_k\}$ converges to a root, the reader may wonder why $|f(x_k)|$ should not be used as a measure for stopping the iterations. The reason is that the residual is not always a reliable measure (although it is often used in practice to decide convergence). We will address this question later in this chapter, when we introduce the concept of *condition number of a root*.

1.3 Newton's Method

Newton's method, also known as *Newton–Raphson method*,[7] is probably the most important root-finding algorithm of numerical mathematics. Its formulation naturally leads to a wide variety of other root-finding algorithms and is easily adaptable to solving systems of nonlinear equations.[8] To formulate this method we will assume that $f(x)$ is differentiable and that the root is close to some *initial guess* x_0. To get a better estimation of the root we assume that, close to x_0, the function $f(x)$ can be approximated by its tangent line $T_0(x)$ at the point $(x_0, f(x_0))$ plotted in gray below. In other words, we locally approximate $f(x)$ by its first order *Taylor expansion* $T_0(x)$ at $x = x_0$:

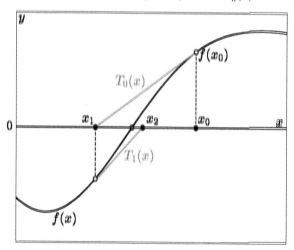

$$T_0(x) \equiv f(x_0) + f'(x_0)(x - x_0), \tag{1.6}$$

---

7 Named after the English mathematicians Sir Isaac Newton (1642–1727) and Joseph Raphson (1648–1715).
8 This will be addressed in the second part of the book.

where $f'(x_0)$ is the value of the derivative of $f(x)$ at x_0. The motivation for approximating $f(x)$ by $T_0(x)$ is simple: if $T_0(x)$ is a reasonably good approximation of $f(x)$, then the solution of $T_0(x) = 0$ should be a reasonable good estimation of the solution of $f(x) = 0$. Let $x = x_1$ be the solution of $T_0(x) = 0$ satisfying

$$f(x_0) + f'(x_0)(x_1 - x_0) = 0, \quad \text{or} \quad x_1 = x_0 - \frac{f(x_0)}{f'(x_0)}. \tag{1.7}$$

The resulting abscissa x_1 where $T_0(x)$ intercepts the x-axis is the new (hopefully more accurate) estimation of the root. The process can be repeated approximating $f(x)$ by its Taylor expansion $T_1(x)$ at x_1 (straight gray line below the curve) in order to obtain an even better estimation x_2 satisfying

$$x_2 = x_1 - \frac{f(x_1)}{f'(x_1)}, \tag{1.8}$$

and so on, leading to the iterative formula:

Newton's Iteration:

$$x_{k+1} = x_k - \frac{f(x_k)}{f'(x_k)} \quad (k \geq 0). \tag{1.9}$$

A simple implementation of Newton's method can be found in the next code, whose structure essentially differs from the one seen in the bisection in two main aspects. First, the code needs an initial guess $x_0 = a$ as starting point, instead of an interval. Second, in addition to the string name corresponding to the M-file function **fun**, the code also needs the one associated with $f'(x)$ in the argument **dfun**. Providing the M-file for the exact derivative entails both analytical differentiation and its subsequent programming, both being error-prone tasks. While this is the correct procedure, it is sometimes advisable to let the computer do the job by approximating the value of $f'(x_k)$ by the ratio

$$f'(x_k) \approx \frac{f(x_k + \delta) - f(x_k)}{\delta}, \tag{1.10}$$

where δ is a small increment ($\delta = 10^{-8}$ in our code). This approximation of the derivative, along with the suitability of the chosen δ, will be properly justified later in the chapter devoted to numerical differentiation.[9] The reader may check

---

9 This approximation is crucial in Newton's method for systems of equations (Part II).

that there are no noticeable differences when using either the exact or the approximate derivative (1.10).

```
% Code 2: Newton's method for solving f(x) = 0
% Input:    1. a: initial guess
%           2. tol: tolerance so that abs(x_k+1 - x_k) < tol
%           3. itmax: maximum number of iterations allowed
%           4. fun: function's name
%           5. dfun: derivative's name
%           (If dfun = '0', then f'(x) is approximated)
% Output:   1. xk: resulting sequence
%           2. res: resulting residuals
%           3. it: number of required iterations
function [xk,res,it] = newton(a,tol,itmax,fun,dfun)
    xk = [a]; fk = feval(fun,xk); res = abs(fk); it = 0;
    tolk = res(1); dx = 1e-8 ;
    while it < itmax & tolk > tol
        if dfun == '0'
            dfk = (feval(fun,xk(end)+dx)-fk)/dx;
        else
            dfk = feval(dfun,xk(end));
        end
        xk = [xk, xk(end) - fk/dfk]; fk = feval(fun,xk(end));
        res = [res abs(fk)]; tolk = abs(xk(end)-xk(end-1));
        it = it + 1;
    end
```

Let us compare the performance of Newton's method with the bisection by applying both algorithms to solve the cubic equation (1.2) starting from the same initial guess $x_0 = 1.5$. The first two columns of Table 1.1 outline the sequences x_k^B and x_k^{NR} resulting from the bisection and Newton–Raphson methods, respectively. While the bisection method requires almost 50 iterations to achieve full accuracy, Newton's method does the same job in just 5. In fact, Newton–Raphson's sequence nearly *doubles* the number of converged digits from one iterate to the next, whereas in the bisection sequence this number grows very slowly (and sometimes even decreases, as seen from $k = 5$ to 6). This is better understood when looking at the third and fourth columns of Table 1.1, where we have included the absolute error corresponding to both methods $\varepsilon_k = |x_k - \alpha|$, where α is the reference value given in the exact expression (1.3), and whose numerical evaluation with Matlab is 1.796 321 903 259 44.

Table 1.1 Iterates resulting from using bisection (x_k^{B}) and Newton–Raphson (x_k^{NR}) when solving (1.2), with added shading of converged digits.

k	Bisection (x_k^{B})	Newton–Raphson (x_k^{NR})	$\varepsilon_k^{\text{B}} = \lvert x_k - \alpha \rvert$	$\varepsilon_k^{\text{NR}} = \lvert x_k - \alpha \rvert$
0	1.5	1.5	3.0×10^{-1}	3.0×10^{-1}
1	1.75	1.869 565 215 355 94	4.6×10^{-2}	7.3×10^{-2}
2	1.875	1.799 452 405 786 30	7.9×10^{-2}	3.1×10^{-3}
3	1.812 5	1.796 327 970 874 37	1.6×10^{-2}	6.1×10^{-6}
4	1.781 25	1.796 321 903 282 30	1.5×10^{-2}	2.3×10^{-11}
5	1.796 875	1.796 321 903 259 44	5.5×10^{-4}	
6	1.789 062 5	1.796 321 903 259 44	7.3×10^{-3}	
7	1.792 968 75		3.4×10^{-3}	
8	1.794 921 875		1.4×10^{-3}	
$\vdots$	$\vdots$		$\vdots$	
46	1.796 321 903 259 45		3.6×10^{-15}	
47	1.796 321 903 259 44			
48	1.796 321 903 259 44			

1.4 Order of a Root-Finding Method

From Table 1.1, it is clear that the Newton–Raphson algorithm converges much faster than the bisection method. In numerical mathematics it is crucial to have a reliable measure of how fast (or slow) a method converges to the desired solution. Some readers may think that we are taking an alarming position since the bisection and Newton Matlab codes provide the solution almost instantaneously, the difference in speed between both methods in terms of computational time being unnoticeable. However, we will see in Part II that the efficiency of the method becomes much more relevant when solving *systems of nonlinear equations*.

A rigorous way of measuring quantitatively the performance of an iterative method is given by the concept of *order of convergence*:

Order of Convergence: A root-finding method providing a sequence $\{x_k\}$ with $\lim\limits_{k \to \infty} x_k = \alpha$ has order of convergence $p \geq 1$ if

$$\lim_{k \to \infty} \frac{\lvert x_{k+1} - \alpha \rvert}{\lvert x_k - \alpha \rvert^p} = C, \tag{1.11}$$

for some positive *bounded* constant C, usually termed as the *asymptotic error constant*. In particular, if $p = 1$, then C must be *smaller than unity* for the criterion to be applicable.

If $p = 1$ (and accordingly $C < 1$) then the sequence $\{x_k\}$ is said to converge *linearly*. If $p > 1$, it is said that the sequence converges *superlinearly*. In particular, for $p = 2$ and $p = 3$, the sequence is said to have *quadratic* or *cubic* convergence, respectively. However, the value of p of a given sequence does not need to be an integer and in general will depend not only on the method used but also on the particular root sought, as we will see later. This implies that in practical situations, we will have to rely on the actual numerical sequence in order to estimate p. Notice that the limit (1.11) can be rewritten in terms of the absolute errors:

$$\lim_{k \to \infty} \frac{\varepsilon_{k+1}}{\varepsilon_k^p} = C, \tag{1.12}$$

which means that for large enough k,

$$\varepsilon_{k+1} \approx C \, \varepsilon_k^p. \tag{1.13}$$

Expression (1.13) is less formal but certainly provides more insight. For example, in the linear case (1.13) implies that $\varepsilon_{k+1} \lesssim \varepsilon_k$, since $C < 1$ for $p = 1$. In other words, the sequence ε_k must be *monotonically decreasing* in order to have linear convergence. According to Table 1.1, $\varepsilon_0^B > \varepsilon_5^B$, and therefore the bisection method does not qualify to have linear convergence in the sense of (1.11)[10]

We can numerically estimate the actual order of convergence p by taking the logarithm of both sides of (1.13) and setting the quantities $X_k \equiv \log_{10} \varepsilon_k$, $Y_k \equiv \log_{10} \varepsilon_{k+1}$, and $D \equiv \log_{10} C$, so that the expression now reads

$$Y_k \approx D + p X_k. \tag{1.14}$$

According to (1.14), the set of points (X_k, Y_k) should follow a linear law with slope p. However, expression (1.13) must be conceived just as a *local law*, i.e. sufficiently close to the converged root. Typically, unless the iteration is started really close to the root, the first iterates of the sequence must be discarded from the analysis. On the other hand, since $\varepsilon_k \to 0$, the quotient $\varepsilon_{k+1}/\varepsilon_k^p$ appearing in (1.12) is affected by numerical cancelation and therefore the last few iterates must also be ruled out from the numerical evaluations.

Figure 1.2a shows the convergence of Newton's method based on the actual sequence obtained in Table 1.1 for the solution of $x^3 - x = 4 = 0$, where the points (X_k, Y_k) seem to align parallel to a straight line of slope $p = 2$ (dashed gray line), as an indication of quadratic convergence in this particular case. Sometimes, the slope of the numerical data (X_k, Y_k) can be clearly identified by simple eye inspection (particularly for $p = 1$ or $p = 2$), although linear regression may also be used if data are more scattered or if the value of p is not an easily identifiable integer.

---

10 However, some texts consider the bisection method to be linearly convergent if one considers an *upper bound* of the error $\tilde{\varepsilon}_k \equiv \dfrac{b-a}{2^k}$, so that $\lim_{k \to \infty} \dfrac{\tilde{\varepsilon}_{k+1}}{\tilde{\varepsilon}_k} = \dfrac{1}{2}$.

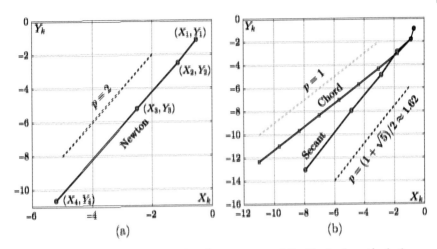

Figure 1.2 (a) Analysis of the order of convergence of the Newton's method: the points (X_k, Y_k) seem to align with a straight line of slope $p = 2$ (dashed line). (b) Same analysis for chord (gray squares) and secant methods (gray circles), showing linear and golden ratio orders, respectively.

We can rigorously prove the quadratic convergence of Newton's method by evaluating the limit (1.11) for $p = 2$

$$\lim_{k \to \infty} \frac{|x_{k+1} - \alpha|}{|x_k - \alpha|^2} = \lim_{k \to \infty} \frac{|e_{k+1}|}{|e_k|^2}, \tag{1.15}$$

Before calculating the limit above, first notice that Newton's iteration $x_{k+1} = x_k - f(x_k)/f'(x_k)$ can be expressed in terms of the errors $e_k = x_k - \alpha$. Since $x_k = \alpha + e_k$ we may write

$$e_{k+1} = e_k - \frac{f(\alpha + e_k)}{f'(\alpha + e_k)} = \frac{e_k f'(\alpha + e_k) - f(\alpha + e_k)}{f'(\alpha + e_k)},$$

and therefore

$$\lim_{k \to \infty} \frac{e_{k+1}}{e_k^2} = \lim_{k \to \infty} \frac{e_k f'(\alpha + e_k) - f(\alpha + e_k)}{e_k^2 f'(\alpha + e_k)},$$

Since $e_k \to 0$ when $k \to \infty$, we may define $z = e_k$ as a continuous variable that approaches zero so that we can rewrite the previous limit as

$$\lim_{z \to 0} \frac{z f'(\alpha + z) - f(\alpha + z)}{z^2 f'(\alpha + z)} = \frac{f''(\alpha)}{2 f'(\alpha)},$$

where we have used L'Hôpital's rule to solve the indeterminate form $0/0$. Therefore we conclude that Newton's method has quadratic convergence with asymptotic error constant

$$\boxed{C_{\mathrm{N}} = \frac{1}{2} \left| \frac{f''(\alpha)}{f'(\alpha)} \right|,} \tag{1.16}$$

1.5 Chord and Secant Methods

One of the drawbacks of Newton's method is that we must supply the derivative of the function at every iteration. As mentioned before, the derivative can be approximated using (1.10) so there is no need for providing a supplementary function for the evaluation $f'(x)$. Either in the exact or approximate version of Newton's method, we need *two* function evaluations per iteration. There are situations where two or more evaluations per iteration may be computationally expensive, such as in the case of extending Newton's method to solve systems of nonlinear equations, as we will address in Part II. Now, let us assume that we have to provide a Newton-like iteration without explicitly appealing to the derivative of $f(x)$ and with just one evaluation of $f(x)$ per iteration.

Assume that we have identified an interval $[a, b]$ such that $f(a)f(b) < 0$. The key point is to provide an estimation q of the slope $f'(x_k)$ of the function at the kth iterate and substitute Newton's iteration by

$$x_{k+1} = x_k - \frac{f(x_k)}{q}. \tag{1.17}$$

Provided that $f(x)$ does not oscillate very much within $[a, b]$, a reasonable estimation of q is the slope $(f(b) - f(a))/(b - a)$ provided by the mean value theorem. In this case, the expression (1.17) leads to what is usually termed as the *chord method*:

Chord Iteration:

$$x_{k+1} = x_k - \frac{b - a}{f(b) - f(a)} f(x_k). \tag{1.18}$$

The chord method can be improved by updating q at every iteration with a new quantity q_k obtained from the values of the jth iterates x_j and their images $f(x_j)$ obtained at the previous stages $j = k - 1$ and $j = k$:

$$q_k = \frac{f(x_k) - f(x_{k-1})}{x_k - x_{k-1}}, \tag{1.19}$$

so that (1.17) now leads to the *secant method*:

Secant Iteration:

$$x_{k+1} = x_k - \frac{x_k - x_{k-1}}{f(x_k) - f(x_{k-1})} f(x_k). \tag{1.20}$$

It is common practice to start the indexing at $k = -1$ by taking the initial values $x_{-1} = a$ and $x_0 = b$, so that the first iterate x_1 is the same as the one obtained with the chord method. It should be clear that both chord and secant methods involve just the *new* evaluation $f(x_k)$ per iteration, the remaining terms being previously computed (and stored) in the past.

Starting from Newton's code as a reference, the reader may easily program chord and secant algorithms and compare their efficiency with Newton's method when solving the test cubic equation $x^3 - x - 4 = 0$ starting from the same initial interval $[1, 2]$, as was done with Newton's and bisection methods. Figure 1.2b shows the resulting convergence history of the two methods. From the two sets of data of Figure 1.2b we may conclude that while the chord method seems to converge linearly, the secant methods converges *superlinearly* with an approximate order $p \approx 1.62$. It can be formally proved that the exact convergence order of the chord method is $p = 1$, whereas for the secant method the order is the *golden ratio* $p = (1 + \sqrt{5})/2 \approx 1.618\,034\ldots$ (see exercises at the end of the chapter).

Practical 1.1: Sliding Particles over Surfaces

A small point-like object of mass m initially at rest is released from the highest point of the surface A of height $f_0 = f(0)$, as shown in the following figure. The object slowly starts to slide under the effects of gravity until it reaches point B at which it loses contact with the surface. The goal of this practical is to determine the abscissa x_B of that point.

First, assuming there are no friction forces, show that the speed of the object at $(x, f(x))$, i.e. still in contact with the surface, is

$$v = (2g)^{1/2}(f_0 - f(x))^{1/2}.$$

Then show that the horizontal velocity $\dfrac{\mathrm{d}x}{\mathrm{d}t}$ is given by the expression in the figure on the left. Imposing that there is no horizontal acceleration at point B (i.e. $\ddot{x}_B = 0$) show that (excluding points where $f' = 0$)

$$1 + (f')^2 + 2(f_0 - f)f'' = 0,$$

at the point where contact is lost.

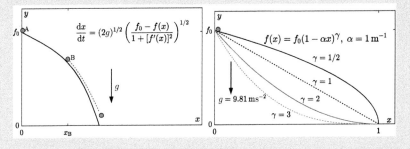

Consider the family of surfaces shown in the figure on the right. Releasing the object from point $(0, 1)$ m, find the abscissas x_B for $\gamma = 0.1, 0.25, 0.5,$ and 0.9.

1.6 Conditioning

As mentioned before, nonlinear equations $f(x) = 0$ usually involve parameters and constants coming from known external data. For example, Eq. (1.5) for the quantum energy levels depend on the well's depth V_0 and width a, as well as on the mass of the particle m. Therefore, it is obvious that changing the value of V_0, a, or m in (1.5) will change the value of the corresponding energy levels $\mathcal{E}_n$. In particular, we may also expect that if the parameters are changed just by a tiny amount, then the changes in the solutions should also be very small.

In numerical mathematics, a problem is said to be *well conditioned* when small changes in the input data always lead to small changes in the output solution. By contrast, a problem is said to be *ill conditioned* when even very small changes in the input data may lead to very large variations in the outcome. In practice it is very important to *quantify* this sensitivity of the solution to small changes in the input data. Let us quantify this in the case of root-finding.

Imagine we are asked to find the abscissa $x = a$ at which the graph $(x, f(x))$ intercepts the ordinate $y = b$; see Figure 1.3a. Mathematically, the sought abscissa a is the root x of the equation $f(x) - b = 0$, where b plays the role of a parameter. If b is slightly changed to $b + \Delta b$ (with $|\Delta b|$ small), the new root will accordingly move to a new value $x = a + \Delta a$. If the conditions of the inverse function theorem are satisfied in a neighborhood of $(x, y) = (a, b)$, the

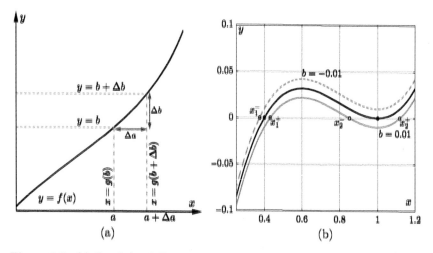

Figure 1.3 (a) Graph $(x, f(x))$ intercepting ordinates $y = b$ and $y = b + \Delta b$ at abscissas $x = a$ and $x = a + \Delta b$, respectively. (b) Roots of the equation $x^3 - 2.4x^2 + 1.8x - 0.4 - b = 0$ for $b = 0$ (solid black curve), $b = -0.01$ (dashed gray), and $b = 0.01$ (solid gray).

equation $f(x) - y = 0$ defines x as a function $g(y)$ with $a = g(b)$, where $g(y)$ is the inverse function of $f(x)$. Accordingly, the derivatives of $f(x)$ at $x = a$ and of $g(y)$ at $y = b$ satisfy the relation

$$\left[\frac{dg}{dy}\right]_{y=b} = \left(\left[\frac{df}{dx}\right]_{x=a}\right)^{-1}, \tag{1.21}$$

or $g'(b) = 1/f'(a)$ for short, where the prime symbols acting on f and g must be understood as derivatives with respect to x and y, respectively. We can now estimate the location of the new root. Since Δb is small,

$$g'(b) \approx \frac{g(b + \Delta b) - g(b)}{\Delta b}, \quad \text{and therefore} \quad g(b + \Delta b) \approx g(b) + \Delta b \, g'(b).$$

Since the new root is precisely located at $x = a + \Delta a = g(b + \Delta b)$,

$$a + \Delta a \approx g(b) + \Delta b \, g'(b).$$

Finally, since $a = g(b)$, we conclude from the last equation that $\Delta a \approx \Delta b \, g'(b)$. Taking the absolute value and recalling that $g'(b) = 1/f'(a)$,

$$\boxed{|\Delta a| \approx \frac{|\Delta b|}{|f'(a)|}.} \tag{1.22}$$

This last expression predicts how far the new root will move in terms of how much we have perturbed b. From (1.22), we see that no matter how tiny $|\Delta b|$ may be, $|\Delta a|$ can potentially be very large if $|f'(a)|$ is very small. As an example, take the equation $x^3 - 2.4x^2 + 1.8x - 0.4 - b = 0$. For $b = 0$, the previous equation has a simple root at $x_1 = 0.4$ and a double root at $x_2 = 1$ (see Figure 1.3b, black curve). For $b = -0.01$ and $b = 0.01$, the simple root x_1 exhibits a slight displacement to $x_1^- \approx 0.375$ and $x_1^+ \approx 0.431$, respectively (see dashed and solid gray curves in Figure 1.3b, respectively). However, for the same values of b, the double root does experience remarkable changes. In particular, for $b = 0.01$, the double root disappears, leading to two simple roots located at $x_2^- \approx 0.851$ and $x_2^+ \approx 1.118$. For $b = -0.01$ the effects are even more drastic since the function has no longer any root near $x = 1$.

From the previous analysis, we can clearly conclude that the double root $x_2 = 1$ is more sensitive (or ill-conditioned) than the simple root $x_1 = 0.4$. This phenomenon could have been predicted in advance just by evaluating the denominator $|f'(a)|$ appearing in (1.22) with $f(x) = x^3 - 2.4x^2 + 1.8x - 0.4$ and $a = 0.4$ or $a = 1$, since $f'(0.4) = 0.36$, whereas $f'(1) = 0$.

In general, for a given numerical problem, it is common practice to quantify its conditioning by the simple relation

$$\varepsilon_{out} = K \, \varepsilon_{in}, \tag{1.23}$$

where ε_{in} and ε_{out} are the size of the variations introduced in the input data and their corresponding deviation effect in the outcome solution, respectively, and K is a positive constant known as the *condition number* of the problem. The quantity ε_{in} may represent uncertainties in the parameters, numerical noise or, within the context of this book, numerical inaccuracies due to limited machine precision. The condition number K must be understood as a *noise amplifier*, which magnifies small uncertainties. A condition number of order 1 is an indication of *well-conditioning*, whereas a problem with $K \gg 1$ is definitely *ill-conditioned*.

By comparing (1.23) with (1.22), we can easily identify $\varepsilon_{\text{out}} = |\Delta a|$ as the displacement exhibited by the root, $\varepsilon_{\text{in}} = |\Delta b|$ as the numerical uncertainty in the evaluation of $f(x)$, and

$$K = \frac{1}{|f'(a)|} \tag{1.24}$$

as the *condition number of the root* $x = a$. As we will see in Section 1.6, the performance of Newton's method can be affected if the root we are looking for is ill-conditioned.

1.7 Local and Global Convergence

Newton's method converges properly only under certain conditions. One required condition is that the initial guess from which the iteration is initiated must be *sufficiently close* to the root, that is, a *local* initial guess. In that sense, it is said that Newton's method has only *local convergence*. Even if the sequence converges to the root, the order may not be always $p = 2$, as in Figure 1.2a.

It is a common misconception that every single method has its associated local order of convergence (Newton's has $p = 2$, secant has $p = (1 + \sqrt{5})/2$, etc.) In actuality, the order also depends on the conditioning of the root to which our sequence approaches. For example, the asymptotic constant from Newton's method C_N appearing in (1.16) is proportional to $|f'(\alpha)|^{-1}$. As a consequence, if α is a double root and, accordingly, $f'(\alpha) = 0$, the convergence criterion (1.11) is no longer valid since C_N is not bounded.[11]

Figure 1.4a shows the result of applying Newton's and secant methods to find the double ill-conditioned root $x_2 = 1$ of the equation $x^3 - 2.4x^2 + 1.8x - 0.4 = 0$ studied in Section 1.6. Newton's iteration is started from $x_0 = 1.2$, whereas the secant has been initialized from the interval $[1.1, 1.2]$ that contains the root. The reader may check that in this case Newton's and secant methods lose

---

11 The asymptotic constants of secant or chord methods have the same problem with double roots (see Exercises 1.3 and 1.4).

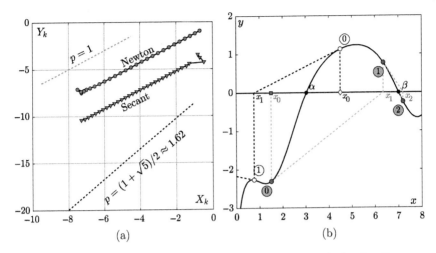

Figure 1.4 (a) Convergence history of Newton's and secant methods when the sequences approach the double root $x_2 = 1$ of equation $x^3 - 2.4x^2 + 1.8x - 0.4 = 0$. The Y_k ordinates corresponding to the secant method (triangles) have been shifted downwards three units to avoid overlap between the two sets of data and help visualize. (b) Newton's method iterates for the solution of $\log x - \exp(\sin x) = 0$.

their quadratic and golden ratio orders, respectively, both exhibiting linear convergence, as shown in Figure 1.4a. Double or ill-conditioned roots appear in physics more frequently than one may expect, particularly in problems where the transcendental equation to be solved is the result of imposing some kind of critical or threshold condition (we refer the reader to Practical 1.2, for example).

In general, root-finding methods converge to the desired solution only if the initial guess x_0 is really close to the sought root, i.e. most of the methods are just *locally convergent*. In practice, a root-finding algorithm starting from an initial guess x_0 moderately far away from the root could easily lead to a sequence x_k that may wander from one point to another of the real axis, eventually diverging to infinity or converging to a solution (not necessarily the sought one). Figure 1.4b illustrates this phenomenon by showing the result of computing the roots of the function $f(x) = \log x - e^{\sin x}$ using Newton's method starting from different initial guesses. The first two roots of $f(x)$ are located at $\alpha \approx 3.04$ and $\beta \approx 7.01$ (black bullets in Figure 1.4b). In this example, we initialize Newton's method from two initial guesses reasonably close (but not too close) to α. To guide the eye, we have indicated the history of each of the two sequences by encircled numbering of their ordinates. The first sequence starts at $x_0 = 1.5$ (gray square) and Newton's first iterate $x_1 \approx 6.3$ is already very close to β, to which the sequence eventually converges (gray dashed lines and symbols). The second sequence starts from $x_0 = 4.5$ (white diamond), unfortunately leading to a location x_1 where Newton's algorithm predicts a negative second iterate x_2, which is not even within the function's domain due to the logarithmic term.

From the previous example we can conclude that forecasting the fate of a Newton's sequence based on the location of the initial guess x_0 is usually impossible. In the first case, we may have naturally expected the sequence to approach α instead of β (because of its initial proximity to the former one). In the second case, we could have also naturally expected the sequence to converge at least to either one of the two roots, but never such a dramatic failure of the iteration. We recommend the reader to explore the complex convergence properties of Newton's method by starting the iteration from a wide range of initial guesses located between α and β. The reader may also repeat the experiment with the secant or chord algorithms to conclude that the behavior of the sequences is also unpredictable when using these methods.

Under particular circumstances, there are certain root-finding methods that always converge to the same root, regardless of the initial guess from which they have been started. In general, for a given function $f(x)$ with a unique root α in the open interval I, a root-finding method is said to be *globally convergent* within I if $\lim_{k \to \infty} x_k = \alpha$ for all initial guesses $x_0 \in I$. The bisection method or the *Regula Falsi*[12] method are examples of globally convergent algorithms. However, these and other univariate globally convergent methods are not always easily adaptable to solve systems of nonlinear equations.

Complementary Reading

For an authoritative study of different root-finding methods for nonlinear scalar equations, I strongly recommend Dahlquist and Björk's *Numerical Methods in Scientific Computing*, Vol. I. The reader will find there detailed mathematical proofs of the convergence of many algorithms, as well as other very important topics that have not been covered in this chapter, such as *Fixed-Point Iteration*, *Minimization* or the solution of algebraic equations and *deflation*. That chapter also addresses the *Termination Criteria* problem (i.e. when to stop the iteration) in depth and with mathematical rigor.

For a different approach to the root-finding problem, Acton's *Numerical Methods that Work* is an alternative. The reader will find in that text very clear geometrical explanations of why different root-finding methods may have convergence problems. Acton's book also provides deep insight into the technical aspects of root-finding algorithms, as well as very useful tips and strategies.

---

12 Also known as the false position method (see Exercise 1.6).

Practical 1.2: Throwing Balls and Conditioning

A ball is thrown from the lowest point of a hill $(x_0, y_0) = (0,0)$m with initial angle $\theta_0 = 67.5°$ and speed 37 m s^{-1} (see the figure below). The height of the hill is given by the expression $y(x) = ax^2 e^{-bx}$, where x and y are measured in meters, $a = 0.15$m^{-1}, and $b = 0.04$ m^{-1}.

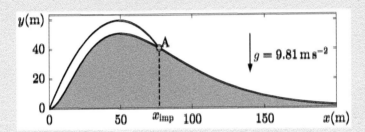

(a) Edit a Matlab .m function corresponding to the equation $F(x; v_0, \theta_0) = 0$ whose solution x is the abscissa where the parabola and the hill's profile intersect (for arbitrary initial speed and angle v_0 and θ_0, respectively).

(b) Using Newton's method, find the abscissa x_{imp} of the point A where the ball impacts with the hill. Provide your result with at least five exact figures.

(c) With the same initial speed, represent the impact abscissa x_{imp} for initial angles within the range $[50°, 70°]$.

(d) Find the minimum initial angle $\theta_{\min}$ that allows the ball to impact beyond $x_{\text{imp}} = 50$ m, i.e. to the right of the hill's peak. Provide your result with at least three exact figures.

(e) Explore the order of convergence of the root-finding method when computing the minimum angle in (d). Do you observe an increase in the number of iterations required to achieve the desired accuracy? If so, explain what may be the reason for that phenomenon.

Problems and Exercises

The symbol (A) means that the problem has to be solved analytically, whereas (N) means to use numerical Matlab codes. Problems with the symbol * are slightly more difficult.

1. (A) Consider the bisection method starting from the interval $I_0 = [a, b]$. Let $\{x_k\}$ be the resulting bisection sequence satisfying $\lim_{k \to \infty} x_k = \alpha$. Let $\varepsilon > 0$ be a given tolerance.

   (a) How many iterations are required in order to satisfy that $|\alpha - x_k| < \varepsilon$?

   (b) Once that tolerance has been reached, how many iterations should be added to have one more digit of precision?

2. (N) Apply the bisection algorithm to solve the equations over the intervals indicated below. Provide the numerical approximation of the root α with at least 10 significant digits and also report how many bisections K were required to achieve that accuracy.

   (a) $x^{-1} - \tan x = 0$, $[0, \pi/2]$.

   (b) $x^{-1} - 2^x = 0$, $[0, 1]$.

   (c) $2^{-x} + e^x + 2 \cos x = 6$, $[1, 3]$.

3. (A) Show that the chord method starting within the interval $[a, b]$ with $f(a)f(b) < 0$ has linear convergence to the root α and obtain its corresponding asymptotic error constant C. If $C < 1$ for linear convergence, what constraints does this condition impose? Will the method converge linearly to a double root?

4. (A)* Show that the secant method converges superlinearly with order of convergence $p = (1 + \sqrt{5})/2$. As a hint, use Taylor's expansion to show that the errors $e_k = x_k - \alpha$ obey (to the lowest order) the recurrence:
   $$e_{k+1} \approx \frac{f''(\alpha)}{2f'(\alpha)} e_k e_{k-1}.$$
   Next, assume that the absolute error obeys the relation $\varepsilon_{k+1} = C\varepsilon^p$, with $p \geq 1$ in order to provide the value of the exponent p as well as the asymptotic error constant C.

5. (N) Using your own codes for the chord and secant methods, solve the equation $x^3 - x - 4 = 0$ starting from the interval $[1, 2]$ and reproduce the results of Figure 1.2b.

6. (N) *Regula Falsi* or *false position* method: this root-finding algorithm is in essence a globally convergent version of the secant method. Starting from the interval $[x_{-1}, x_0]$ satisfying $f(x_{-1})f(x_0) < 0$, proceed with the iteration

$$x_{k+1} = x_k - \frac{x_k - x_\ell}{f(x_k) - f(x_\ell)} f(x_\ell),$$

where x_ℓ is the last iterate satisfying $f(x_\ell)f(x_k) < 0$, that is,

$$\ell = \max\{-1, 0, 1, \ldots, k-1\} \quad \text{such that} \quad f(x_\ell)f(x_k) < 0.$$

You need to modify your secant method code by storing the history of the iterates in order to include the condition above.

(a) Solve again the cubic $x^3 - x - 4 = 0$ starting from the interval $[1, 2]$ and compute its local order of convergence based on the sequence obtained. What order can you identify?

(b) Solve $\log x - e^{\sin x} = 0$ (see Figure 1.4b) starting from initial intervals $[x_{-1}, x_0]$ containing either α or β and convince yourself that the method is globally convergent within those intervals. In particular, verify for a few values of k that $[x_k, x_{k+1}] \subset [x_{-1}, x_0]$.

7. (A) Applying Newton's method to a certain equation leads to the iteration $x_{k+1} = 2x_k - zx_k^2$, with $z \neq 0$. Find the original equation or, equivalently, find the purpose of such iteration and its corresponding limit $\lim_{k \to \infty} x_k$.

8. (A) We apply Newton's method to solve $x^2 - 1 = 0$ starting from $x_0 = 10^{10}$. How many iterations are required to obtain the root with an error of 10^{-8}? Solve the problem *analytically*. After that, verify your estimation numerically.

9. (A) For $p > 0$, find the value of the quantity $x = \sqrt{p + \sqrt{p + \sqrt{p + \cdots}}}$. Hint: x is the limit of the sequence

$$x_1 = \sqrt{p}, \quad x_2 = \sqrt{p + \sqrt{p}} = \sqrt{p + x_1}, \ldots$$

10. (A–N) One technique to accelerate Newton's method for roots of multiplicity $m > 1$ consists of replacing the original equation $f(x) = 0$ by the auxiliary one $u(x) = 0$, where $u(x) = \dfrac{f(x)}{f'(x)}$.

(a) If α is a root of $f(x) = 0$ with multiplicity $m > 1$, find the limit

$$\lim_{x \to \alpha} \frac{u(x)}{x - \alpha}.$$

(b) If the modified Newton's iteration using the auxiliary equation reads
$x_{n+1} = x_n - s_n$, find s_n.

(c) If we use the original Newton's method to solve $\cos^2 x = 0$ starting from $x_0 = 2$, find the minimum number of iterations n such that $|x_n - \pi/2| < 10^{-4}$.

(d) Repeat (c) using the auxiliary equation.

11. (A-N) *Steffensen's* root-finding method is given by the iteration

$$x_{n+1} = x_n - \frac{f(x_n)}{g(x_n)}, \quad \text{where } g(x_n) \equiv \frac{f(x_n + f(x_n)) - f(x_n)}{f(x_n)}.$$

(a) Starting from the initial guess $x_0 = 1.2$, apply the method to solve $x^3 - x - 4 = 0$. What order of convergence p can you identify?

(b) Solve $x^3 - 2.4x^2 + 1.8x - 0.4 = 0$ starting also from $x_0 = 1.2$. Do you observe the same exponent p as in part (a)?

(c) (*) Find the exact order of convergence p of the method for simple roots, as well as its corresponding asymptotic constant.

12. (A-N) *Halley's*[13] root-finding method is an algorithm that improves Newton's quadratic convergence when solving $f(x) = 0$. In this case, the method solves the auxiliary equation $F(x) = 0$, where $F(x) \equiv f(x)/\sqrt{f'(x)}$ and $f'(x)$ is assumed to be positive and differentiable within the domain containing the root.

(a) Find the resulting iterative formula in terms of $f_n \equiv f(x_n)$, $f'_n \equiv f'(x_n)$, and $f''_n \equiv f''(x_n)$.

(b) Find the root of $f(x) \equiv -\cos x$ within $[0, \pi]$ using Halley's method starting from $x_0 = 1$. Can you identify the order of convergence p?

13. (A-N) *Dynamics and Kinematics*: A trapeze artist is at rest on top of a platform of height $a \equiv 10$ m (hollow circle A in the following figure) holding a rigid rod of length a and negligible mass that can rotate around the fixed point O. The performer starts falling under the effect of gravity (dashed gray curve), passing through point $(0, 0)$ at maximum speed without touching the ground. Later, the acrobat releases the rod when the angle it forms with the vertical axis is θ_0 (point B). From point B onward, the artist traces a parabolic trajectory, finally touching the ground at point C, whose abscissa is $x = G(\theta_0)$. Neglect friction forces.

---

13 Edmond Halley, English astronomer who predicted in 1705 the orbital periodicity of the comet that now bears his name.

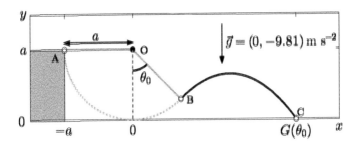

(a) Show that $G(\theta_0) \equiv a\{(1 + 2\cos^2\theta_0)\sin\theta_0 + 2\sqrt{\cos^3\theta_0(1 - \cos^3\theta_0)}\}$.

(b) Find the angle θ_0 for which $G(\theta_0) = 20$ m.

(c) What is the angle θ_0 so that the acrobat maximizes the range G? Provide your numerical answers with at least four exact figures. Advice: for parts (b) and (c), consider the alternative equation

$$F(z) \equiv a\left\{(1 + 2z^2)\sqrt{1 - z^2} + 2\sqrt{z^3(1 - z^3)}\right\}, \quad \text{where} \quad z \equiv \cos\theta_0.$$

2

Polynomial Interpolation

2.1 Function Approximation

Physical laws are expressed in terms of mathematical functional relations between dependent and independent variables. Such is the case of Hooke's law, $F(x) = -kx$, which relates the force F exerted by a spring of constant k when it is stretched through an elongation x from its equilibrium position. Hooke's law happens to be a simple *linear* relation, but many other physical laws have a more complicated mathematical character, such as Planck's law of black-body radiation $B_\nu(T) = \dfrac{2h\nu^3}{c^2} \dfrac{1}{e^{h\nu/k_B T} - 1}$, relating the spectral radiance density B_ν of frequency ν as a function of the temperature T. There are situations where the functional relation between physical quantities is a priori unknown, and we only have access to a set of experimental values from which we expect to infer an underlying law. Whatever the functional dependence is, physicists and engineers need to operate mathematically with these functions in order to extract relevant information. The two more common mathematical operations are *differentiation* and *integration*. On many occasions, differentiation is analytically possible (if the function is known and simple), whereas finding the primitive of arbitrary functions in closed form is in general a very difficult (sometimes impossible) task.

In its most abstract form, physical laws are often formulated using the language of *differential equations*. Such is the case for Maxwell's equations of electromagnetism, Schrödinger's equation of quantum mechanics, or Navier–Stokes equations of fluid dynamics, to mention just a few. These are complicated equations involving the unknown functions (electric and magnetic fields, for example) and their corresponding derivatives with respect to space and time. In general, solving the differential equations arising in mathematical physics or engineering is a complicated task, only feasible in a limited number of cases.

In numerical mathematics, it is common practice to approximate derivatives and integrals of arbitrary functions by an indirect method. The first

Fundamentals of Numerical Mathematics for Physicists and Engineers, First Edition. Alvaro Meseguer.
© 2020 John Wiley & Sons, Inc. Published 2020 by John Wiley & Sons, Inc.

step consists of approximating the function by means of polynomials. These polynomial combinations are intended to resemble the function to the highest possible accuracy within either a given domain (global approximation) or in the neighborhood of a point on the real axis (local approximation). The second step consists of differentiating or integrating these polynomials instead of the original function. If these polynomials are accurate approximations of the original function, their differentiation and integration are expected to be also very good approximations of the derivative and integral of the original function, respectively. The motivation for using polynomials is that they are easy to differentiate or integrate and, if they are suitably used, they also provide unbeatable accuracy. However, we will see in Part II that functions can also be approximated using more general formulations.

There are different ways of approximating functions using polynomials, and in this chapter we will address one of them: *polynomial interpolation*. As we will see in Chapters 3 and 4, polynomial interpolation lies at the very heart of the most commonly used rules to approximate derivatives and integrals (quadratures) of functions. In Part II, we will also address how to apply polynomial interpolation to approximate the solution of ordinary differential equations in its two possible frameworks: initial and boundary value problems.

2.2   Polynomial Interpolation

We are all familiar with the fundamental algebraic problem of determining the expression $y(x) \equiv a_0 + a_1 x$ of the straight line that passes through two given points (x_0, y_0) and (x_1, y_1). To obtain the coefficients a_0 and a_1, we simply impose the conditions $y(x_0) \equiv y_0$ and $y(x_1) \equiv y_1$ that lead to the linear system of equations

$$a_0 + a_1 x_0 \equiv y_0$$
$$a_0 + a_1 x_1 \equiv y_1. \tag{2.1}$$

Assuming that $x_0 \neq x_1$, we can solve system (2.1) and provide the requested coefficients a_0 and a_1 so that the sought straight line is

$$y(x) \equiv \frac{y_0 x_1 - y_1 x_0}{x_1 - x_0} + \frac{y_1 - y_0}{x_1 - x_0}\, x. \tag{2.2}$$

Since the solution above is a polynomial of degree 1, a natural question is whether this procedure can be applied to find polynomials of suitable degree passing through an arbitrary number of points. The answer is yes, and the mathematical technique that provides such polynomials is called *polynomial interpolation*.

Suppose we are given a set of distinct $n+1$ abscissas $\{x_0, x_1, x_2, \ldots, x_n\}$ and $n+1$ arbitrary ordinates $\{y_0, y_1, y_2, \ldots, y_n\}$. Imagine that we want to find a polynomial $\Pi_n(x)$ of degree n that passes through the $n+1$ points $\{(x_0, y_0), (x_1, y_1), \ldots, (x_n, y_n)\}$, as shown in the figure below.

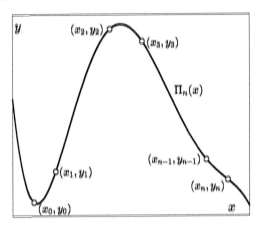

Let this polynomial be

$$\Pi_n(x) = a_0 + a_1 x + a_2 x^2 + \cdots + a_n x^n, \tag{2.3}$$

with a_j $(j = 0, \ldots, n)$ being $n+1$ unknown coefficients. As we did for the case of the straight line, we can identify the coefficients a_j by imposing

$$\Pi_n(x_j) = a_0 + a_1 x_j + a_2 x_j^2 + \cdots + a_n x_j^n = y_j \quad (j = 0, 1, 2, \ldots, n), \tag{2.4}$$

that is, the polynomial must pass through the given points (x_j, y_j). The reader may easily check that equations (2.4) lead to a linear system for the undetermined coefficients a_j whose matrix form is

$$
\begin{bmatrix}
1 & x_0 & x_0^2 & \cdots & x_0^n \\
1 & x_1 & x_1^2 & \cdots & x_1^n \\
1 & x_2 & x_2^2 & \cdots & x_2^n \\
\vdots & \vdots & \vdots & & \vdots \\
1 & x_n & x_n^2 & \cdots & x_n^n
\end{bmatrix}
\begin{bmatrix}
a_0 \\
a_1 \\
a_2 \\
\vdots \\
a_n
\end{bmatrix}
=
\begin{bmatrix}
y_0 \\
y_1 \\
y_2 \\
\vdots \\
y_n
\end{bmatrix}. \tag{2.5}
$$

The square matrix V with elements $V_{ij} \equiv x_i^j$ appearing on the left-hand side of (2.5) is usually referred to as the *Vandermonde's matrix*[1] associated with

---

[1] Alexandre Théóphile Vandermonde (1735–1796), French mathematician better known for his remarkable contributions to the theory of determinants.

the abscissas $\{x_0, x_1, \ldots, x_n\}$, and whose determinant is (see Problem 2.1)

$$
\begin{aligned}
\det V &= (x_n - x_{n-1})(x_n - x_{n-2}) \cdots (x_n - x_0) \cdots \\
&\quad (x_{n-1} - x_{n-2})(x_{n-1} - x_{n-3}) \cdots (x_{n-1} - x_0) \cdots \\
&\quad \cdots \\
&\quad (x_3 - x_2)(x_2 - x_1) \\
&= \prod_{0 \le i < j \le n} (x_j - x_i)
\end{aligned}
\tag{2.6}
$$

Since the abscissas x_j are assumed to be different, $\det V \ne 0$. Consequently, system (2.5) has always a *unique* solution. Therefore, the resulting polynomial $\Pi_n(x)$ is also unique. Since $\deg \Pi_n(x) \le n$, this polynomial belongs to $\mathbb{R}_n[x]$, that is, the $(n+1)$-dimensional vector space of polynomials of degree less than or equal to n. What has just been described is usually known as the *method of undetermined coefficients* to obtain the polynomial that goes exactly through the given $n + 1$ distinct points (x_j, y_j).

As mentioned in the introduction, one of the main applications of interpolation is to approximate functions on a given domain. In what follows, we will assume that $f(x)$ is a continuous function within the interval $[a, b]$. Let $\{x_0, x_1, x_2, \ldots, x_n\}$ be a set of distinct abscissas with $a \le x_j \le b$, and let $f_j = f(x_j)$ be the values of $f(x)$ at these abscissas x_j, henceforth termed as *nodes of interpolation*. To approximate $f(x)$, we search for the interpolatory polynomial of degree n, here denoted as $\Pi_n f(x)$, passing through the coordinates $\{(x_0, f_0), \ldots, (x_n, f_n)\}$. This polynomial must satisfy the conditions

$$
\Pi_n f(x_j) = a_0 + a_1 x_j + a_2 x_j^2 + \cdots + a_n x_j^n = f_j \quad (j = 0, 1, 2, \ldots, n), \tag{2.7}
$$

which lead exactly to the same linear system as in (2.4) for $y_j = f_j$ $(j = 0, \ldots, n)$. The polynomial $\Pi_n f(x)$ resulting from solving system (2.7) is usually known as the *interpolant* of $f(x)$ at the nodes $\{x_0, x_1, \ldots, x_n\}$.

We have always assumed that the degree of the interpolant is n (with $n + 1$ being the number of interpolation nodes) for system (2.7) to be determinate. However, there are many other interpolatory techniques based on different types of constraints to be satisfied by the interpolant. For example, we may obtain a more accurate polynomial by imposing $\Pi f(x_j) = f(x_j)$ as well as $[\Pi f(x)]'|_{x_j} = f'(x_j)$ at the nodes x_j, where $'$ stands for differentiation with respect to x. That is, in this particular interpolation, usually known as *oscula-tory*[2] interpolation, both the polynomial and its derivative must coincide with $f(x)$ and $f'(x)$ at the nodes, respectively. Since the number of constraints to be satisfied by the polynomial has been doubled, we may accordingly look for polynomials of degree $2n + 1$.

---

2 Also known as *Hermite–Birkhoff* interpolation. See references at the end of this chapter for details.

2.3 Lagrange's Interpolation

In Section 2.2, we saw how to determine the interpolant

$$\Pi_n f(x) = a_0 + a_1 x + a_2 x^2 + \cdots + a_n x^n \tag{2.8}$$

of an arbitrary function $f(x)$ that takes the values $f_j = f(x_j)$ on a given set of $n+1$ distinct nodes x_j ($j = 0, 1, \ldots, n$). In practice, it is sometimes convenient to express the interpolant appearing in (2.8) in a different way. This new form will be particularly useful in the forthcoming chapters on numerical differentiation and integration.

We start with an example by interpolating linearly (i.e. with an interpolant of degree 1) a function that takes the values $f_0 = f(x_0)$ and $f_1 = f(x_1)$ at the nodes x_0 and x_1. According to expression (2.2) for the straight line passing through the points (x_0, y_0) and (x_1, y_1), this polynomial is

$$\Pi_1 f(x) = \frac{f_0 x_1 - f_1 x_0}{x_1 - x_0} + \frac{f_1 - f_0}{x_1 - x_0} x, \tag{2.9}$$

where we have just replaced y_0 and y_1 by f_0 and f_1 in (2.2), respectively. Another way of obtaining this interpolant would be to look for a suitable linear combination of polynomials in $\mathbb{R}_1[x]$ so that $\Pi_1 f(x_0) = f_0$ and $\Pi_1 f(x_1) = f_1$. This can be done by considering the linear combination

$$\Pi_1 f(x) = f_0 \ell_0(x) + f_1 \ell_1(x), \tag{2.10}$$

where $\ell_0(x)$ and $\ell_1(x)$ are the two linear polynomials

$$\ell_0(x) = \frac{x - x_1}{x_0 - x_1} \tag{2.11}$$

and

$$\ell_1(x) = \frac{x - x_0}{x_1 - x_0}, \tag{2.12}$$

shown in the figure below.

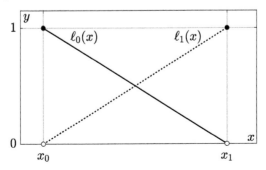

Since the polynomials $\ell_0(x)$ and $\ell_1(x)$ are the two straight lines satisfying

$$\ell_0(x_0) \equiv 1, \quad \ell_0(x_1) \equiv 0, \quad \ell_1(x_0) \equiv 0, \quad \text{and} \quad \ell_1(x_1) \equiv 1, \tag{2.13}$$

the conditions $\Pi_1 f(x_0) \equiv f_0$ and $\Pi_1 f(x_1) \equiv f_1$ required by the interpolant are automatically satisfied. We would like to emphasize that the interpolant appearing in (2.10) with $\ell_0(x)$ and $\ell_1(x)$ given by (2.11) and (2.12), respectively, is the same polynomial previously shown in (2.9), as can be checked by formal substitution and rearrangement of the factors:

$$\Pi_1 f(x) \equiv f_0 \frac{x - x_1}{x_0 - x_1} + f_1 \frac{x - x_0}{x_1 - x_0} \equiv \left(\frac{f_0}{x_0 - x_1} + \frac{f_1}{x_1 - x_0} \right) x + \frac{f_0 x_1 - f_1 x_0}{x_1 - x_0}.$$

The reader may wonder how the functions $\ell_0(x)$ and $\ell_1(x)$ were designed in order to satisfy the conditions (2.13). To better illustrate their origin, let us expand this method for the case $n \equiv 2$. Suppose we have to find the interpolant $\Pi_2 f(x)$ of $f(x)$ passing through the points (x_0, f_0), (x_1, f_1), and (x_2, f_2). Let us write the polynomial as the linear combination

$$\Pi_2 f(x) \equiv f_0 \ell_0(x) + f_1 \ell_1(x) + f_2 \ell_2(x), \tag{2.14}$$

where $\ell_0(x)$, $\ell_1(x)$, and $\ell_2(x)$ are suitable quadratic polynomials such that $\Pi_2 f(x_j) \equiv f_j$ for $0 \leq j \leq 2$. We can see that these three constraints are automatically satisfied if $\ell_i(x_j) \equiv 1$ if $i \equiv j$ and $\ell_i(x_j) \equiv 0$ if $i \neq j$. For example, $\ell_0(x)$ can be easily constructed from the product of the binomials $(x - x_1)$ and $(x - x_2)$

$$\ell_0(x) \equiv C \, (x - x_1)(x - x_2),$$

so that it cancels at the nodes x_1 and x_2. The constant C is a suitable normalization factor so that $\ell_0(x_0) \equiv 1$. In this case, this factor has the value

$$C \equiv \frac{1}{(x_0 - x_1)(x_0 - x_2)}.$$

Proceeding similarly with $\ell_1(x)$ and $\ell_2(x)$, we obtain the polynomials

$$\ell_0(x) \equiv \frac{(x - x_1)(x - x_2)}{(x_0 - x_1)(x_0 - x_2)}, \tag{2.15}$$

$$\ell_1(x) \equiv \frac{(x - x_0)(x - x_2)}{(x_1 - x_0)(x_1 - x_2)}, \tag{2.16}$$

and

$$\ell_2(x) \equiv \frac{(x - x_0)(x - x_1)}{(x_2 - x_0)(x_2 - x_1)}, \tag{2.17}$$

depicted in the figure below.

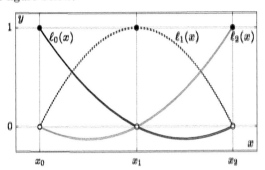

As an example, let us find the interpolant of $f(x) \equiv \cos\left(\frac{\pi}{2}x\right)$ at the nodes $x_0 \equiv -1$, $x_1 \equiv 0$, and $x_2 = 1$, where the function takes the values $f_0 \equiv 0$, $f_1 \equiv 1$, and $f_2 \equiv 0$, respectively. The interpolant in this case is

$$\Pi_2 f(x) \equiv 0 \cdot \ell_0(x) + 1 \cdot \ell_1(x) + 0 \cdot \ell_2(x) \equiv \ell_1(x),$$

and, since $\ell_1(x) \equiv (x - x_0)(x - x_2)(x_1 - x_0)^{-1}(x_1 - x_2)^{-1} \equiv -(x + 1)(x - 1)$,

$$\Pi_2 f(x) \equiv 1 - x^2.$$

The reader may easily check that this polynomial could also have been obtained using the method of undetermined coefficients described in Section 2.2 for the interpolant $\Pi_2 f(x) \equiv a_0 + a_1 x + a_2 x^2$. These coefficients are obtained by solving system (2.5), which in this case reads

$$\begin{bmatrix} 1 & -1 & 1 \\ 1 & 0 & 0 \\ 1 & 1 & 1 \end{bmatrix} \begin{bmatrix} a_0 \\ a_1 \\ a_2 \end{bmatrix} = \begin{bmatrix} 0 \\ 1 \\ 0 \end{bmatrix},$$

and whose solution is $a_0 = 1$, $a_1 \equiv 0$, and $a_2 \equiv -1$, as expected.

For an arbitrarily given number of $n + 1$ distinct nodes $\{x_0, \ldots, x_n\}$ and corresponding ordinates $\{f_0, \ldots, f_n\}$, we can generalize this new form of factorization by expressing the interpolant as the linear combination

$$\Pi_n f(x) \equiv f_0 \ell_0(x) + f_1 \ell_1(x) + \cdots + f_n \ell_n(x) \equiv \sum_{i=0}^{n} f_i \ell_i(x) \in \mathbb{R}_n[x], \tag{2.18}$$

where $\ell_i(x)$ is the polynomial of degree n given by

$$\ell_i(x) = \frac{(x - x_0)(x - x_1) \cdots (x - x_{i-1})(x - x_{i+1}) \cdots (x - x_n)}{(x_i - x_0)(x_i - x_1) \cdots (x_i - x_{i-1})(x_i - x_{i+1}) \cdots (x_i - x_n)}, \quad (2.19)$$

or, in a more compact form,

$$\ell_i(x) = \prod_{\substack{k=0 \\ (k \neq i)}}^{n} \frac{x - x_k}{x_i - x_k} \quad (i = 0, \dots, n). \tag{2.20}$$

The polynomials $\ell_i(x)$ satisfy

$$\ell_i(x_j) = \delta_{ij} = \begin{cases} 1 & (i = j), \\ 0 & (i \neq j), \end{cases} \tag{2.21}$$

where we have used *Kronecker's* symbol δ_{ij}. Using property (2.21), the interpolant can be expressed in terms of the $\ell_i(x)$ polynomials, in a form that is usually known as the

---

Lagrange's Interpolation Formula:

$$\Pi_n f(x) = \sum_{i=0}^{n} f_i \ell_i(x), \quad \text{with } \ell_i(x) = \prod_{\substack{k=0 \\ (k \neq i)}}^{n} \frac{x - x_k}{x_i - x_k} \quad (i = 0, \dots, n). \tag{2.22}$$

This formula was published in 1794 by the French mathematician Joseph-Louis Lagrange (1736–1813), also known by his contributions to the field of variational calculus and analytical mechanics.[3] For the same reason, the polynomials $\ell_i(x)$ appearing in (2.19) or in (2.20) are usually known as *Lagrange's cardinal polynomials*.

As we will see, Lagrange's formula (2.22) will be extremely useful to obtain many important properties of interpolants. However, in practice, formula (2.22) is rarely used since it is computationally *very expensive*. To better understand why, let us estimate the number of products required to evaluate $\Pi_n f(x)$. At a given arbitrary abscissa x, the numerator of each one of the terms $f_i \ell_i(x)$ appearing inside the sum of (2.22) requires to perform the n products $f_i(x - x_0)(x - x_1) \cdots (x - x_{i-1})(x - x_{i+1}) \cdots (x - x_n)$ involving the numerator of (2.19). Since the sum includes $n + 1$ terms of that type, we approximately

---

[3] However, it was Edward Waring, Lucasian Professor of Mathematics at the University of Cambridge, who first published this polynomial factorization in 1779.

need n^2 (neglecting n in front of n^2) products for its computation. The amount of work required for the computation of the products appearing in the denominators of (2.19) is the same as for the numerators. However, if we want to evaluate $\ell_i(x)$ at a different value of x (which is precisely one of the main purposes of interpolation), we *do not* need to compute the denominators again; they only need to be computed once and stored for future evaluations. For this reason, the computational cost of the denominators is usually neglected. As a result, it is said that Lagrange's formula has a quadratic computational cost, or $O(n^2)$, i.e. doubling the number of nodes quadruples the number of products required for its computation. In Section 2.4, we will see that there are other factorizations of the interpolant requiring just $O(n)$ operations for its evaluation.

2.3.1 Equispaced Grids

Previously, we interpolated the function $\cos(\pi x/2)$ at the set of *equidistant* nodes $\{x_0, x_1, x_2\} = \{-1, 0, 1\}$, that is, x_1 is at the same distance from x_0 as it is from x_2. In general, a set of nodes $\{x_0, x_1, \ldots, x_n\}$ satisfying $x_{j+1} - x_j = h$, $(j = 0, 1, \ldots, n-1)$ for some positive distance h is said to be an *equispaced grid*. For a generic closed interval $[a, b]$, with $x_0 = a$, $x_n = b$, and $h = (b-a)/n$, the corresponding equispaced distribution of nodes is explicitly given by the expression

$$x_j = a + \frac{b-a}{n}j = a + hj, \quad (j = 0, 1, \ldots, n), \tag{2.23}$$

as shown in Figure 2.1. Equispaced grids for a small or even moderate[4] number of nodes appear quite frequently in different areas of numerical mathematics, such as in the design of differentiation and quadrature rules for approximating derivatives and integrals, respectively, as we will see in the next two chapters.

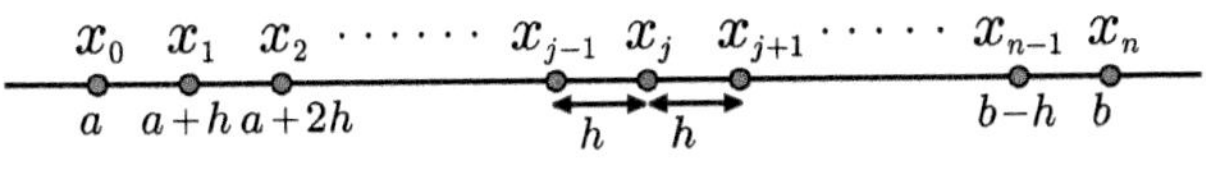

Figure 2.1 Equispaced grid $x_j = a + hj$, $(j = 0, 1, \ldots, n)$ with $h = (b-a)/n$.

---

4 The reason for this limitation will be explained in the next section.

```
% Code 3: Lagrange cardinal polynomials (equispaced nodes)
function [P,xn,z] = cardpolequi(a,b,n,m)
% Input: 1. a & b: interval [a,b]
%        2. n: with x_0 = a and x_n = b (n+1 interp nodes)
%        3. m: number of points in dense grid
%
% Output: 1. P: matrix of card. polynomials.
%         2. xn: equispaced interpolation nodes
%         3. z: dense grid
xn = a + [0:n]'*(b-a)/n; z = linspace(a,b,m)';

% Matrix containing cardinal polynomials
P = ones(m,n+1) ;
for jj = 0:n
    knj = [0:jj-1 jj+1:n];
    for kk = knj
    P(:,jj+1) = P(:,jj+1).*(z-xn(kk+1))/(xn(jj+1)-xn(kk+1));
    end
end
% If fn is a column vector containing f(x) at the nodes xn,
% then P*fn provides the interpolant P_n f(z) on the dense
% grid z (see Practical 3 for details)
```

For a given interval $[a, b]$ and two positive integers n and m, with $m \gg n$, Code 3 provides the Lagrange cardinal polynomials $\ell_i(x)$ defined in (2.20) associated with the equispaced set of nodes $\{x_0, x_1, \ldots, x_n\}$ introduced in (2.23). These $n + 1$ polynomials are evaluated at a denser grid of equidistant abscissas $\{z_1, z_2, \ldots, z_m\}$. This second grid is just intended for numerical inspection of the behavior of the cardinal polynomials between neighboring nodes x_j and $x_{j \pm 1}$. Code 3 returns the vectors **xn** and **z** containing the set of $n + 1$ interpolation nodes $\{x_j\}$ and m inspection abscissas $\{z_k\}$, for $j \equiv 0, 1, \ldots, n$ and $k \equiv 1, 2, \ldots, m$, respectively. Code 3 also returns an $m \times (n + 1)$ matrix **P** containing the evaluated cardinal polynomials at the denser grid $\{z_k\}$. Practical 2.1 explains how this matrix can be used to evaluate the interpolant $\Pi_n f(x)$ of a given function $f(x)$ at the inspection grid.

Practical 2.1: Lagrange Equispaced Interpolation

The main goal of this practical is to compute the Lagrange's interpolant $\Pi_n f(x)$ of an arbitrary function $f(x)$ within the interval $[-1, 1]$ using its values $f_j = f(x_j)$ at the nodes $x_j = -1 + 2j/n$, for $j = 0, 1, \ldots, n$ (gray circles in the figure below.) To explore the behavior of the interpolation, we need to evaluate the interpolant on a denser grid (hollow circles) of equally spaced points $\{z_1, \ldots, z_m\} \in [-1, 1]$ ($m = 200$ is enough).

(a) First build the $m \times (n + 1)$ matrix whose columns are the cardinal polynomials evaluated at the abscissas z_k.

$$\mathbb{P} = \begin{bmatrix} \ell_0(z_1) & \ell_1(z_1) & \cdots & \ell_n(z_1) \\ \ell_0(z_2) & \ell_1(z_2) & \cdots & \ell_n(z_2) \\ \vdots & \vdots & \ddots & \vdots \\ \ell_0(z_m) & \ell_1(z_m) & \cdots & \ell_n(z_m) \end{bmatrix}, \quad \text{with } \ell_i(z) = \prod_{\substack{k=0 \\ (k \neq i)}}^{n} \frac{z - x_k}{x_i - x_k}.$$

(b) For $n = 3$, 6, and 9, plot the polynomials $\ell_i(z)$ on the grid z_k and check that $\ell_i(x_j) = \delta_{ij}$.

The matrix $\mathbb{P}$ is particularly useful to evaluate the interpolant $\Pi_n f(x)$ on the denser grid z_k:

$$\begin{bmatrix} \Pi_n f(z_1) \\ \Pi_n f(z_2) \\ \vdots \\ \Pi_n f(z_m) \end{bmatrix} = f_0 \begin{bmatrix} \ell_0(z_1) \\ \ell_0(z_2) \\ \vdots \\ \ell_0(z_m) \end{bmatrix} + f_1 \begin{bmatrix} \ell_1(z_1) \\ \ell_1(z_2) \\ \vdots \\ \ell_1(z_m) \end{bmatrix} + \cdots + f_n \begin{bmatrix} \ell_n(z_1) \\ \ell_n(z_2) \\ \vdots \\ \ell_n(z_m) \end{bmatrix} = \mathbb{P} \begin{bmatrix} f_0 \\ f_1 \\ \vdots \\ f_n \end{bmatrix}.$$

(c) For $n = 4, 8$, and 16, interpolate $f(x) = e^x$ and plot the interpolant $\Pi_n f(z_k)$ on the finer grid.

2.4 Barycentric Interpolation

In the Sections 2.2 and 2.3, we saw two different ways of computing the interpolant $\Pi_n f(x)$ resulting from the evaluation of a given function $f(x)$ on a set of distinct nodes $\{x_0, \ldots, x_n\}$. The first interpolatory formula (2.3) consists of a linear combination of the monomial basis $\{1, x, x^2, \ldots, x^n\}$ with undetermined coefficients that can only be obtained if we solve Vandermonde's system of linear equations (2.5), whose solution is sometimes extremely inaccurate (as we will see in the practicals).[5] The second method leads to Lagrange's formula (2.22) consisting of a linear combination of the cardinal basis $\{\ell_0(x), \ell_1(x), \ldots, \ell_n(x)\}$ whose evaluation requires approximately n^2 operations. In what follows, we look for numerically reliable and computationally more efficient factorizations of the interpolant $\Pi_n f(x)$. In the process of obtaining this alternative interpolatory formula, we will introduce new concepts and obtain intermediate results that will be very useful in the forthcoming chapters.

We start our analysis by reviewing some mathematical properties of polynomials. According to the fundamental theorem of algebra, an arbitrary polynomial $p(x) \in \mathbb{R}_n[x]$ has at most n roots (if they are counted with their multiplicities). As a consequence, a real polynomial $p(x)$ of degree less than or equal to n will vanish, at most, at n different locations on the real axis. Therefore, if a polynomial $p(x) \in \mathbb{R}_n[x]$ vanishes at more than n different abscissas on the real axis, this polynomial must necessarily be the zero polynomial, i.e. $p = 0$.

Consider two polynomials $p(x)$ and $q(x)$ in $\mathbb{R}_n[x]$ that take the same values on a set of distinct nodes $\{x_0, x_1, \ldots, x_n\}$ on the real axis, that is,

$$p(x_j) = q(x_j) \quad (j = 0, \ldots, n).$$

Since the difference polynomial $r(x) = p(x) - q(x) \in \mathbb{R}_n[x]$ vanishes at $n + 1$ distinct points, i.e. $r(x_j) = 0 \quad (j = 0, \ldots, n)$, this polynomial must be the zero polynomial $r = 0$ or, equivalently, p and q must be the same polynomial.

The properties just seen before lead to the apparently irrelevant conclusion,

$$\Pi_n p(x) = p(x), \quad \forall p(x) \in \mathbb{R}_n[x]. \tag{2.24}$$

In other words, the interpolant $\Pi_n p(x)$ of any polynomial $p(x)$ in $\mathbb{R}_n[x]$ is itself. The reason is that by construction, $\Pi_n p(x_j) = p(x_j)$ at the $n + 1$ distinct real nodes $\{x_0, x_1, \ldots, x_n\}$ and, therefore, $\Pi_n p(x)$ and $p(x)$ must be the same polynomial. In particular, if we interpolate the unit polynomial $p(x) = 1$, we should obtain $\Pi_n p(x) = 1$ or, in terms of Lagrange's interpolatory formula (2.22) with

---

5 In Part II of this book, we will provide a proper explanation of why the solution of large Vandermonde's linear systems may be numerically inaccurate.

$$f_j = p(x_j) = 1,$$

$$\sum_{i=0}^{n} \ell_i(x) = 1. \tag{2.25}$$

Equation (2.25) is usually known as Cauchy's identity and is a general property satisfied by the cardinal polynomials associated with any arbitrary set of $n+1$ distinct nodes. We will numerically explore this property in Practical 2.2.

The Lagrange's cardinal polynomials,

$$\ell_i(x) = \frac{(x - x_0)(x - x_1) \cdots (x - x_{i-1})(x - x_{i+1}) \cdots (x - x_n)}{(x_i - x_0)(x_i - x_1) \cdots (x_i - x_{i-1})(x_i - x_{i+1}) \cdots (x_i - x_n)},$$

can be compactly written by introducing two new quantities. The first quantity is the normalizing factor appearing in the denominator of $\ell_i(x)$

$$\lambda_i = \frac{1}{\displaystyle\prod_{\substack{k=0 \\ (k \neq i)}}^{n} (x_i - x_k)}, \tag{2.26}$$

usually known as *barycentric weight*, whereas the second is the $(n+1)$-th degree polynomial appearing in the numerator

$$\ell(x) = (x - x_0)(x - x_1) \cdots (x - x_n) = \prod_{k=0}^{n} (x - x_k) \in \mathbb{R}_{n+1}[x], \tag{2.27}$$

known as *nodal polynomial* associated with the set of distinct nodes $\{x_0, x_1, \ldots, x_n\}$. Since the numerator of $\ell_i(x)$ appearing in (2.19) contains all the binomials $(x - x_k)$ for $k \neq i$, we can write the cardinal function $\ell_i(x)$ as

$$\ell_i(x) = \lambda_i \frac{\ell(x)}{x - x_i}. \tag{2.28}$$

Substituting (2.28) in (2.22) and factorizing $\ell(x)$ outside of the sum, we obtain the so-called

First Barycentric Interpolation Formula:

$$\Pi_n f(x) = \ell(x) \sum_{i=0}^{n} \frac{f_i \lambda_i}{x - x_i}, \quad \text{with } \lambda_i = \frac{1}{\displaystyle\prod_{\substack{k=0 \\ (k \neq i)}}^{n} (x_i - x_k)}. \tag{2.29}$$

This interpolatory form cannot be evaluated at any of the nodes x_j (where the polynomial precisely takes the known values f_j). The computational cost of (2.29) is $O(n)$, since evaluating the nodal polynomial $\ell(x)$ requires n products and the sum also involves $n + 1$ products and divisions. There is an even more efficient factorization that does not require evaluation of the nodal polynomial. First of all, notice that formal substitution of the cardinal polynomial (2.28) in Cauchy's identity (2.25) leads to

$$\sum_{i=0}^{n} \frac{\lambda_i \ell(x)}{x - x_i} \equiv \ell(x) \sum_{i=0}^{n} \frac{\lambda_i}{x - x_i} \equiv 1, \tag{2.30}$$

and therefore the nodal polynomial can be expressed as $\ell(x) = \left[\sum_{i=0}^{n} \frac{\lambda_i}{x - x_i} \right]^{-1}$.

Introducing this last expression in (2.29) leads to what is usually known as

Second Barycentric Interpolation Formula:

$$\Pi_n f(x) \equiv \frac{\displaystyle\sum_{i=0}^{n} \frac{f_i \lambda_i}{x - x_i}}{\displaystyle\sum_{i=0}^{n} \frac{\lambda_i}{x - x_i}}. \tag{2.31}$$

The reader interested in more details regarding the origins, accuracy, and other aspects of the barycentric formulas (2.29) and (2.31) is referred to the recommended readings at the end of the chapter.

For a given distribution of distinct nodes $\{x_0, x_1, \ldots, x_n\}$, their associated $n + 1$ barycentric weights $\{\lambda_0, \lambda_1, \ldots, \lambda_n\}$ can be obtained using the definition (2.26), which generally requires $O(n)$ products per weight. Fortunately, the barycentric weights can sometimes be calculated explicitly for some particular sets of nodes. In particular, for nodes equally spaced in $[a, b]$ according to (2.23), the barycentric weights are (see Problem 5)

$$\lambda_j \equiv \frac{n^n}{(b - a)^n} \frac{(-1)^{n-j}}{n!} \binom{n}{j}. \tag{2.32}$$

Interpolation Forms (Summary)

For a given set of $n+1$ distinct nodes $\{x_0, x_1, \ldots, x_n\}$, and corresponding ordinates of the function $f(x)$ at these nodes $f_j \equiv f(x_j)$, the unique interpolant $\Pi_n f(x)$ of degree n that takes the values f_j at the nodes x_j can be expressed in any of the following forms:

$$\Pi_n f(x) \equiv \begin{cases} a_0 + a_1 x + a_2 x^2 + \cdots + a_n x^n & \text{(monomial)} \\[2mm] \sum_{i=0}^{n} f_i \ell_i(x) & \text{(Lagrange)} \\[2mm] \ell(x) \sum_{i=0}^{n} f_i \lambda_i (x - x_i)^{-1} & \text{(first barycentric)} \\[2mm] \left[\sum_{i=0}^{n} f_i \lambda_i (x - x_i)^{-1} \right] \left[\sum_{i=0}^{n} \lambda_i (x - x_i)^{-1} \right]^{-1} & \text{(second barycentric)} \end{cases}$$

where the vector of coefficients $[a_0\ a_1\ a_2\ \cdots\ a_n]^{\mathrm{T}}$ is the solution of the linear system (2.5) with $y_j = f_j$ ($j = 0, 1, \ldots, n$), and where

$$\lambda_i^{-1} = \prod_{\substack{k=0 \\ (k \neq i)}}^{n} (x_i - x_k) \quad \text{(barycentric weights)}$$

$$\ell(x) = \prod_{k=0}^{n} (x - x_k) \quad \text{(nodal polynomial)}$$

and

$$\ell_i(x) = \lambda_i \prod_{\substack{k=0 \\ (k \neq i)}}^{n} (x - x_k) \quad \text{(cardinal polynomials)}$$

Barycentric forms cannot be evaluated at any of the interpolation nodes x_j, where the interpolant precisely adopts the value f_j.

2.5 Convergence of the Interpolation Method

So far, we have just seen constructive methods to devise interpolants of functions in different ways, without mentioning how accurate these interpolants are when compared with the original function they approximate. For example, in Section 2.3 we interpolated the function $f(x) \equiv \cos(\pi x/2)$ using the nodes $\{-1, 0, 1\}$. Figure 2.2a shows the resulting interpolant $\Pi_2 f(x) \equiv 1 - x^2$, which is a fair qualitative approximation of the trigonometric function. Figure 2.2b shows on a semi-log plot the discrepancy between $f(x)$ and

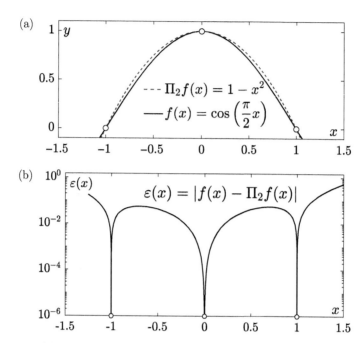

Figure 2.2 (a) Trigonometric function $f(x) = \cos(\pi x/2)$ (black solid curve) and the interpolant $\Pi_2 f(x) = 1 - x^2$ (dashed curve) at the nodes $\{-1, 0, 1\}$. (b) Pointwise discrepancy between $f(x)$ and $\Pi_2 f(x)$.

$\Pi_2 f(x)$, measured by the absolute error function $\varepsilon(x) = |f(x) - \Pi_2 f(x)|$. As expected, the function $\varepsilon(x)$ is zero at the nodes, and it grows as we move away from them.

This section deals with two fundamental questions of polynomial approximation of functions. The first question is, *can a function be always approximated using polynomials?* If the answer is affirmative, then the second question is, *how close is the interpolant $\Pi_n f(x)$ to the original function $f(x)$ it interpolates?* In this part of the book we will simply state, without proof, two fundamental results from the mathematical theory of approximation. The first result is as follows:

Result I: Weiertrass' Approximation Theorem

Let $f(x)$ be a continuous function over the interval $[a, b]$. Then, for any $\epsilon > 0$ there exists an integer $n = n(\epsilon)$ and a polynomial $p(x) \in \mathbb{R}_n[x]$ such that

$$|f(x) - p(x)| < \epsilon,$$

for all x in $[a, b]$.

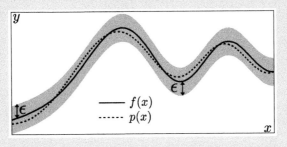

This theorem states that for any prescribed tolerance ϵ, we can always find a polynomial $p(x)$ of suitable degree n (dashed curve in the figure above) that differs, in absolute value, from the original function (solid curve) by less than ϵ (gray band of thickness 2ϵ). The second result describes the accuracy of our interpolants:

Result II: Cauchy's Remainder for Interpolation

Let $f(x)$ have an $(n+1)$th derivative, $f^{(n+1)}(x)$, in the interval $I = [a, b]$, and let $\Pi_n f(x)$ be the interpolant of $f(x)$ on the distinct nodes $\{x_0, x_1, \ldots, x_n\} \in I$. Then, for all $x \in I$ there exists a point $\xi(x)$ within the interval $[\min\{x_0, \ldots, x_n, x\}, \max\{x_0, \ldots, x_n, x\}]$ such that

$$R_n(x) \equiv f(x) - \Pi_n f(x) = \frac{f^{(n+1)}(\xi)}{(n+1)!} \ell(x), \tag{2.33}$$

where $\ell(x)$ is the nodal polynomial $\ell(x) = (x - x_0)(x - x_1) \cdots (x - x_n)$.

The quantity $R_n(x)$ is known as *Cauchy's remainder*. Equation (2.33) provides a quantitative measure of the discrepancy between the function and the interpolant, sometimes referred to as the *pointwise interpolation error*. Obviously, the error is zero at the nodes since the nodal polynomial vanishes precisely at those points. But Cauchy's remainder formula also reveals two more aspects that are worth mentioning. The first is the factorial term $(n+1)!$

appearing in the denominator, thus indicating that the error should quickly decrease when increasing the number of nodes used in the interpolation. The second is the factor $f^{(n+1)}(\xi)$ appearing on the numerator that, if large in absolute value, might potentially deteriorate the accuracy of our interpolation.

For an equispaced node distribution $\{x_0, x_1, \ldots, x_n\} \in I$ satisfying $|x_{j+1} - x_j| = h$, it can be shown (see Problem 3) that $|\ell(x)| \leq n!h^{n+1}/4$, so that the absolute value of the pointwise interpolation error (2.33) is bounded by

$$|f(x) - \Pi_n f(x)| \leq \max_{\xi \in I}|f^{(n+1)}(\xi)| \frac{h^{n+1}}{4(n+1)}. \tag{2.34}$$

This bound (2.34) tells us that linear interpolations ($n = 1$) lead to errors that are proportional to h^2, whereas quadratic interpolations ($n = 2$) lead to errors proportional to h^3, for example. That is, halving h reduces the error by a factor of 4 and a factor of 8 in linear and quadratic interpolations, respectively. In fact, for an arbitrary equispaced grid (2.23) over the interval $[a, b]$, the spacing is $h = (b - a)/n$, that is, $h = O(n^{-1})$. As a consequence, when n tends to infinity, a simple calculation shows that the interpolation error (2.34) should tend to zero as $O(n^{-(n+2)})$, that is, polynomial interpolation should converge very quickly when increasing n. However, we have so far ignored the effect of the term appearing in (2.34) corresponding to the maximum absolute derivative $|f^{(n+1)}(\xi)|$. For functions where this term is bounded, polynomial interpolation converges rapidly for moderate values of n. For functions where this term grows significantly with n, the convergence may be at stake, as the following counterexample shows.

2.5.1 Runge's Counterexample

Consider the interpolation of the function

$$f(x) \equiv \frac{1}{1 + 25x^2} \tag{2.35}$$

within the interval $[-1, 1]$ using equally spaced sets of nodes given by (2.23):

$$x_j \equiv -1 + \frac{2j}{n}, \quad (j \equiv 0, 1, \ldots, n). \tag{2.36}$$

The resulting interpolants for $n \equiv 7, 14$, and 28 are shown in Figure 2.3. For $n = 7$ (Figure 2.3a), we can see that the interpolant fails to reproduce the flat shape of the original function near the edges of the interval. Doubling n (Figure 2.3b) remarkably improves the interpolation in the interior region of the interval, where the differences between the interpolant and the function are not visible to the naked eye. However, the interpolant still exhibits large oscillations close to the boundaries of the interval. Increasing n further just worsens things (see Figure 2.3c for $n \equiv 28$) because the oscillations of the interpolant at the

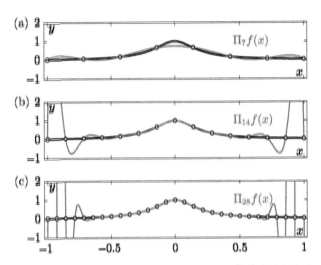

Figure 2.3 Interpolation of the function $f(x) \equiv (1 + 25x^2)^{-1}$ (solid black curve) using equispaced nodes (2.36) for $n = 7$ (a), $n = 14$ (b), and $n = 28$ (c). The resulting interpolants $\Pi_n f(x)$ are the gray solid curves.

edges, far from being mitigated, exhibit even larger overshootings between the nodes. The reader may check that these oscillations exhibited by the interpolant keep growing when n is increased further, thus indicating that

$$\lim_{n \to \infty} |f(x) = \Pi_n f(x)| \neq 0 \tag{2.37}$$

near the boundaries of the interpolation interval $I \equiv [-1, 1]$. Summing up, whereas increasing n makes the interpolant to converge pointwise to the function in the central region of the interval, it also leads to a systematical divergence in some regions near the boundaries. This unexpected behavior is usually known as *Runge's phenomenon* or *Runge's counterexample*, as it was described in 1910 by the German mathematician Carl David Tolmé Runge.[6]

The origin of the lack of convergence is that Runge's function (2.35) has high order derivatives with *extremely* large absolute values. Table 2.1 contains these large values for the three cases plotted in Figure 2.3. Table 2.1 also includes the factor $h^{n+1}/4(n+1)$ arising from the bound of the absolute nodal polynomial $|\ell(x)| \leq n! h^{n+1}/4$ for equispaced nodes. A simple inspection of the orders of magnitude of the numbers shown in Table 2.1 clearly shows that the high absolute derivatives overweigh the factor obtained from the nodal polynomial bound, therefore leading to a divergence of the interpolating polynomial.

---

6 Also known by the Runge–Kutta method for the numerical integration of differential equations, or by the Laplace–Runge–Lenz vector of the two-body problem of classical mechanics.

Table 2.1 Runge's counterexample.

n	7	14	28
$\max\lvert f^{(n+1)}(\xi)\rvert$	1.6×10^{10}	3.7×10^{22}	1.6×10^{51}
$h^{n+1}/4(n+1)$	1.4×10^{-6}	3.5×10^{-15}	5×10^{-36}

a) Divergence of equispaced polynomial interpolation of function (2.35) with $h = 2/n$.

From this example it is logical to think that polynomial interpolation is not a reliable tool for approximating functions. However, we must emphasize that the divergent interpolatory polynomials seen in Runge's counterexample have been obtained in a very particular way – *using equidistant nodes*. Although an equispaced distribution of nodes may seem the natural choice, it leads to a nodal polynomial $\ell(x)$ with an absolute bound that cannot mitigate large values of $\lvert f^{(n+1)}(\xi)\rvert$. In Section 2.7, we will see how to solve this problem simply by using suitable *non-equispaced* distributions of nodes with associated nodal polynomials that have optimal bounds, resulting in an efficient damping effect on large values of $\lvert f^{(n+1)}(\xi)\rvert$ when increasing n. However, it is worth mentioning that equispaced interpolation applied on many other functions *does not always* lead to catastrophic behavior such as the one shown in Figure 2.3. For example, equispaced interpolation of the function $(1 + x^2)^{-1}$ converges perfectly on the same interval, without any trace of wild oscillations. In general, for *moderate* values of n, equispaced interpolation works perfectly at the center of the interval. For this reason, equispaced interpolation is widely used in many applications of numerical mathematics, as we will see in Chapters 3 and 4.

Because of Runge's counterexample, polynomial interpolation has traditionally had a very bad reputation as a reliable numerical method to approximate functions. The main reason is that, for many decades, polynomial interpolation has frequently been mistakenly preconceived as *equispaced polynomial interpolation*. As a consequence, scientific computing literature often advises not to use polynomial interpolation when n is large, mainly motivated by Runge's counterexample. Misleading warnings such as *"Polynomial interpolation should be carried out with the fewest feasible number of data points," "The issues with Lagrange interpolation are numerous (...) The procedure is also highly unstable numerically, especially for datasets with sizes over 20 points,"* or *" Although the interpolation is exact at the given data points, a high-order polynomial can vary wildly between the specified points,"* still appear in many surveys. However, as we will see later in this chapter, we can perfectly work with very large number of nodes and high degree polynomials that do not exhibit fluctuations and that converge to any continuous function.

2.6 Conditioning of an Interpolation

The divergence of the interpolants described in the Section 2.5 is not related to a lack of numerical accuracy. In exact arithmetic, there are functions for which the interpolation error tends to zero when increasing n and there are also functions (such as Runge's counterexample) that have divergent interpolation error. In practical computations, however, polynomial interpolation can also be very sensitive to inaccuracies of the input data, i.e. the sampling values f_j. This section will be focused on the measurement of this sensitivity known as *conditioning of an interpolation*.

In what follows, we consider the interpolation of an arbitrary continuous function $f(x)$ within the interval $I = [-1, 1]$ using a set of distinct nodes $\{x_0, x_1, \ldots, x_n\} \in I$. For simplicity, assume that $|f(x)| \leq 1$ within the interpolation interval.[7] Recall that Lagrange's interpolation form (2.22) of the interpolant of $f(x)$ based on the values $f_j = f(x_j)$ at the nodes is

$$\Pi_n f(x) = \sum_{j=0}^{n} f_j \ell_j(x). \tag{2.38}$$

For our analysis, we need a couple of definitions:

Lebesgue Function and Lebesgue Constant: the Lebesgue function associated with the nodes $\{x_0, x_1, \ldots, x_n\} \in [-1, 1]$ is

$$\lambda_n(x) = \sum_{j=0}^{n} |\ell_j(x)|. \tag{2.39}$$

The Lebesgue constant corresponding to $\lambda_n(x)$ is

$$\Lambda_n = \max_{x \in [-1,1]} \lambda_n(x). \tag{2.40}$$

Since $|f(x)| \leq 1$ for all x in the interval I, the Lebesgue function is an upper bound on the absolute value of the interpolant (2.38):

$$|\Pi_n f(x)| = \left| \sum_{j=0}^{n} f_j \ell_j(x) \right| \leq \sum_{j=0}^{n} |f_j| |\ell_j(x)| \leq \sum_{j=0}^{n} |\ell_j(x)| = \lambda_n(x). \tag{2.41}$$

---

7 If not, normalize the function with respect to its maximum absolute value $\max_{x \in [-1,1]} |f(x)|$.

Lagrange's form (2.38) assumes that the values of the function at the nodes $f_j \equiv f(x_j)$ are *exact*. However, the quantity f_j is always inaccurate due to limited computer precision or errors in data measurement, so that its actual value is

$$\widetilde{f}_j = f_j + \Delta f_j, \quad (j = 0, 1, \ldots, n), \tag{2.42}$$

where f_j is the exact value $f(x_j)$ and Δf_j its deviation due to error. For simplicity, we will assume that the absolute deviations are bounded, i.e. $|\Delta f_j| < \varepsilon$, where $\varepsilon > 0$ is some positive constant representing the accuracy of our input data.[8] The Lagrange interpolant associated with the inaccurate data $\widetilde{f}_j$ is

$$\widetilde{\Pi}_n f(x) = \sum_{j=0}^{n} \widetilde{f}_j \ell_j(x) = \sum_{j=0}^{n} (f_j + \Delta f_j) \ell_j(x). \tag{2.43}$$

Assuming ε to be a very small quantity, the question is whether the discrepancy between the perturbed $\widetilde{\Pi}_n f(x)$ and the exact $\Pi_n f(x)$ interpolants is also small. In other words, we wonder about the sensitivity of the interpolant with respect to very small perturbations in the data values. The absolute difference between the exact and inexact interpolants can be bounded as follows:

$$|\widetilde{\Pi}_n f(x) - \Pi_n f(x)| = \left| \sum_{j=0}^{n} (f_j + \Delta f_j) \ell_j(x) - \sum_{j=0}^{n} f_j \ell_j(x) \right|$$

$$= \left| \sum_{j=0}^{n} \Delta f_j \ell_j(x) \right| \leq \sum_{j=0}^{n} |\Delta f_j| |\ell_j(x)|$$

$$\leq \varepsilon \sum_{j=0}^{n} |\ell_j(x)| = \varepsilon \lambda_n(x). \tag{2.44}$$

By virtue of (2.40), an upper bound of the maximum discrepancy between both polynomials is given by the inequality

$$\max_{x \in I} |\widetilde{\Pi}_n f(x) - \Pi_n f(x)| \leq \varepsilon \Lambda_n, \tag{2.45}$$

therefore implying that small perturbations in the input data f_j can potentially be amplified if the Lebesgue constant Λ_n is large, thus leading to large discrepancies between the interpolants (2.38) and (2.43). In other words, the condition

---

8 This constant is usually taken as the machine epsilon in double precision $= \varepsilon \approx 10^{-16}$.

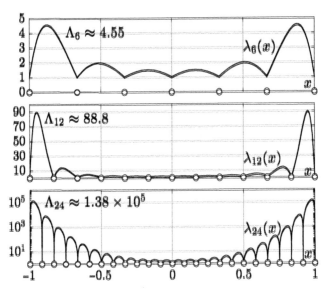

Figure 2.4 Lebesgue functions $\lambda_n(x)$ associated with sets of equispaced nodes (2.36) within the interval $[-1, 1]$.

number of an interpolation procedure based on a set of nodes $\{x_0, x_1, \ldots, x_n\}$ is its associated Lebesgue constant given by (2.40), that is,

$$K \equiv \Lambda_n.$$
(2.46)

In order to understand the numerical implications of these theoretical results, Figure 2.4 shows the Lebesgue functions associated with sets of equispaced nodes (2.36) within the interval $[-1, 1]$. Figure 2.4 also includes the corresponding Lebesgue constant for each case, showing a clear exponential growth of this constant when increasing n (notice the logarithmic scale used for $n \equiv 24$). For equally spaced points, the Lebesgue constant grows following the asymptotic law[9]

$$\Lambda_n^{\text{Equi.}} \propto \frac{2^n}{n \log n}.$$
(2.47)

Figure 2.5 shows the result of interpolating $f(x) \equiv e^x$ within the interval $[-1, 1]$ using equispaced interpolations up to $n \equiv 60$. In particular, Figure 2.5a shows

---

9 We refer the reader to the Complementary Reading section at the end of this chapter for details.

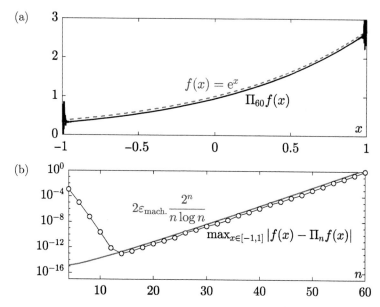

(a)

$f(x) = e^x$

$\Pi_{60} f(x)$

x

(b)

$2\varepsilon_{\text{mach}} \cdot \dfrac{2^n}{n \log n}$

$\max_{x \in [-1,1]} |f(x) - \Pi_n f(x)|$

n

Figure 2.5 Interpolation of $f(x) = e^x$ with equispaced nodes. (a) $f(x)$ (solid black) and interpolant $\Pi_n f(x)$ (dashed gray) for $n = 60$ (since both curves graphically overlap, the interpolant has deliberately shifted slightly down to ease visualization). (b) Maximum discrepancy of $\varepsilon(x) = |f(x) - \Pi_n f(x)|$ as a function of n.

the original function (gray dashed) and the resulting interpolant for $n = 60$ (solid black)[10] exhibiting large errors due to ill-conditioning near the edges of the interval. Figure 2.5b shows the maximum discrepancy between the original function and the interpolant (hollow circles) as a function of n, compared with the asymptotic law (2.47) using twice the machine epsilon as proportionality factor (solid gray). The reader can easily check that the discrepancy curve extrapolated backwards for $n = 4$ intercepts the vertical axis at an ordinate whose order of magnitude is precisely 10^{-16}. The solid gray curve corresponding to the asymptotic law has been extended for low values of n in order to guide the eye to the aforementioned interception point.

---

10 The interpolant curve has deliberately shifted slightly down for a better visualization.

Practical 2.2:   Conditioning of Equispaced Nodes

Consider the Lagrange matrix $\mathbb{P}$ previously obtained in Practical 2.1 for the same set of equispaced nodes and dense grid z_k.

(a) For $n = 8, 16, 24$, and 32, evaluate and plot the functions

$$u(z) = \sum_{j=0}^{n} \ell_j(z) \quad \text{and} \quad \lambda_n(z) = \sum_{j=0}^{n} |\ell_j(z)|$$

on the denser grid and confirm property (2.25), that is, confirm that $u(z_k) = 1$. For each n, reproduce Figure 2.4 and estimate the Lebesgue constants using the approximation $\Lambda_n \approx \max_{0 \le k \le m} \lambda_n(z_k)$.

(b) For $n = 60$, interpolate $f(x) = e^x$ and plot the interpolant $\Pi_n f(z_k)$ and the original function e^{z_k} both evaluated on the finer grid (Figure 2.5a). On a different semi-log plot, represent the pointwise error $\varepsilon_n(z_k) = |\Pi_n f(z_k) - e^{z_k}|$.

(c) For $n = 4, 6, 8, 10, \ldots, 60$, interpolate $f(x) = e^x$ as in part (b) and, for each value of n, compute the maximum pointwise error

$$\varepsilon_n = \max_{0 \le k \le m} |\Pi_n f(z_k) - e^{z_k}|,$$

and compare it with the asymptotic law $\sim 2^n / n \log(n)$ as in Figure 2.5b.

Optional: failure of the monomial basis using equispaced grids
Interpolate $f(x) = e^x$ on the same equispaced grid solving Vandermonde's system (2.5)

$$\begin{bmatrix} 1 & x_0 & x_0^2 & \cdots & x_0^n \\ 1 & x_1 & x_1^2 & \cdots & x_1^n \\ \vdots & \vdots & \vdots & & \vdots \\ 1 & x_n & x_n^2 & \cdots & x_n^n \end{bmatrix} \begin{bmatrix} a_0 \\ a_1 \\ \vdots \\ a_n \end{bmatrix} = \begin{bmatrix} f_0 \\ f_1 \\ \vdots \\ f_n \end{bmatrix}.$$

to obtain the coefficients a_j. Notice Matlab's warning message about the *ill-conditioning* of the matrix. The monomial basis using equispaced grids leads to inaccurate results.[11]

---

[11] Ill-conditioned matrices will be studied in detail in Part II.

2.7 Chebyshev's Interpolation

In Section 2.6, we saw that the main issue with using equispaced interpolation for large n is the order of magnitude of the Lebesgue constant Λ_n and, in particular, the extreme values that the Lebesgue function $\lambda_n(x)$ takes close to the edges of the interpolation interval. It is important to emphasize that this ill-conditioning is *independent* of the function to be interpolated, and *only* caused by the way the nodes are distributed. A natural question is therefore whether there are node distributions whose associated Lebesgue functions are better behaved, i.e. with low Lebesgue constants. The answer is affirmative and not unique, that is, there are *many* node distributions that have particularly well-behaved associated Lebesgue functions with very low or moderate Lebesgue constants. In this section we will describe just one of these distributions: the *Chebyshev*[12] nodes. The Chebyshev nodes[13] are the abscissas given by the expression:

$$x_j = \cos\left(\frac{j\pi}{n}\right), \quad (j = 0, 1, \ldots, n). \tag{2.48}$$

Geometrically, these nodes are the horizontal projections of the set of points located along the unit circle at the equidistant angles $\theta_j \equiv j\pi/n$, $(j = 0, 1, \ldots, n)$, as shown in Figure 2.6. Note that these nodes are sorted from right to left on the x-axis, since $x_0 \equiv \cos(0) \equiv 1$ and $x_n = \cos(\pi) \equiv -1$. Occasion-

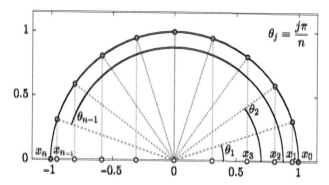

Figure 2.6 Chebyshev nodes (hollow circles) as the horizontal projection of equally distributed points along the unit circle (gray circles).

---

12 Pafnuty Chebyshev (1821–1894), Russian mathematician known for his contributions to probability and number theory.
13 Also known as Chebyshev nodes of *second kind* or Chebyshev *practical abscissas*.

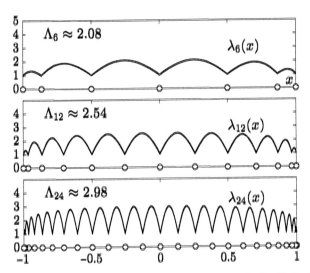

Figure 2.7 Lebesgue functions and constants associated with the Chebyshev nodes distribution (2.48), to be compared with the equispaced case shown in Figure 2.4 for the same values of n.

ally these nodes are defined as $x_j \equiv -\cos(j\pi/n)$ so that $x_j > x_i$, $(j > i)$, but it is not a standard practice.

In order to study the conditioning of the Chebyshev nodes, we have computed their associated Lebesgue functions and constants for the same increasing values of n as in Section 2.6 for the equispaced case. Figure 2.7 outlines the results. It can be observed that the Lebesgue constants are much smaller than in the equispaced case, and actually for n as large as 24 they still are of the order of unity. It can be shown that for the Chebyshev node distribution (2.48), the Lebesgue constant grows following the logarithmic asymptotic law:

$$\Lambda_n^{\text{Cheb.}} \propto \frac{2}{\pi} \log n. \tag{2.49}$$

Another advantage of the Chebyshev nodes is that their barycentric weights are particularly simple:

$$\lambda_j \equiv \begin{cases} (-1)^j \dfrac{2^{n-1}}{n} & (j = 1, 2, \ldots, n-1) \\[2ex] (-1)^j \dfrac{2^{n-2}}{n} & (j = 0, n), \end{cases} \tag{2.50}$$

which greatly simplifies the second barycentric formula (2.31):

$$\Pi_n f(x) = \left[\sum_{j=0}^{n}{}'(-1)^j f_j (x - x_j)^{-1} \right] \left[\sum_{j=0}^{n}{}'(-1)^j (x - x_j)^{-1} \right]^{-1}, \qquad (2.51)$$

where the primes on the summation signs indicate that the terms $j = 0$ and $j = n$ must be halved.

As a final test, we reproduce the interpolation of Runge's function appearing in Figure 2.3, but in this case using Chebyshev nodes. For $n = 28$ (bottom panel of Figure 2.8) the interpolant and the original function are already indistinguishable to the naked eye. To have an accurate measure of the discrepancy between the function and the interpolant, Figure 2.9a shows the pointwise absolute error $|f(x) - \Pi_{28} f(x)|$ on a semi-log scale. Simultaneously, Figure 2.9b shows the maximum pointwise error as a function of n revealing a very neat convergence. This result is clear evidence of the well-conditioning of Chebyshev's interpolation, regardless of the degree of the interpolants, which in this case is as large as $n = 240$ without any traces of numerical instabilities.

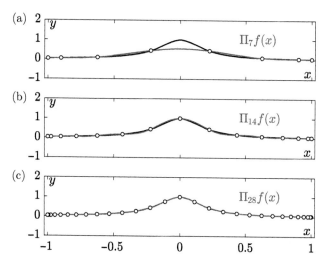

Figure 2.8 Interpolation of the function $f(x) = (1 + 25x^2)^{-1}$ (solid black curve) using Chebyshev nodes (2.48) for $n = 7$ (a), $n = 14$ (b), and $n = 28$ (c). The resulting interpolants $\Pi_n f(x)$ are the gray solid curves.

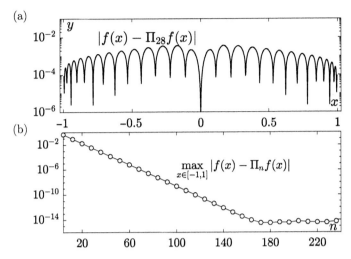

Figure 2.9 Interpolation of the function $f(x) = (1 + 25x^2)^{-1}$ using Chebyshev nodes. (a) Pointwise error for $n = 28$ and (b) maximum error as a function of n.

The log-linear scale graph used in Figure 2.9b clearly reveals that the maximum pointwise error decays *exponentially* as a function of n (in essence, the number of nodes used or the degree of the interpolant). This is one of the most important advantages of using a well-conditioned interpolation: *exponential convergence*. However, it should be mentioned that exponential convergence can be lost if the function is not sufficiently smooth. The rule of thumb is as follows:

Convergence Rate of Chebyshev's Interpolation

If the k-th derivative of the function $f(x)$ is bounded within the interval $[-1, 1]$, then the interpolant $\Pi_n f(x)$ based on the Chebyshev nodes (2.48) satisfies

$$\max_{x \in [-1,1]} |f(x) - \Pi_n f(x)| = O(n^{-k}). \tag{2.52}$$

In general, a convergence rate of the form $O(n^{-k})$ is said to be *algebraic*, with *algebraic index of convergence* k. In particular, if $f(x) \in C^\infty[-1, 1]$, k is unbounded so we may expect a much faster convergence (faster than any algebraic index),[14] such as the one exhibited in Figure 2.9b. We will exploit this property in Chapters 3 and 4, when approximating derivatives and integrals, respectively.

In general, to obtain a well-conditioned interpolant within an arbitrary interval $[a, b]$, we need to evaluate the function at the Chebyshev nodes previously transformed according to the linear map[15]

$$x_j \equiv a + \frac{b-a}{2}\left[1 + \cos\left(\frac{\pi}{n}j\right)\right],\tag{2.53}$$

for $0 \leq j \leq n$. It is straightforward to see that expression (2.51) remains valid since the change in the barycentric weights is a common proportionality factor that simplifies from its numerator and denominator. Code 4 takes advantage of this property.

```
% Code 4: Chebyshev Barycentric Interpolation
% Chebyshev nodes of second kind (practical abscissas)
  function ff = barifun(xn,fn,z)
% Input:   1. xn = interpolation Chebyshev nodes
%             mapped to [a,b] (column vector):
%             x_j = a + (b-a)*(1+cos(j*pi/n))/2; j=0,1,...,n
%          2. fn = function at nodes xn (column vector)
%          3. z = evaluation grid (z =/= xn; column vector)
% Output: ff interpolated function at z

  n = length(xn); ff = 0*z; num = ff ; den = ff ;
  s = (-1).^[0:n-1]';
  sn = s.*[fn(1)/2; fn(2:end-1); fn(end)/2];
  sd = [s(1)/2; s(2:end-1); s(end)/2];

  for ii = 1:length(num);
  num = ((z(ii) - xn).^(-1))'*sn;
  den = ((z(ii) - xn).^(-1))'*sd;
  ff(ii) = num/den;
  end
```

---

[14] Known as *exponential convergence*. For a more rigorous treatment, we refer the reader to the Complementary Reading section at the end of the chapter.

[15] Remember that due to the reverse order of Chebyshev nodes, $x_0 \equiv b$ and $x_n \equiv a$.

Practical 2.3: Barycentric-Chebyshev Interpolation

Interpolate the function

$$f(x) = \tanh\{20\sin(12x)\} + \frac{1}{50}e^{3x}\sin(300x),$$

within the interval $[0,1]$ using Chebyshev nodes $z_j = \cos\left(\frac{j\pi}{n}\right)$, for $j = 0, 1, \ldots, n$, along with the transformation $x_j = \frac{1}{2}(1 + z_j)$.

(a) Show that the barycentric form of the interpolant is exactly the same as in (2.51):

$$\Pi_n f(x) = \left[\sum_{j=0}^{n}{}'(-1)^j f_j (x - x_j)^{-1}\right] \left[\sum_{j=0}^{n}{}'(-1)^j (x - x_j)^{-1}\right]^{-1},$$

the reason being that the barycentric weights have a common factor of proportionality that cancels out from numerator and denominator.

(b) Plot the original function on a dense grid consisting of $m = 12\,000$ points.

(c) Interpolate the function using $n = 10^2$ and 10^3 Chebyshev nodes and compare the plots of the resulting interpolants with the original function.

(d) For $n = 10^3$, plot a semi-log graph of the pointwise absolute error $\varepsilon(x) = |\Pi_n f(x) - f(x)|$ on the dense grid. Find a suitable value of n such that $\varepsilon(x) < 10^{-10}$, $\forall x \in [0,1]$.

Complementary Reading

One of the classical and authoritative references is the book by Davis, *Interpolation and Approximation*, mainly written for a mathematical audience with a slightly advanced background. Nevertheless, polynomial interpolation is such an important subject that almost every monograph on numerical analysis devotes at least one chapter to it. A formal and also very accessible approach to the subject can be found in Kincaid and Cheney's *Numerical Analysis* text, which provides very elegant proofs for the existence and uniqueness of the interpolatory polynomial, as well as for remainder formula.

The term *barycentric* to describe interpolation formula (2.31) was allegedly used for the first time by H. Rutishauser in his monograph

(Continued)

Lectures on Numerical Mathematics, due to its apparent similarity to the formulas used to determine the center of gravity (barycenter) of a system of particles. For a very nice exposition of the concept of conditioning of an interpolation, I strongly recommend Trefethen's *Approximation Theory and Approximation Practice* (the test function used in Practical 2.3 is taken from this book). For a very detailed analysis of Runge's phenomenon I suggest having a look at Epperson's book *An Introduction to Numerical Methods and Analysis*. For a detailed characterization of all types of convergence rates (algebraic, exponential, etc.), I particularly like Boyd's *Chebyshev and Fourier Spectral Methods*.

Topics such as *divided differences*, *Newton's interpolation formula*, *Hermite's interpolation* or *Bernstein polynomials* (among many others) have not been covered. For a comprehensive and updated survey including these and many other aspects of interpolation and approximation, at a more advance level, I personally like Chapter 4 of Bjork and Dahlquist's *Numerical Methods in Scientific Computing*. The reader will find there a lot of material, including proofs of the Weiertrass theorem and remainder formula, as well as references for further study.

Problems and Exercises

The symbol (A) means that the problem has to be solved analytically, whereas (N) means to use numerical Matlab codes. Problems with the symbol * are slightly more difficult.

1. (A) Show that the determinant of Vandermonde's matrix $V_{ij} = x_i^j$, $(i, j = 0, 1, \ldots, n)$ shown in (2.5) is

$$\det V = \prod_{0 \le i < j \le n} (x_j - x_i).$$

2. (AN) Show that the linear interpolation error of an arbitrary twice differentiable function $f(x)$ within the interval (a, b) satisfies

$$\varepsilon(x) \le |x - a||x - b| \max_{t \in (a,b)} \left| \frac{f''(t)}{2} \right|.$$

Apply this result to estimate the error of approximating $\arcsin(x)$ at the center of the interval $[0.5330, 0.5340]$ using the linear interpolant based on the values of $f(x)$ at the ends of the interval.

3. (A) Show that for an equispaced interpolation with nodes $x_i = x_0 + i\,h$ $(h > 0,\ i = 0, 1, \ldots, n)$, the absolute value of the nodal polynomial satisfies

$$|\ell(x)| \le \frac{n!}{4} h^{n+1}, \quad \forall x \in (x_0, x_n).$$

Use this result to show that the interpolation error of a sufficiently differentiable function satisfies

$$\max_{x \in I} |f(x) - \Pi_n f(x)| \le \max_{x \in I} |f^{(n+1)}(x)| \frac{h^{n+1}}{4(n+1)}.$$

4. (A*) Consider a set of distinct nodes $\{x_0, x_1, \ldots, x_n\}$ and its corresponding nodal polynomial $\ell(x)$ given by expression (2.27). Show that the first derivative of the nodal polynomial evaluated at the j-th node is

$$\ell'(x_j) = \prod_{k \ne j}(x_j - x_k).$$

Using the property above, show that the cardinal polynomials can be expressed as

$$\ell_j(x) = \frac{\ell(x)}{\ell'(x_j)(x - x_j)}.$$

5. (A) Determine the barycentric weights λ_j associated with the set of equispaced nodes $x_j = -1 + 2j/n$, $(j = 0, 1, \ldots, n)$ within the interval $[-1, 1]$. Generalize your result for the equispaced distribution $x_j = a + (b - a)j/n$, for $j = 0, 1, \ldots, n$, within an arbitrary interval $[a, b]$ and recover expression (2.32).

6. (A*) Consider a set of distinct nodes $\{x_0, x_1, \ldots, x_n\}$ and their associated cardinal polynomials $\ell_j(x)$ given by (2.22). Use the result obtained in Problem 2.4 to evaluate the quantities $\ell'_j(x_i)$, $(i, j = 0, 1, \ldots, n)$, i.e. the derivatives of the cardinal polynomials at the nodes.

7. (A) We want to interpolate $f(x) = 4^x$ using the nodes $\{x_0, x_1, x_2\} = \{-1/2, 0, 1/2\}$ in order to provide a rational approximation (i.e. of the form m/n, with $m, n \in \mathbb{N}$) of $\sqrt[3]{4}$.

(a) Obtain the interpolant in monomial form $\Pi_n f(x) = \sum_{j=0}^{n} a_j x^j$.

(b) Obtain the interpolant in Lagrange form $\Pi_n f(x) = \sum_{j=0}^{n} f_j \ell_j(x)$.

(c) Evaluate the interpolant at a suitable point to approximate $\sqrt[3]{4}$.

8. (A) (*Osculatory interpolation*): We aim to interpolate a differentiable function $f(x)$ using its values at the nodes x_0 and $x_1 = h$ (with $h > 0$), that is, $f_0 = f(0)$ and $f_1 = f(h)$. In order to improve the accuracy of the interpolant, we want also to involve the values of the derivative of $f(x)$, $f_0' = f'(0)$ and $f_1' = f(h)$ on the same nodes. Therefore, we consider a cubic polynomial of the form

$$H(x) = a_0 + a_1 x + a_2 x^2 + a_3 x^3,$$

and impose $H(x)$ and $H'(x)$ to take the same values as $f(x)$ and $f'(x)$, respectively, at the nodes.
(a) Show that $H(x)$ exists and is unique.
(b) Obtain the coefficients a_j as a function of f_0, f_1, f_0', and f_1'.
(c) Obtain the interpolant of $f(x) \equiv \cos x$ using the nodes $x_0 = 0$ and $x_1 = \pi$ and find its roots. Is $\pi/2$ a root of $H(x)$?

9. (A) For an arbitrary set of distinct nodes $\{x_0, x_1, \ldots, x_n\}$, show that their associated barycentric weights λ_j and cardinal polynomials $\ell_j(x)$ satisfy
(a) $\lambda_0 + \lambda_1 + \cdots + \lambda_n = 0$ (*).
(b) $\displaystyle\sum_{j=0}^{n} \ell_j'(x) \equiv 0.$

10. (A) We want to interpolate within the interval $[-1, 1]$ any continuous function $f(x)$ satisfying $f(-1) \equiv f(1) \equiv 0$. In what follows, assume that the values of the function $f_j \equiv f(x_j)$, $(j = 0, 1, \ldots, n)$ at given internal $n + 1$ nodes,

$$-1 < x_0 < x_1 < \cdots < x_{n-1} < x_n < 1, \quad (n > 1),$$

are known. Let $\ell_j(x)$, $(j = 0, 1, \ldots, n)$ be the Lagrange cardinal polynomials associated with the previous internal nodes and the interpolant of $f(x)$ expressed as

$$\Pi f(x) = \sum_{k=0}^{n} g_k\, m_k(x) = \sum_{\ell=0}^{n} a_\ell x^\ell$$

where $m_k(x) \equiv (x^2 - 1)\ell_k(x)$, $(k = 0, 1, \ldots, n)$.
(a) Find the coefficients g_k as a function of the values f_j.
(b) For $n \equiv 1$, $x_0 \equiv -1/2$, and $x_1 \equiv 1/2$, calculate the coefficients a_0 and a_1 of the monomial form in terms of the values f_0 and f_1. Interpolate $f(x) \equiv \cos(\pi x/2)$ and obtain $\Pi f(x)$.

3

Numerical Differentiation

3.1 Introduction

Let us start by reviewing the mathematical concept of derivative. The derivative of a continuous function $f(x)$ at a given abscissa x is defined by the limit

$$\frac{\mathrm{d}f}{\mathrm{d}x} \equiv f'(x) = \lim_{h \to 0} \frac{f(x+h) - f(x)}{h}. \tag{3.1}$$

To carry out the previous limit, it is assumed that we have access to the values of $f(x)$ at *any* given abscissa x. Numerically, taking the limit $h \to 0$ is not feasible because computers work with finite precision so, in practice, we cannot evaluate $f(x)$ at any given abscissa x or $x + h$. However, we may have access to the values of the function $f_j \equiv f(x_j)$, $0 \le j \le n$, on a prescribed *discrete* set of nodes $\{x_0, x_1, \ldots, x_n\}$. Under this circumstance, would it be possible to obtain useful approximations of $f'(x)$ at the nodes where the function is evaluated and only using those sampled values?

To understand the problem better, we start with an example. Figure 3.1a shows the graph of a function (black dashed curve) $f(x)$ over the interval $[0.5, 2]$. Let us assume that the ordinates of this function are only known on a set of equispaced[1] abscissas $\{x_0, x_1, \ldots, x_n\}$, with $x_{j+1} = x_j + h$, where it takes the values $\{f_0, f_1, \ldots, f_n\}$ (black bullets in Figure 3.1a), respectively. Figure 3.1b shows the first derivative of the function (dashed gray curve) $f'(x)$. The main goal of numerical differentiation is to obtain approximations of $f'(x)$ at the same nodes where $f(x)$ was originally sampled and using only information from that sampling. Henceforth, we will denote u_j as the approximation of $f'(x)$ at the jth node, that is,

$$u_j \approx \left(\frac{\mathrm{d}f}{\mathrm{d}x}\right)_{x = x_j} \equiv f'(x_j), \tag{3.2}$$

for $0 \le j \le n$.

1 We will consider non-equispaced distributions later on.

Fundamentals of Numerical Mathematics for Physicists and Engineers, First Edition. Alvaro Meseguer.
© 2020 John Wiley & Sons, Inc. Published 2020 by John Wiley & Sons, Inc.

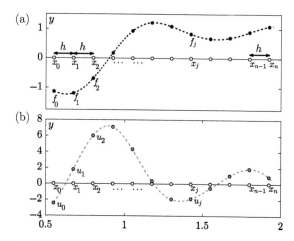

Figure 3.1 Numerical differentiation: starting from the known values $f_j = f(x_j)$ of a given function evaluated at some nodes x_j, as depicted in (a), obtain numerical approximations u_j of $f'(x_j)$, depicted in (b).

Numerical Differentiation

For a given function $f(x)$ with known ordinates $\{f_0, f_1, \ldots, f_n\}$ at the abscissas $\{x_0, x_1, \ldots, x_n\}$, respectively, provide numerical approximations $\{u_0, u_1, \ldots, u_n\}$ of the derivative of the function at those abscissas $\{f'(x_0), f'(x_1), \ldots, f'(x_n)\}$, respectively, using the sampled values $\{f_0, f_1, \ldots, f_n\}$.

A possibility is to consider the approximation

$$u_j = \frac{f_{j+1} - f_j}{h} = \frac{f(x_j + h) - f(x_j)}{h} \quad (0 \leq j \leq n-1). \tag{3.3}$$

Since this formula takes values of $f(x)$ at the next node $x_j + h$, it is usually known as *Forward Difference*. Expression (3.3)[2] is clearly similar to (3.1) and, if h is small (compared with the order of magnitude of $|x_j|$), its numerical value should be close to $f'(x_j)$. We will study the accuracy of this and other similar formulas in Section 3.4. We could also provide an approximation to the derivatives using the alternative formula

$$u_j = \frac{f_j - f_{j-1}}{h} = \frac{f(x_j) - f(x_j - h)}{h} \quad (1 \leq j \leq n), \tag{3.4}$$

usually termed as *Backward Difference*, for obvious reasons. The similarity of this formula with (3.1) is less evident but it is clear that in the limit for $h \to 0$

---

2 The reader may have noticed that this formula was already used in Code 2 (Newton's method) for the solution of transcendental equations in order to approximate $f'(x_k)$.

it also provides the value of the derivative $f'(x_j)$. The reader may have already noticed that (3.3) cannot be used at x_n (since f_{n+1} is not available) or (3.4) at x_0 (since f_{-1} is not available either). To illustrate how these formulas are used in practice, let us approximate the derivative of $f(x) = \sin x$ at $x = 1$ with $h = 0.01$. The forward difference approximation formula (3.3) can be applied by setting $x_j = 1$ and $x_{j+1} = 1.01$ to obtain

$$u_j = \frac{\sin(1.01) - \sin(1)}{0.01} = 0.536\ 085\ 981\ 011\ 869,$$

where only the black figures are correct[3] with an approximate absolute error $|f'(x_j) - u_j| \approx 4 \times 10^{-3}$. The backward difference formula (3.4) can be used similarly by setting $x_j = 1$ and $x_{j-1} = 0.99$ to obtain

$$u_j = \frac{\sin(1) - \sin(0.99)}{0.01} = 0.544\ 500\ 620\ 737\ 598,$$

also with an approximate error $|f'(x_j) - u_j| \approx 4 \times 10^{-3}$. In order to improve the accuracy we can simply reduce h. However, this can also be accomplished keeping the same value of h and using more accurate formulas instead. One example is the *Centered Difference* given by the expression

$$u_j = \frac{1}{h}\left(\frac{1}{2}f_{j+1} - \frac{1}{2}f_{j-1}\right) = \frac{1}{2h}(f(x_j + h) - f(x_j - h)), \tag{3.5}$$

for $1 \leq j \leq n - 1$. By setting $x_j = 1$, $x_{j-1} = 0.99$, and $x_{j+1} = 1.01$ we obtain an approximation $u_j = 0.540\ 293\ 300\ 874\ 733$ with a remarkably reduced absolute error $|f'(x_j) - u_j| \approx 10^{-5}$. Alternatively, we could also use the formula

$$u_j = \frac{1}{h}\left(-\frac{1}{2}f_{j+2} + 2f_{j+1} - \frac{3}{2}f_j\right) = \frac{1}{2h}(-f(x_j + 2h) + 4f(x_j + h) - 3f(x_j)), \tag{3.6}$$

for $0 \leq j \leq n - 2$. By setting $x_j = 1$, $x_{j+1} = 1.01$, and $x_{j+2} = 1.02$ we obtain $u_j = 0.540\ 320\ 104\ 950\ 42$, also with a similar absolute error $|f'(x_j) - u_j| \approx 2 \times 10^{-5}$. Expressions (3.3), (3.4), (3.5), and (3.6) are particular instances of *finite difference formulas*. Figure 3.2 indicates graphically the function values used in these particular cases, where the arrows indicate the nodes required by each formula to evaluate u_j.

The reader may wonder what the origin of formulas such as (3.5) or (3.6) is, how have they been devised, what their accuracy is, etc. In the rest of the chapter, we will try to answer these and other questions.

---

3 Compared with the exact reference value $\cos(1) = 0.540\ 302\ 305\ 868\ 1$.

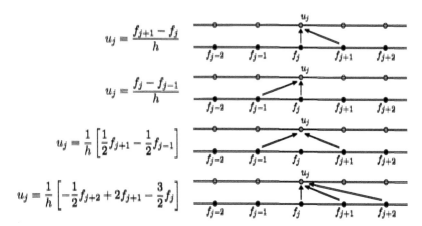

$$u_j \equiv \frac{f_{j+1} - f_j}{h}$$

$$u_j \equiv \frac{f_j - f_{j-1}}{h}$$

$$u_j \equiv \frac{1}{h}\left[\frac{1}{2}f_{j+1} - \frac{1}{2}f_{j-1}\right]$$

$$u_j \equiv \frac{1}{h}\left[-\frac{1}{2}f_{j+2} + 2f_{j+1} - \frac{3}{2}f_j\right]$$

Figure 3.2 From top to bottom, finite difference formulas (3.3), (3.4), (3.5), and (3.6), respectively.

3.2 Differentiation Matrices

As promised, we start by explaining the origin of the formulas seen in Section 3.1, in particular formulas (3.5) and (3.6). Recall the interpolatory polynomial of an arbitrary function $f(x)$ on a set of three equidistant nodes located at $\{x_0, x_1, x_2\} \equiv \{0, h, 2h\}$ given by

$$\Pi_2 f(x) \equiv f_0 \ell_0(x) + f_1 \ell_1(x) + f_2 \ell_2(x), \qquad (3.7)$$

where the $\ell_j(x)$ cardinal polynomials are

$$\ell_0(x) \equiv \frac{1}{2h^2}(x - h)(x - 2h), \;\; \ell_1(x) \equiv -\frac{1}{h^2}x(x - 2h), \;\; \text{and} \;\; \ell_2(x) \equiv \frac{1}{2h^2}x(x - h).$$

Since the interpolant $\Pi_2 f(x)$ is a reasonably good approximation of the function $f(x)$ within the interval $[0, 2h]$, we expect its derivative, $\Pi_2' f(x)$, to be a good approximation of the exact derivative $f'(x)$ as well. Therefore, we approximate the derivative of the function $f(x)$ at the nodes by the derivative of the corresponding interpolant at those nodes. First we differentiate the interpolant:

$$\frac{\mathrm{d}}{\mathrm{d}x}\Pi_2 f(x) \equiv \Pi_2' f(x) \equiv f_0 \ell_0'(x) + f_1 \ell_1'(x) + f_2 \ell_2'(x),$$

which in this particular case is

$$\Pi_2' f(x) \equiv f_0 \frac{1}{2h^2}(2x - 3h) + f_1\left(-\frac{1}{h^2}\right)(x - 2h) + f_2 \frac{1}{2h^2}(2x - h).$$

As a result, the values u_0, u_1, and u_2 that approximate $f'(x)$ at the nodes x_0, x_1, and x_2, respectively, are given by

$$u_j \equiv \Pi_2' f(x_j) \equiv f_0 \ell_0'(x_j) + f_1 \ell_1'(x_j) + f_2 \ell_2'(x_j),$$

for $j = 0, 1, 2$. After a bit of algebra, the reader may check that the evaluation of the derivative of the interpolant at the nodes $\{x_0, x_1, x_2\} = \{0, h, 2h\}$ provides the quantities sought:

$$u_0 = f_0 \ell'_0(0) + f_1 \ell'_1(0) + f_2 \ell'_2(0) = \frac{1}{h}\left(-\frac{3}{2}f_0 + 2f_1 - \frac{1}{2}f_2\right),$$

$$u_1 = f_0 \ell'_0(h) + f_1 \ell'_1(h) + f_2 \ell'_2(h) = \frac{1}{h}\left(-\frac{1}{2}f_0 + \frac{1}{2}f_2\right),$$

and

$$u_2 = f_0 \ell'_0(2h) + f_1 \ell'_1(2h) + f_2 \ell'_2(2h) = \frac{1}{h}\left(\frac{1}{2}f_0 - 2f_1 + \frac{3}{2}f_2\right).$$

The previous expressions give the answer to the main question in numerical differentiation: to provide approximations of $f'(x)$ at the nodes in terms of the sampled values of $f(x)$. The three previous equations can be synthesized in one matrix-vector product

$$\begin{bmatrix} u_0 \\ u_1 \\ u_2 \end{bmatrix} = \begin{bmatrix} \Pi'_2 f(x_0) \\ \Pi'_2 f(x_1) \\ \Pi'_2 f(x_2) \end{bmatrix} = \frac{1}{h}\begin{bmatrix} -3/2 & 2 & -1/2 \\ -1/2 & 0 & 1/2 \\ 1/2 & -2 & 3/2 \end{bmatrix}\begin{bmatrix} f_0 \\ f_1 \\ f_2 \end{bmatrix}.$$

The reader can easily identify the first row of the matrix above as formula (3.6) for $j = 0$, and second row as (3.5) for $j = 1$.

The previous analysis can be generalized for an arbitrary set of distinct nodes (not necessarily equispaced). First, interpolate a given function $f(x)$ on the grid $\{x_0, x_1, \ldots, x_n\}$ to obtain the interpolant $\Pi_n f(x)$. Second, differentiate this interpolant in its Lagrange form

$$\Pi'_n f(x) \equiv \sum_{j=0}^{n} f_j \ell'_j(x). \tag{3.8}$$

Third and last, approximate the derivatives $\{f'(x_0), f'(x_1), \ldots, f'(x_n)\}$ of the function at the nodes by evaluating the derivative of the interpolant at each x_j:

$$f'(x_i) \approx u_i = \Pi'_n f(x_i) = \sum_{j=0}^{n} f_j \ell'_j(x_i), \tag{3.9}$$

for $0 \leq i \leq n$. Rewriting (3.9) as a matrix-vector product, it becomes

$$\begin{bmatrix} u_0 \\ u_1 \\ \vdots \\ u_n \end{bmatrix} = \begin{bmatrix} \ell'_0(x_0) & \ell'_1(x_0) & \cdots & \ell'_n(x_0) \\ \ell'_0(x_1) & \ell'_1(x_1) & \cdots & \ell'_n(x_1) \\ \vdots & \vdots & & \vdots \\ \ell'_0(x_n) & \ell'_1(x_n) & \cdots & \ell'_n(x_n) \end{bmatrix}\begin{bmatrix} f_0 \\ f_1 \\ \vdots \\ f_n \end{bmatrix}, \tag{3.10}$$

where the $n+1$ components of the vector $[u_0 \; u_1 \; \cdots \; u_n]^{\mathrm{T}}$ appearing in (3.10) are explicitly given by the expressions

$$u_i = \sum_{j=0}^{n} \mathbf{D}_{ij} f_j, \tag{3.11}$$

for $0 \leq i \leq n$. The quantities $\mathbf{D}_{ij}$ are the elements of the $(n+1) \times (n+1)$ matrix appearing on the right-hand side of (3.10):

$$\boxed{\mathbf{D}_{ij} = \ell'_j(x_i), \quad 0 \leq i, j \leq n.} \tag{3.12}$$

The matrix $\mathbf{D} \in \mathrm{M}_{n+1}(\mathbb{R})$ appearing in (3.12) is known as the *differentiation matrix* associated with the nodes $\{x_0, x_1, \ldots, x_n\}$, and its elements are the derivatives of the Lagrange's cardinal polynomials associated with the nodes and evaluated precisely at each one of them.[4] Matrix (3.12) is extremely important since many numerical methods arising in the discretization of ordinary and partial differential equations are essentially derived from it in one way or another. Understanding the origin and properties of this matrix is crucial, in order to understand the advantages, accuracy, and limitations of many numerical methods appearing in the literature on numerical differential equations. From a theoretical point of view, the concept of differentiation matrix is also of extreme relevance since it is a finite dimensional representation of the differential operator $\dfrac{\mathrm{d}}{\mathrm{d}x}$. We will revisit these concepts in the second half of this book, within the framework of boundary and initial value problems of ordinary differential equations.

We can find explicit expressions of the matrix elements $\mathbf{D}_{ij}$ in terms of the barycentric weights λ_j associated with the nodes x_j. For example, in order to obtain the diagonal terms $\mathbf{D}_{jj}$ of the differentiation matrix, take the logarithm of $\ell_j(x)$

$$\log \ell_j(x) = \log \left[\lambda_j \prod_{k \neq j} (x - x_k) \right] = \log \lambda_j + \sum_{k \neq j} \log(x - x_k),$$

and differentiate it to obtain

$$\frac{\ell'_j(x)}{\ell_j(x)} = \sum_{k \neq j} \frac{1}{x - x_k}.$$

Evaluating this last expression at x_j leads to

$$\boxed{\mathbf{D}_{jj} = \ell'_j(x_j) = \sum_{\substack{k=0 \\ (k \neq j)}}^{n} \frac{1}{x_j - x_k} \quad (0 \leq j \leq n).} \tag{3.13}$$

---

4 $\mathrm{M}_n(\mathbb{R})$ denotes the linear space of $n \times n$ real matrices.

To obtain the off-diagonal elements (that is, $\mathbf{D}_{ij}$, with $i \neq j$) we first need to write the jth Lagrange's cardinal polynomial in a different way. If we take the logarithm of the nodal polynomial

$$\log \ell(x) = \log \prod_{k=0}^{n} (x - x_k) = \sum_{k=0}^{n} \log(x - x_k),$$

and differentiate, we obtain

$$\frac{\ell'(x)}{\ell(x)} = \sum_{k=0}^{n} \frac{1}{x - x_k},$$

or

$$\ell'(x) = \sum_{k=0}^{n} \frac{\ell(x)}{x - x_k} = \prod_{k \neq 0} (x - x_k) + \prod_{k \neq 1} (x - x_k) + \cdots + \prod_{k \neq j} (x - x_k) + \cdots + \prod_{k \neq n} (x - x_k).$$

Evaluating $\ell'(x)$ at x_j cancels the products above except the one with $k \neq j$:

$$\ell'(x_j) = \prod_{k \neq j} (x_j - x_k) = \lambda_j^{-1}.$$

Therefore, we can write

$$\ell_j(x) = \lambda_j \prod_{\substack{k=0 \\ (k \neq j)}}^{n} (x - x_k) = \frac{1}{\ell'(x_j)} \prod_{\substack{k=0 \\ (k \neq j)}}^{n} (x - x_k) = \frac{\ell(x)}{\ell'(x_j)(x - x_j)}.$$

Differentiation of this new factorization of $\ell_j(x)$ leads to the expression

$$\ell_j'(x) = \frac{1}{\ell'(x_j)} \frac{\ell'(x)(x - x_j) - \ell(x)}{(x - x_j)^2},$$

which can be evaluated at $x_i \neq x_j$. As a result,

$$\boxed{\mathbf{D}_{ij} = \ell_j'(x_i) = \frac{\lambda_j}{\lambda_i} \frac{1}{x_i - x_j} \qquad (0 \leq i, j \leq n, \quad i \neq j).} \qquad (3.14)$$

Substituting the barycentric weights (2.32) associated with the equidistant nodes $x_j = jh$ ($0 \leq j \leq n$) in (3.13) and (3.14) leads to the differentiation matrix

$$\mathbf{D}_{ij} = \begin{cases} \dfrac{1}{h} \displaystyle\sum_{\substack{k=0 \\ (k \neq j)}}^{n} \dfrac{1}{j - k} & (0 \leq i = j \leq n), \\[4mm] \dfrac{1}{h} \dfrac{(-1)^{i-j}}{i - j} \dfrac{i! \, (n - i)!}{j! \, (n - j)!} & (0 \leq i, j \leq n, \quad i \neq j). \end{cases} \qquad (3.15)$$

The reader may check that for $n = 1$ and $n = 2$, these matrices are

$$\frac{1}{h}\begin{bmatrix} -1 & 1 \\ -1 & 1 \end{bmatrix} \tag{3.16}$$

and

$$\frac{1}{h}\begin{bmatrix} -3/2 & 2 & -1/2 \\ -1/2 & 0 & 1/2 \\ 1/2 & -2 & 3/2 \end{bmatrix}, \tag{3.17}$$

respectively. For $n = 1$, the product of matrix (3.16) with a vector $[f_0 \ f_1]^{\mathrm{T}} = [f(x_0) \ f(x_1)]^{\mathrm{T}}$ leads to formulas (3.3) and (3.4):

$$\begin{bmatrix} u_0 \\ u_1 \end{bmatrix} = \frac{1}{h}\begin{bmatrix} -1 & 1 \\ -1 & 1 \end{bmatrix}\begin{bmatrix} f_0 \\ f_1 \end{bmatrix} = \frac{1}{h}\begin{bmatrix} -f_0 + f_1 \\ -f_0 + f_1 \end{bmatrix}.$$

In particular, the first row of the last identity is simply (3.3) for $j = 0$, whereas the second row is (3.4) for $j = 1$. Similarly, the product of matrix (3.17) with the vector $[f_0 \ f_1 \ f_2]^{\mathrm{T}} = [f(x_0) \ f(x_1) \ f(x_n)]^{\mathrm{T}}$ containing the values of a function $f(x)$ at the nodes $\{x_0, x_1, x_2\}$ leads to another vector containing formulas (3.6) and (3.5) for $j = 0$ and $j = 1$ in its first and second components, respectively:

$$\begin{bmatrix} u_0 \\ u_1 \\ u_2 \end{bmatrix} = \frac{1}{h}\begin{bmatrix} -3/2 & 2 & -1/2 \\ -1/2 & 0 & 1/2 \\ 1/2 & -2 & 3/2 \end{bmatrix}\begin{bmatrix} f_0 \\ f_1 \\ f_2 \end{bmatrix} = \frac{1}{h}\begin{bmatrix} -\dfrac{3}{2}f_0 + 2f_1 - \dfrac{1}{2}f_2 \\ -\dfrac{1}{2}f_0 + \dfrac{1}{2}f_2 \\ \dfrac{1}{2}f_0 - 2f_1 + \dfrac{3}{2}f_2 \end{bmatrix}.$$

Finally, the third row provides the backward difference formula

$$u_2 = \frac{1}{h}\left(\frac{1}{2}f_0 - 2f_1 + \frac{3}{2}f_2\right), \tag{3.18}$$

which can be used to approximate the derivative at x_2.

Code 5A provides the differentiation matrix for an arbitrary set of distinct nodes $\{x_0, x_1, \ldots, x_n\}$ using expressions (3.13) and (3.14).

In particular, Code 5A can provide the coefficients of the differentiation matrix (3.15) corresponding to equispaced node distributions. For example, to obtain the rational coefficients (for $h = 1$) of the differentiation matrices (3.16) and (3.17) for $n = 1$ or $n = 2$, just type in Matlab:

```
>> format rat ; n = 1 ; x = [0:n] ; D = diffmat(x)
```

or

```
>> format rat; n = 2; x = [0:n]; D = diffmat(x),
```

```
% Code 5A: Differentiation matrix
function D = diffmat(x)
% Input:  vector of nodes x= [x_0; x_1; ...; x_n]
% Output: differentiation matrix D

  N = length(x); lamb = zeros(N,1); D = zeros(N,N);
  for jj = 1:N
    lamb(jj) = 1/prod(x(jj) - x([1:jj-1 jj+1:N]));
  end

  for ii = 1:N
    for jj = 1:N
      if jj == ii
        D(ii,jj) = sum(1./(x(ii)-x([1:ii-1 ii+1:N])));
      else
        D(ii,jj) = lamb(jj)/((x(ii)-x(jj))*lamb(ii));
      end
    end
  end
```

respectively. For arbitrary n, Code 5A helps provide the coefficients of an arbitrary backward, forward, and centered differentiation formula such as (3.3), (3.4), (3.5), (3.6), or (3.18). For even n, the first, central, and last rows of the matrix provided by Code 5A are the classical forward, centered, and backward finite difference formulas that frequently appear in the literature on numerical differentiation. As an example, for $n = 4$, we obtain the *five-point stencil* (central row):

```
>> format rat ; n = 4 ; x = [0:n] ; D = diffmat(x)+
D=
```

-25/12	4	-3	4/3	-1/4
-1/4	-5/6	3/2	-1/2	1/12
1/12	-2/3	0	2/3	-1/12
-1/12	1/2	-3/2	5/6	1/4
1/4	-4/3	3	-4	25/12 .

Summary: Differentiation Matrix

The differentiation matrix associated with a set of distinct nodes $\{x_0, x_1, \ldots, x_n\}$ with barycentric weights $\{\lambda_0, \lambda_1, \cdots, \lambda_n\}$ given by (2.26) is

$$\mathbf{D}_{ij} = \ell_j'(x_i) = \begin{cases} \displaystyle\sum_{\substack{k=0 \\ (k \neq j)}}^{n} \frac{1}{x_j - x_k}, & (0 \leq j \leq n; \ i = j), \\[2em] \displaystyle\frac{\lambda_j}{\lambda_i} \frac{1}{x_i - x_j}, & (0 \leq i,j \leq n; \ i \neq j). \end{cases} \qquad (3.19)$$

The product of $\mathbf{D}$ with the vector $[f_0 \ f_1 \ \cdots f_n]^{\mathrm{T}}$ containing the ordinates $f_j = f(x_j)$ of the function $f(x)$ at the nodes provides the vector $[u_0 \ u_1 \ \cdots u_n]^{\mathrm{T}}$ whose components are $u_j = \Pi_n' f(x_j)$, that is, the derivative of the interpolant of $f(x)$ evaluated at the nodes:

$$\begin{bmatrix} \mathbf{D}_{00} & \mathbf{D}_{01} & \cdots & \mathbf{D}_{0n} \\ \mathbf{D}_{10} & \mathbf{D}_{11} & \cdots & \mathbf{D}_{1n} \\ \vdots & \vdots & \cdots & \vdots \\ \mathbf{D}_{n0} & \mathbf{D}_{n1} & \cdots & \mathbf{D}_{nn} \end{bmatrix} \begin{bmatrix} f_0 \\ f_1 \\ \vdots \\ f_n \end{bmatrix} = \begin{bmatrix} u_0 \\ u_1 \\ \vdots \\ u_n \end{bmatrix}.$$

3.3 Local Equispaced Differentiation

If we want to approximate the derivative of a function $f(x)$ on a given set of abscissas $\{x_0, x_1, \ldots, x_m\}$, we can simply use expression (3.19) with $n = m$. In particular, if the points are equidistant, it is a temptation to apply (3.15). However, if m is large, we expect our approximation to be numerically ill-conditioned (we are implicitly using an interpolant generated from an equispaced grid). In this section, we show one possible way of solving the problem: *local finite differences*.

For large sets of uniformly distributed points $x_{j+1} = x_j + h$ (for some small spacing $h > 0$), it is common practice to proceed as follows. First, calculate the differentiation matrix $\mathbf{D}$ using (3.15) with the prescribed spacing h and a low or moderate value of n (typically $n < 10$). Second, extract from $\mathbf{D}$ one of its rows to generate a formula that is applied on clusters of $n + 1$ adjacent abscissas.

To better illustrate this idea, Figure 3.3 shows an example in which we use the centered formula (3.5) to approximate the derivative of the function $f(x)$ whose values are known at the abscissas $\{x_0, \ldots, x_m\}$. The second row of the

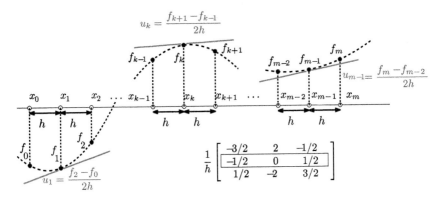

Figure 3.3 Centered difference formula (3.5) applied on a set of equidistant points. The parabolas (black dashed curves) are the local interpolants used to approximate the derivative at the center of each triad of adjacent abscissas.

differentiation matrix (boxed in Figure 3.3), corresponding to the equispaced case $n = 2$ with spacing h, is used as a stencil or template that is replicated to be applied on clusters of three adjacent nodes (formulas in gray). This formula approximates the derivative of $f(x)$ at the central node by computing the slope u_k of the tangent line (solid gray line) to the local interpolant $\Pi_2 f(x)$ of degree $n = 2$ (dashed parabola). It is important to notice that by replicating (3.5) at different triads $\{x_{k-1}, x_k, x_{k+1}\}$, we implicitly use a *different local interpolant* $\Pi_2 f(x)$ to obtain the approximated derivative u_k at the central nodes x_k. This is the reason why finite differences are considered as a *local* differentiation method.

The reader may have noticed that the centered formula used in the previous example cannot be applied to the abscissa x_0 or to x_m, since f_{-1} and f_{m+1} are not available, respectively. However, this local differentiation procedure can also be represented as the matrix–vector product

$$
\underbrace{\begin{pmatrix} u_1 \\ u_1 \\ u_2 \\ \vdots \\ u_{m-1} \end{pmatrix}}_{(m-1)\times 1} = \frac{1}{h} \underbrace{\begin{pmatrix} -\frac{1}{2} & 0 & \frac{1}{2} & 0 & \cdots & & 0 \\ 0 & -\frac{1}{2} & 0 & \frac{1}{2} & 0 & \cdots & \vdots \\ \vdots & 0 & \ddots & \ddots & \ddots & \ddots & \\ \vdots & & \ddots & \ddots & \ddots & \ddots & \vdots \\ \vdots & & & \ddots & -\frac{1}{2} & 0 & \frac{1}{2} & 0 \\ 0 & \cdots & & \cdots & 0 & -\frac{1}{2} & 0 & \frac{1}{2} \end{pmatrix}}_{(m-1)\times(m+1)} \cdot \underbrace{\begin{pmatrix} f_0 \\ f_1 \\ f_2 \\ \vdots \\ f_{m-1} \\ f_m \end{pmatrix}}_{(m+1)\times 1}
$$

$$(3.20)$$

The $(m-1) \times (m+1)$ matrix appearing in (3.20) makes use of the $m+1$ abscissas to approximate the derivatives just at the $m-1$ interior points. Although this is not a differentiation matrix in the sense of (3.19), it performs a similar job by producing approximations of the derivative of $f(x)$ at the inner abscissas $\{x_1, \ldots, x_{m-1}\}$. This matrix is usually known as the *second order centered finite difference matrix*. The matrix and its corresponding finite difference formula are simply referred to as CD2, where the acronym CD stands for *centered difference* and the number 2 signifies that their origin is the derivative of second degree interpolants obtained from triads of adjacent abscissas.

Similarly, replicating forward difference stencil (3.3) on pairs of adjacent points leads to the matrix-vector product

$$
\underbrace{\begin{pmatrix} u_0 \\ u_1 \\ u_2 \\ \vdots \\ u_{m-1} \end{pmatrix}}_{m \times 1} = \frac{1}{h} \underbrace{\begin{pmatrix} -1 & 1 & 0 & \cdots & & \cdots & 0 \\ 0 & -1 & 1 & 0 & & & \vdots \\ \vdots & 0 & \ddots & \ddots & \ddots & & \\ & & \ddots & \ddots & \ddots & \ddots & \vdots \\ \vdots & & & \ddots & -1 & 1 & 0 \\ 0 & \cdots & & \cdots & 0 & -1 & 1 \end{pmatrix}}_{m \times (m+1)} \underbrace{\begin{pmatrix} f_0 \\ f_1 \\ f_2 \\ \vdots \\ f_{m-1} \\ f_m \end{pmatrix}}_{(m+1) \times 1} , \tag{3.21}
$$

The $m \times (m+1)$ matrix appearing in (3.21) is the *first order forward finite difference* matrix associated with (3.3), and is usually known as FD1, since it originates from a first degree local interpolant. In this case, matrix (3.21) can only approximate the derivatives at the abscissas $\{x_0, x_1, \ldots, x_{m-1}\}$. Backwards difference formula (3.4), known as BD1, leads to the same matrix as in (3.21),

$$
\underbrace{\begin{pmatrix} u_1 \\ u_1 \\ u_2 \\ \vdots \\ u_m \end{pmatrix}}_{m \times 1} = \frac{1}{h} \underbrace{\begin{pmatrix} -1 & 1 & 0 & \cdots & & \cdots & 0 \\ 0 & -1 & 1 & 0 & & & \vdots \\ \vdots & 0 & \ddots & \ddots & \ddots & & \\ & & \ddots & \ddots & \ddots & \ddots & \vdots \\ \vdots & & & \ddots & -1 & 1 & 0 \\ 0 & \cdots & & \cdots & 0 & -1 & 1 \end{pmatrix}}_{m \times (m+1)} \underbrace{\begin{pmatrix} f_0 \\ f_1 \\ f_2 \\ \vdots \\ f_{m-1} \\ f_m \end{pmatrix}}_{(m+1) \times 1} , \tag{3.22}
$$

However, the reader must notice that this stencil cannot provide u_0, for obvious reasons.

Finite difference formulas are extremely important as they are widely used in many branches of mathematical physics and engineering. They are also very convenient from the point of view of computational efficiency. On the one hand, the associated matrices are *sparse* (i.e. a matrix in which most of the elements are zero). In particular, finite difference matrices have a *banded* structure (with a few nonzero super and subdiagonals), which, as we will see in Part II, is very convenient from the point of view of numerical conditioning. In general, the discretization of differential equations using finite differences eventually leads to linear systems of equations that are relatively easy to solve precisely because of this banded structure. On the other hand, since the approximation of the derivative u_j at x_j only requires operating with the values of the function at neighboring nodes (f_j and $f_{j\pm1}$, for example), this allows for a very efficient *divide and conquer* computational strategy where these operations can independently (and simultaneously) be performed by different processors.

However, the main drawback of finite difference formulas is their accuracy. If one requires accurate values of the derivative, a very small grid spacing h is usually needed. More specifically, let us approximate the derivative of a function within a given physical domain $[a, b]$ using a set of equidistant points $x_j \equiv a + jh$ ($0 \leq j \leq m$), where $h \equiv (b-a)/m$. For a given stencil such as CD2, the only way to improve the accuracy is therefore to increase m in order to reduce h, so the number of abscissas (and therefore the number of function evaluations) could potentially be quite large, increasing the computational cost. There are different strategies available to improve the accuracy in numerical differentiation. First, let us have a look at the error of the previous finite difference formulas.

3.4 Accuracy of Finite Differences

In Chapter 2, we studied the accuracy of polynomial interpolation. Cauchy's remainder formula (2.33) provided the upper bound (2.34) of the pointwise interpolation error for equidistant node distributions. For a grid spacing h, (2.34) establishes that

$$|f(x) - \Pi_n f(x)| \equiv O(h^{n+1}). \tag{3.23}$$

Since finite difference formulas are extracted from evaluating the derivative of the interpolant $\Pi_n f(x)$ at the equispaced distributed nodes, we are interested in the measure of the discrepancy between the quantities $u_j = \Pi'_n f(x_j)$ and $f'(x_j)$, similar to (3.23). This measure is given by the following result[5]:

Nodal Error of Interpolatory Differentiation

Let $\Pi'_n f(x)$ be the derivative of the interpolant obtained from the ordinates $\{f_0, f_1, \ldots, f_n\}$ of the function $f(x)$ evaluated at the $n+1$ distinct nodes $\{x_0, x_1, \ldots, x_n\}$ within the interval I. Then,

$$|f'(x_j) - \Pi'_n f(x_j)| \leq \frac{1}{(n+1)!} \max_{\xi \in I} |f^{(n+1)}(\xi)| \prod_{\substack{k=0 \\ k \neq j}}^{n} |x_j - x_k|. \tag{3.24}$$

For equidistant node distributions of the form $x_j = a + hj$ $(0 \leq j \leq n)$, where $h = (b-a)/n$, (3.24) leads to the upper bound

$$|f'(x_j) - \Pi'_n f(x_j)| \leq \frac{1}{n+1} \max_{\xi \in I} |f^{(n+1)}(\xi)| \frac{h^n}{\binom{n}{j}}. \tag{3.25}$$

In other words, the error of the derivative of the nth degree interpolant $\Pi'_n f(x)$ evaluated at any of the nodes $\{x_0, x_1, \ldots, x_n\}$ satisfies

$$\boxed{|f'(x_j) - \Pi'_n f(x_j)| = O(h^n).} \tag{3.26}$$

By comparing Cauchy's remainder formula (2.34) for equispaced interpolation with (3.26), we can immediately realize that during the process of differentiation we have lost one order of accuracy, from $O(h^{n+1})$ in the original interpolant, to $O(h^n)$ in its derivative. At this point, the reader may have guessed that further differentiation may deteriorate the accuracy of our original interpolation. This is in general true, and it has important implications since we will need higher degree interpolants in order to approximate higher order derivatives; otherwise, we may have accuracy problems (even if we diminish h). Equations involving second or even fourth order derivatives are common in mathematical physics and in many areas of engineering such as elasticity theory, for example.

The accuracy of a finite difference formula can also be studied in terms of Taylor expansions. For example, consider CD2 formula (3.5) for the approximation of $f'(x)$:

$$f'(x) \approx \frac{1}{2h}(f(x+h) - f(x-h)). \tag{3.27}$$

---

5 For details and proof, see recommended reading at the end of this chapter.

If the difference $f(x + h) - f(x - h)$ on the right-hand side of (3.27) is expanded in its Taylor series in a neighborhood of x, we obtain

$$f(x + h) - f(x - h) = f(x) + hf'(x) + \frac{h^2}{2}f''(x) + \frac{h^3}{6}f'''(x) + \frac{h^4}{24}f^{IV}(x) + \cdots$$

$$- \left[f(x) - hf'(x) + \frac{h^2}{2}f''(x) - \frac{h^3}{6}f'''(x) + \frac{h^4}{24}f^{IV}(x) + \cdots \right]$$

$$= 2hf'(x) + \frac{h^3}{3}f'''(x) + \cdots .$$

Introducing the last quantity into the right-hand side of (3.27) leads to

$$\frac{1}{2h}(f(x + h) - f(x - h)) = f'(x) + \frac{h^2}{6}f'''(x) + \cdots = f'(x) + O(h^2).$$

$$(3.28)$$

That is, the discrepancy between CD2 and the actual derivative $f'(x)$ is

$$\left| \frac{1}{2h}(f(x + h) - f(x - h)) - f'(x) \right| = O(h^2),$$

verifying (3.26) for $n = 2$.

We end this section with a study of the accuracy of finite differences formulas for the approximation of second order derivatives. Following the same procedure as in Section 3.2, we consider the second derivative of the interpolant $\Pi_2 f(x)$ of (3.7) based at the nodes $\{x_0, x_1, x_2\} = \{0, h, 2h\}$:

$$\Pi_2'' f(x) = f_0 \ell_0''(x) + f_1 \ell_1''(x) + f_2 \ell_2''(x),$$

$$(3.29)$$

where $\ell_j(x)$ are the Lagrange cardinal polynomials

$$\ell_0(x) = \frac{1}{2h^2}(x - h)(x - 2h), \quad \ell_1(x) = -\frac{1}{h^2}x(x - 2h), \text{ and } \ell_2(x) = \frac{1}{2h^2}x(x - h).$$

A simple calculation yields

$$\Pi_2'' f(x) = \frac{1}{h^2}f_0 - \frac{2}{h^2}f_1 + \frac{1}{h^2}f_2.$$

$$(3.30)$$

Let $[v_0\ v_1\ v_2]^T = [\Pi_2'' f(x_0)\ \Pi_2'' f(x_1)\ \Pi_2'' f(x_2)]^T$ be the approximated second derivative based on the evaluation of $\Pi_2'' f(x)$. According to (3.30), this vector is given by

$$\begin{bmatrix} v_0 \\ v_1 \\ v_2 \end{bmatrix} = \frac{1}{h^2} \begin{bmatrix} 1 & -2 & 1 \\ 1 & -2 & 1 \\ 1 & -2 & 1 \end{bmatrix} \begin{bmatrix} f_0 \\ f_1 \\ f_2 \end{bmatrix} .$$

We can rewrite the last identity as

$$v_j \equiv \sum_{k=0}^{2} \mathbf{D}_{jk}^{(2)} f_k, \qquad (3.31)$$

for $j = 0, 1, \ldots, n$, where $\mathbf{D}^{(2)}$ is the *second order differentiation matrix* associated with the nodes $\{x_0, x_1, x_2\}$ and whose elements are

$$\boxed{\mathbf{D}_{ij}^{(2)} = \ell_j''(x_i), \quad 0 \le i, j \le n,} \qquad (3.32)$$

for $n = 2$. In this case, $\mathbf{D}^{(2)}$ is a 3×3 matrix that is simply the square of the first order differentiation matrix $\mathbf{D}$ obtained in the Section 3.2:

$$\mathbf{D}^{(2)} = \frac{1}{h^2} \begin{bmatrix} 1 & -2 & 1 \\ 1 & -2 & 1 \\ 1 & -2 & 1 \end{bmatrix}$$

$$= \frac{1}{h} \begin{bmatrix} -3/2 & 2 & -1/2 \\ -1/2 & 0 & 1/2 \\ 1/2 & -2 & 3/2 \end{bmatrix} \frac{1}{h} \begin{bmatrix} -3/2 & 2 & -1/2 \\ -1/2 & 0 & 1/2 \\ 1/2 & -2 & 3/2 \end{bmatrix} = \mathbf{D}^2.$$

As an example, let us approximate the second derivative of $\sin x$ at $x = 1$ using the forward formula given by the first row of the previous expression

$$f''(x) \approx \frac{1}{h^2}(f(x) - 2f(x+h) + f(x+2h)).$$

For $h = 0.01$, the nodes are $\{x_0, x_1, x_2\} = \{1.00, 1.01, 1.02\}$ so we obtain

$$\sin''(1) \approx \frac{\sin(1) - 2\sin(1.01) + \sin(1.02)}{(0.01)^2} \approx -0.846\,824\,787\,709\,144,$$

compared to the exact value $-0.841\,470\,984\,807\,897$, with a remarkably deteriorated local error $|f''(x_0) - v_0| \approx 5 \times 10^{-3}$. As an exercise, the reader may confirm that

$$\frac{1}{h^2}(f(x) - 2f(x+h) + f(x+2h)) \equiv f''(x) + \frac{2}{3}hf'''(x) + \cdots,$$

so that the discrepancy is $O(h)$ and the order 2 of the error of FD2 for the first derivative has been lost.

Practical 3.1: Equispaced (Local) Differentiation

Finite difference matrices are of *Toeplitz* type in which each descending diagonal from left to right is constant. Matlab's **toeplitz** command builds up a matrix from the components of two vectors $u = (u_0, u_1, \ldots, u_m)$ and $v = (v_0, v_1, \ldots, v_m)$, with $v_0 = u_0$, placing them as shown below:

$$\begin{pmatrix} u_0 & v_1 & v_2 & v_3 & \cdots & v_{m-1} & v_m \\ u_1 & u_0 & v_1 & v_2 & \cdots & v_{m-2} & v_{m-1} \\ u_2 & u_1 & u_0 & v_1 & \cdots & v_{m-3} & v_{m-2} \\ u_3 & u_2 & u_1 & u_0 & \cdots & v_{m-4} & v_{m-3} \\ \vdots & \vdots & \vdots & \vdots & & \vdots & \vdots \end{pmatrix}.$$

For instance, type in Matlab's prompt:
toeplitz([1 3 5 7],[1 2 3 4]).

(a) Using **toeplitz**, build the matrices

$$\begin{pmatrix} -1 & 1 & 0 & 0 & \ldots & 0 & 0 \\ 0 & -1 & 1 & 0 & \ldots & 0 & 0 \\ 0 & 0 & -1 & 1 & \ldots & 0 & 0 \\ \vdots & \vdots & \vdots & \vdots & & \vdots & \vdots \\ 0 & 0 & 0 & 0 & \ldots & 1 & 0 \\ 0 & 0 & 0 & 0 & \ldots & -1 & 1 \end{pmatrix} \quad \text{and} \quad \begin{pmatrix} -\frac{1}{2} & 0 & \frac{1}{2} & 0 & \ldots & 0 & 0 & 0 & 0 \\ 0 & -\frac{1}{2} & 0 & \frac{1}{2} & \ldots & 0 & 0 & 0 & 0 \\ 0 & 0 & -\frac{1}{2} & 0 & \ldots & 0 & 0 & 0 & 0 \\ \vdots & \vdots & \vdots & \vdots & & \vdots & \vdots & \vdots & \vdots \\ 0 & 0 & 0 & 0 & \ldots & -\frac{1}{2} & 0 & \frac{1}{2} & 0 \\ 0 & 0 & 0 & 0 & \ldots & 0 & -\frac{1}{2} & 0 & \frac{1}{2} \end{pmatrix},$$

and generate FD1 and CD2 finite difference matrices formulas (3.3) and (3.5), respectively, dividing by h (check their dimension):
toeplitz([-1 zeros(1,m-1)],[-1 1 zeros(1,m-1)])/h
toeplitz([-1/2 zeros(1,m-2)],[-1/2 0 1/2 zeros(1,m-2)])/h.

(b) Evaluate the function $f(x) \equiv \sin(\pi x)$ on the equispaced grid $x_j = 2j/m$, $(j = 0, 1, \ldots, m)$, with $m = 5, 20, 40$, and 80. Apply the previous matrices on the vector of ordinates $[f_0 \; f_1 \; \cdots \; f_m]^{\mathrm{T}}$ in order to obtain approximations of $f'(x_j) \equiv \pi \cos(\pi x_j)$, $(j = 0, 1, \ldots, m)$. On a semi-log graph, plot the errors $|f'(x_j) - u_j^{\mathrm{FD1}}|$ and $|f'(x_i) - u_i^{\mathrm{CD2}}|$.

(c) For $m = 100, 200, 300, 400, \ldots, 2000$, compute $\varepsilon_m \equiv \max|f'(x_j) - u_j|$ for FD1 and CD2. On a log–log graph, plot ε_m versus h and identify the slopes.

3.5   Chebyshev Differentiation

In the Section 3.3, we have seen that finite difference formulas such as (3.5) or (3.6) are extracted from particular rows of the general equispaced differentiation matrix (3.15) that is generated from the derivative of nth degree interpolants obtained from $n+1$ equidistant nodes. The remainder formula (3.26) tells us that the local error in the approximation of the derivative is $O(h^n)$, where h is the grid spacing.

In order to improve the accuracy of our approximation of the derivative of an arbitrary given function $f(x)$ within the interval $[a, b]$ we have two different strategies. The first one is what we have just seen in the Section 3.3:

Reducing h (Constant n): Local Differentiation

First, sample the function on an equidistant set of $m+1$ abscissas $x_j = a + hj$, $(j = 0, 1, \ldots, m)$, with $h = (b-a)/m$. Second, apply any of the finite difference stencils of low or moderate order n on neighboring clusters of abscissas (this can be done since the points are equidistant). Finally, to reduce the local error, simply reduce h by increasing the number of points, that is, increasing m (using the same finite difference stencil, i.e. keeping n constant). Since h is proportional to m^{-1} and the local error of the finite difference stencil is $O(h^n)$, the error of the approximate derivative at the points tends to zero following the rule

$$\boxed{O(m^{-n}),} \tag{3.33}$$

that is, the error approaches zero *algebraically*, with the order n of the finite difference stencil used.

---

or

Reducing the Error by Increasing n: Global Differentiation

Use a unique interpolant based on the Chebyshev nodes

$$x_j = a + \frac{b-a}{2}\left[1 + \cos\left(\frac{\pi}{n}j\right)\right],$$

for $0 \le j \le n$. If the function has k bounded derivatives, (2.52) guarantees that the error of the interpolant decreases like $O(n^{-k})$, i.e. *exponentially* if $f(x) \in C^{\infty}[a, b]$. Accordingly, the error in the derivative of the interpolant should also decrease with similar order. This motivates the use of the differentiation matrix (3.19) to evaluate the approximations of the derivative at each node.

In this current section we follow this second approach. First, recall the Chebyshev nodes within the interval $[-1, 1]$,

$$x_j = \cos\left(\frac{\pi}{n}j\right),\tag{3.34}$$

for $j = 0, 1, \ldots, n$, along with their corresponding barycentric weights

$$\lambda_j = \begin{cases} (-1)^j \dfrac{2^{n-1}}{n} & (j = 1, 2, \ldots, n-1) \\[2mm] (-1)^j \dfrac{2^{n-2}}{n} & (j = 0, n). \end{cases}\tag{3.35}$$

Formal substitution of (3.34) and (3.35) in the general expression (3.19) leads to the differentiation matrix associated with the Chebyshev nodes, usually termed as the *Chebyshev Differentiation Matrix*. As an example, the reader may use Code 5A to compute this matrix for any moderate value of n. For example, to obtain the Chebyshev matrix for $n = 16$ simply type

```
>> n = 16; x = cos([0:n]'*pi/n) ; D = diffmat(x);
```

However, when n becomes large, neighboring Chebyshev nodes x_j and x_k may become highly clustered near the edges of the interval so that the expression $x_j - x_k$ appearing in (3.19) may lead to the subtraction of two nearly identical quantities, resulting in remarkable round-off errors that may deteriorate the accuracy of the matrix elements $\mathbf{D}_{ij}$. If we define the coefficients $\delta_0 = \delta_n = 1/2$ and $\delta_1 = \delta_2 = \cdots = \delta_{n-1} = 1$, numerically more accurate explicit relations of the matrix elements of (3.19) for the Chebyshev nodes (3.34) are given by the following:

Chebyshev Differentiation Matrix

$$\mathbf{D}_{ij} = \begin{cases} (-1)^{i+j} \dfrac{\delta_j}{\delta_i(x_i - x_j)}, & (i \neq j), \\[4mm] \dfrac{(-1)^{i+1}}{\delta_i} \displaystyle\sum_{k=0\ (k\neq i)}^{n} (-1)^k \dfrac{\delta_k}{x_i - x_k}, & (i = j). \end{cases}\tag{3.36}$$

For small or even moderate values of n, applying (3.19) or (3.36) does not produce noticeable differences. However, for large n, relations appearing in (3.36) are remarkably more accurate.

```
% Code 5B: Chebyshev Differentiation matrix
% Input: n
% Output: differentiation matrix D and Chebyshev nodes
function [D,x] = chebdiff(n)
x =cos([0:n]'*pi/n); d = [.5;ones(n-1,1);.5];
D = zeros(n+1,n+1);
for ii = 0:n
 for jj = 0:n
  ir = ii + 1; jc = jj + 1;
  if ii == jj
   kk = [0:ii-1 ii+1:n]'; num = (-1).^kk.*d(kk+1);
   D(ir,jc) =((-1)^(ir)/d(ir))*sum(num./(x(ir)-x(kk+1)));
  else
   D(ir,jc) = d(jc)*(-1)^(ii+jj)/((x(ir)-x(jc))*d(ir));
  end
 end
end
```

Code 5B is a simple adaptation of Code 5A for the particular case of Chebyshev nodes, incorporating numerically stable formulas (3.36). We recommend the reader to explore the numerical differences between the Chebyshev differentiation matrix obtained using Codes 5A and 5B.

In order to approximate the derivative of an arbitrary function $f(x)$ within the interval $x \in [a, b]$, we proceed as in Chapter 2 by considering the change of variable

$$x = x(z) = a + \frac{b - a}{2}(z + 1),$$

where $z \in [-1, 1]$. If we define $g(z) \equiv f(x(z))$, then

$$f'(x) = \frac{\mathrm{d}}{\mathrm{d}x}f(x) = \frac{\mathrm{d}z}{\mathrm{d}x}\frac{\mathrm{d}}{\mathrm{d}z}g(z) = \frac{2}{b - a}g'(z).$$

The Chebyshev discretization of the function is carried out within the interval $[-1, 1]$ using the node distribution $z_j = \cos(j\pi/n)$, $(j = 0, 1, \ldots, n)$ so that $f_j \equiv f(x_j)$ is actually $f_j \equiv g(z_j) \equiv g_j$, for $j = 0, 1, \ldots, n$. The approximation u_j of the exact derivative $f'(x_j)$ is therefore

$$f'(x_j) = \frac{2}{b - a}g'(z_j) \approx \frac{2}{b - a}\sum_{k=0}^{n}\mathrm{D}_{jk}g(z_k) = \frac{2}{b - a}\sum_{k=0}^{n}\mathrm{D}_{jk}f_k. \qquad (3.37)$$

In Part II, we will approximate second order derivatives by squaring (3.36) so that $\mathbf{D}^{(2)} = \mathbf{D}^2$.

Practical 3.2: Local Versus Global Numerical Differentiation

To compare the convergence properties of local finite difference formulas versus global differentiation matrices, consider within the interval $[-1,3]$ the function $f(x) = e^{-x} \sin^2(2x)$ and its exact derivative $f'(x) = e^{-x} \sin(2x)(4\cos(2x) - \sin(2x))$.

(a) Consider the set of equidistant points $x_j = -1 + 4j/n$ $(j = 0, 1, \ldots, n)$, with $n = 4, 8, 12, \ldots, 64$. Approximate the derivative $f'(x_j)$ at the interior points of the interval using the finite difference formula

$$u_j = \frac{1}{12h}[-f(x_j + 2h) + 8f(x_j + h) - 8f(x_j - h) + f(x_j - 2h)],$$

for $j = 1, 2, \ldots, n - 1$. For each value of n, compute the maximum error

$$\varepsilon_n = \max_{1 \leq j \leq n-1} |f'(x_j) - u_j|.$$

(b) Repeat (a) using the general differentiation matrix provided by Code 5A applied to the same set of equispaced nodes. Do you observe any anomaly for large n?

(c) Repeat (b) using the general differentiation matrix provided by Code 5A applied to the Chebyshev distribution

$$x_j = a + \frac{b-a}{2}\left[1 + \cos\left(\frac{\pi}{n}j\right)\right],$$

for $a = -1$, $b = 3$, and $0 \leq j \leq n$. For larger values of n, you may alternatively use expression (3.36) combined with the factor $2/(b - a)$ appearing in (3.37).

On a linear-log graph, plot the quantities ε_n registered in each case. According to the plot, for a given tolerance δ, say $\delta = 10^{-3}$, which method provides the derivative with such accuracy requiring less function evaluations, i.e. which of them satisfies $|\varepsilon_n| < \delta$ for a smaller n?

(d) For $n = 48$, approximate $f''(x)$ by squaring Chebyshev differentiation matrix and explore its accuracy by plotting $|\sum_{j=0}^n \mathbf{D}_{ij}^{(2)} f(x_j) - f''(x_i)|$ on a semi-logarithmic scale.

Complementary Reading

Almost every treatise on numerical analysis devotes a chapter to the concept of numerical differentiation. Rigorous treatments of the concept of interpolatory differentiation error can be found in classical texts such as Burden and Faires' *Numerical Analysis* or in Isaacson and Keller's *Analysis of Numerical Methods*, which extends the analysis to arbitrary order of differentiation. For a more general and updated approach, I particularly like Quarteroni, Sacco, and Saleri's monograph *Numerical Mathematics*, where the reader will find modern topics such as *compact finite differences*, not covered in this book.

The concept of differentiation matrix is frequently absent in classical numerical analysis texts and it is more likely found in treatises specialized on numerical methods for partial differential equations (which require a more advanced mathematical background). Alternative algorithms to obtain the coefficients of finite difference stencils of arbitrary order can be found in Fornberg's monograph *A Practical Guide to Pseudospectral Methods*, where the reader will find a very useful survey on central and forward–backward formulas.

For readers with a slightly higher mathematical background, a *must-read* discussion about *algebraic* versus *exponential-geometric* convergence, along with their corresponding graphical interpretations, can be found in Chapter 2 of Boyd's monograph *Chebyshev and Fourier Spectral Methods*. On these lines, I also recommend to have a look at Table 2.4-1 (p. 13) of the aforementioned Fornberg's book.

Problems and Exercises

The symbol (A) means that the problem has to be solved analytically, whereas (N) means to use numerical Matlab codes. Problems with the symbol * are slightly more difficult.

1. (A) Consider the differentiation matrix for the equally spaced nodes $x_j = x_0 + jh$, with $j = 0, 1, 2, 3, 4$:

$$\mathbf{D} = \frac{1}{h} \begin{pmatrix} -25/12 & 4 & -3 & 4/3 & -1/4 \\ -1/4 & -5/6 & 3/2 & -1/2 & 1/12 \\ 1/12 & -2/3 & 0 & 2/3 & -1/12 \\ -1/12 & 1/2 & -3/2 & 5/6 & 1/4 \\ 1/4 & -4/3 & 3 & -4 & 25/12 \end{pmatrix}$$

(a) Observe that the third row of $\mathbf{D}$ is the centered finite difference formula

$$f'(x) \approx \frac{1}{12h}[-f(x+2h) + 8f(x+h) - 8f(x-h) + f(x-2h)].$$

Show that this formula has a local error of order $O(h^4)$.

(b) Obtain the second order differentiation matrix for the same set of nodes by squaring $\mathbf{D}$:

$$\mathbf{D}^2 = \frac{1}{h^2} \begin{pmatrix} 35/12 & -26/3 & 19/2 & -14/3 & 11/12 \\ 11/12 & -5/3 & 1/2 & \cdots & \cdots \\ \cdots & \cdots & \cdots & \cdots & \cdots \\ \cdots & \cdots & \cdots & \cdots & \cdots \\ \cdots & \cdots & \cdots & \cdots & \cdots \end{pmatrix},$$

and obtain the centered difference formula for the approximation of the second derivative

$$f''(x) \approx \frac{1}{12h^2}[-f(x+2h) + 16f(x+h) - 30f(x) + 16f(x-h) - f(x-2h)].$$

What is the order of the error in this case?

2. (A) Which of the following two finite difference formulas is more accurate?

$$f'''(x) \approx \frac{1}{h^3}[f(x+3h) - 3f(x+2h) + 3f(x+h) - f(x)];$$

$$f'''(x) \approx \frac{1}{2h^3}[f(x+2h) - 2f(x+h) + 2f(x-h) - f(x-2h)].$$

3. (A) Calculate the error in the following finite difference formulas:

$$f^{(4)}(x) \approx \frac{1}{h^4}[f(x+4h) - 4f(x+3h) + 6f(x+2h) - 4f(x+h) + f(x)];$$

$$f^{(4)}(x) \approx \frac{1}{h^4}[f(x+2h) - 4f(x+h) + 6f(x) - 4f(x-h) + f(x-2h)].$$

4. Consider the node distribution

$$x_j = \frac{1}{j+1}$$

for $j = 0, 1, 2, \ldots, n$, over the interval $(0, 1]$. For $n = 2$, obtain the cardinal Lagrange polynomials and the differentiation matrix associated with these nodes.

4

Numerical Integration

4.1 Introduction

According to the *Fundamental Theorem of Calculus*, if $F(x)$ is the *antideriva-tive* function of $f(x)$,[1] then

$$\int_a^b f(x)\ \mathrm{d}x \equiv F(b) - F(a). \tag{4.1}$$

In a few situations, these integrals may be calculated exactly because the integrand is relatively simple and the antiderivative function is known. In general, however, we need to approximate these integrals numerically.

Integrals appear regularly in many branches of physics and engineering. We have a clear example in statistical thermodynamics, where Debye theory predicts that the heat capacity of a crystal at absolute temperature T is proportional to

$$\int_0^{\Theta_\mathrm{D}/T} \frac{x^4 \mathrm{e}^x}{(\mathrm{e}^x - 1)^2}\ \mathrm{d}x, \tag{4.2}$$

where Θ_D is the Debye temperature. There is no closed form for the antiderivative of the integrand appearing in (4.2), and therefore that integral must be approximated numerically. In this chapter, we will address how to approximate definite integrals such as (4.2) whenever possible. We will also study how to deal with apparent singularities (improper integrals) of the integrand such as the one exhibited in (4.2) for $x \equiv 0$, or how to compute integrals over infinite or semi-infinite domains, such as in the limit $T \to 0$ for very low temperatures in (4.2).

---

1 That is, if $F'(x) \equiv f(x)$.

Fundamentals of Numerical Mathematics for Physicists and Engineers, First Edition. Alvaro Meseguer.
© 2020 John Wiley & Sons, Inc. Published 2020 by John Wiley & Sons, Inc.

4.2 Interpolatory Quadratures

In Chapter 2, we approximated derivatives of a function $f(x)$ by differentiating its corresponding interpolant obtained from sampling $f(x)$ over a set of prescribed nodes. The underlying concept in numerical integration is essentially the same: to approximate the integral of a function by means of the integration of its corresponding interpolant. For an arbitrary continuous function $f(x)$ in $[a, b]$, consider the integral

$$I(f) = \int_a^b f(x)\, \mathrm{d}x. \tag{4.3}$$

Let

$$\Pi_n f(x) = \sum_{j=0}^n f_j\, \ell_j(x) \in \mathbb{R}_n[x] \tag{4.4}$$

be the Lagrange form of the nth degree interpolant of $f(x)$ obtained from the ordinates $f_j = f(x_j)$ of the function evaluated on a given set of distinct nodes $\{x_0, x_1, \ldots, x_n\} \in [a, b]$. If $\Pi_n f(x)$ is a good approximation of $f(x)$, it seems therefore legitimate to approximate $I(f)$ in (4.3) by the integral of the interpolant, that is,

$$I(\Pi_n f) \approx I(f). \tag{4.5}$$

If we define $I_n(f) \doteq I(\Pi_n f)$, the approximated integral is

$$
\begin{aligned}
I_n(f) &= \int_a^b \left[\sum_{j=0}^n f_j\, \ell_j(x) \right] \mathrm{d}x \\
&= \int_a^b \{ f_0 \ell_0(x) + f_1 \ell_1(x) + \cdots + f_n \ell_n(x) \}\, \mathrm{d}x.
\end{aligned} \tag{4.6}
$$

Integrating each term of the sum above leads to

$$I_n(f) = f_0 \int_a^b \ell_0(x)\, \mathrm{d}x + f_1 \int_a^b \ell_1(x)\, \mathrm{d}x + \cdots + f_n \int_a^b \ell_n(x)\, \mathrm{d}x. \tag{4.7}$$

If we define the quantities

$$w_j \doteq \int_a^b \ell_j(x)\, \mathrm{d}x \quad (j = 0, 1, \ldots, n), \tag{4.8}$$

then (4.7) can be expressed as

$$I_n(f) = f_0 w_0 + f_1 w_1 + \cdots + f_n w_n = \sum_{j=0}^n f_j w_j. \tag{4.9}$$

Expression (4.9) is usually known as an *interpolatory quadrature rule*.[2] The quantities (4.8) are termed as *quadrature weights*, and they are *independent* of the integrand $f(x)$. The integrals appearing in (4.8) simply involve the cardinal Lagrange polynomials in $\mathbb{R}_n[x]$ and they are relatively easy to calculate for low values of n.

Let us start from the simplest case: $n = 0$. We approximate $f(x)$ using the 0th degree interpolant $\Pi_0 f(x) = f_0$ obtained from the ordinate of the function at some prescribed abscissa $x_0 \in [a, b]$, which is usually taken at the center $x_0 = (a + b)/2$ of the domain of integration. Since $\ell_0(x) = 1$, in this case the quadrature formula reduces to

$$I_0(f) = f_0 w_0 = f\left(\frac{a+b}{2}\right) \int_a^b 1 \ \mathrm{d}x = (b-a)f\left(\frac{a+b}{2}\right). \tag{4.10}$$

The approximation (4.10) is usually known as the *midpoint quadrature formula* and its associated quadrature weight is $w_0 = b - a$. Similarly, for $n = 1$, the interpolant is $\Pi_1 f(x) = f_0 \ell_0(x) + f_1 \ell_1(x)$, where the nodes are $x_0 = a$ and $x_1 = b$ so that

$$I_1(f) = f_0 w_0 + f_1 w_1 = f(a) \int_a^b \ell_0(x) \ \mathrm{d}x + f(b) \int_a^b \ell_1(x) \ \mathrm{d}x$$

$$= f(a) \int_a^b \frac{x-b}{a-b} \ \mathrm{d}x + f(b) \int_a^b \frac{x-a}{b-a} \ \mathrm{d}x = \frac{b-a}{2}[f(a) + f(b)]. \tag{4.11}$$

This quadrature is known as the *trapezoidal rule*, with associated quadrature weights $w_0 = w_1 = (b - a)/2$. As an exercise, the reader may repeat the same analysis for $n = 2$, $x_0 = a$, $x_1 = (a + b)/2$, and $x_2 = b$ in order to obtain the so-called *Simpson's rule*,

$$I_2(f) = (b-a)\left[\frac{1}{6}f(a) + \frac{2}{3}f\left(\frac{a+b}{2}\right) + \frac{1}{6}f(b)\right], \tag{4.12}$$

with associated weights $w_0 = w_2 = (b - a)/6$ and $w_1 = 2(b - a)/3$. Figure 4.1 shows a geometrical representation of the three quadrature rules for a particular positive function $f(x)$, where the numerical value of $I(f)$ is the area bounded between the x-axis and the curve $y = f(x)$. The midpoint formula shown in Figure 4.1a provides a very crude approximation of $I(f)$ based on the rectangle of area $(b - a)f_0$. This can be slightly improved by the trapezoidal rule shown in Figure 4.1b based on the straight line passing through the points of coordinates (x_0, f_0) and (x_1, f_1). Simpson's rule shown in Figure 4.1c based on a quadratic interpolant is remarkably more accurate than the previous two quadratures.

For a given arbitrary set of $n + 1$ distinct nodes $\{x_0, x_1, \dots, x_n\} \in [a, b]$ (not necessarily equidistant), calculating the weights w_j introduced in (4.8) requires

---

2 From the Latin word *quadrum* (square).

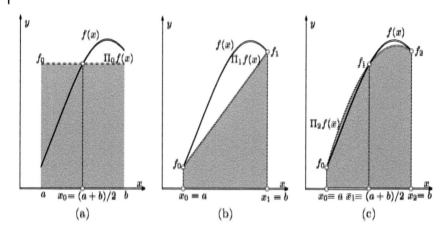

Figure 4.1 Geometrical representation of simple interpolatory quadratures for the approximation of $I(f)$. (a) Midpoint, (b) trapezoidal, and (c) Simpson. The original function f (solid curve) is interpolated and the resulting interpolant (dashed curve) integrated afterwards. The shaded regions represent the approximated area in each case.

the integration of the Lagrange cardinal polynomials $\ell_j(x)$ associated with the nodes x_j. An alternative way of computing the quadrature weights is as follows. Recall from identity (2.24) that $\Pi_n p(x) = p(x)$ for any polynomial $p(x)$ in $\mathbb{R}_n[x]$. An immediate consequence of (2.24) is that

$$I_n(x^k) \equiv I(\Pi_n x^k) \equiv I(x^k) \equiv \frac{1}{k+1}\left[b^{k+1} - a^{k+1}\right], \qquad (4.13)$$

for $k = 0, 1, \ldots, n$. In other words, the quadrature formula $I_n(f)$ must provide the *exact* value $I(f)$ if $f(x)$ is any of the monomials $\{1, x, x^2, \ldots, x^n\} \in \mathbb{R}_n[x]$. Therefore, the quadrature weights must satisfy the relations

$$I_n(x^k) \equiv x_0^k \cdot w_0 + x_1^k \cdot w_1 + \cdots + x_n^k \cdot w_n \equiv \frac{1}{k+1}\left[b^{k+1} - a^{k+1}\right], \quad (4.14)$$

for $k = 0, 1, \ldots, n$. The $n + 1$ equations (4.14) are known as *exactness conditions* of quadrature formula (4.9), and they lead to a linear system for the quadrature weights:

$$
\begin{aligned}
1 \cdot w_0 + 1 \cdot w_1 + \cdots + 1 \cdot w_n &\equiv b - a, \\
x_0 \cdot w_0 + x_1 \cdot w_1 + \cdots + x_n \cdot w_n &\equiv \frac{1}{2}[b^2 - a^2], \\
x_0^2 \cdot w_0 + x_1^2 \cdot w_1 + \cdots + x_n^2 \cdot w_n &\equiv \frac{1}{3}[b^3 - a^3], \\
&\;\;\vdots \\
x_0^n \cdot w_0 + x_1^n \cdot w_1 + \cdots + x_n^n \cdot w_n &\equiv \frac{1}{n+1}[b^{n+1} - a^{n+1}],
\end{aligned}
\qquad (4.15)
$$

or expressed in vector form,

$$
\begin{bmatrix}
1 & 1 & \cdots & 1 \\
x_0 & x_1 & \cdots & x_n \\
x_0^2 & x_1^2 & \cdots & x_n^2 \\
\vdots & \vdots & & \vdots \\
x_0^n & x_1^n & \cdots & x_n^n
\end{bmatrix}
\begin{bmatrix}
w_0 \\
w_1 \\
w_2 \\
\vdots \\
w_n
\end{bmatrix}
\equiv
\begin{bmatrix}
b - a \\
(b^2 - a^2)/2 \\
(b^3 - a^3)/3 \\
\vdots \\
(b^{n+1} - a^{n+1})/(n + 1)
\end{bmatrix} .
\tag{4.16}
$$

The reader must find the matrix above familiar, since it is simply the transposed Vandermonde matrix introduced in Part I; see expression (2.5). Since the nodes are distinct, this matrix is non-singular and therefore system (4.16) always provides a unique set of quadrature weights that makes formula (4.9) exact for any polynomial of degree less than or equal to n, that is, such that $I_n(f) \equiv I(f)$, $\forall f \in \mathbb{R}_n[x]$. In some textbooks, this method is referred to as the *method of undetermined coefficients*, similar to the one used to obtain interpolating polynomials using the natural basis of monomials x^k.

As an illustration, consider the case $n \equiv 1$ with $x_0 \equiv a$ and $x_1 \equiv b$, where (4.16) reduces to the system

$$
\begin{bmatrix}
1 & 1 \\
a & b
\end{bmatrix}
\begin{bmatrix}
w_0 \\
w_1
\end{bmatrix}
\equiv
\begin{bmatrix}
b - a \\
(b^2 - a^2)/2
\end{bmatrix} .
\tag{4.17}
$$

A simple calculation shows that the solution is $w_0 \equiv w_1 \equiv (b - a)/2$, i.e. the trapezoidal rule (4.11), as expected. The reader may also check that the vector of weights corresponding to Simpson's rule (4.12)

$$
[w_0\ w_1\ w_2]^{\mathrm{T}} \equiv \left[\frac{b - a}{6}\ \ \frac{2(b - a)}{3}\ \ \frac{b - a}{6} \right]^{\mathrm{T}}
$$

solves the linear system

$$
\begin{bmatrix}
1 & 1 & 1 \\
a & (a + b)/2 & b \\
a^2 & (a + b)^2/4 & b^2
\end{bmatrix}
\begin{bmatrix}
w_0 \\
w_1 \\
w_2
\end{bmatrix}
\equiv
\begin{bmatrix}
b - a \\
(b^2 - a^2)/2 \\
(b^3 - a^3)/3
\end{bmatrix} ,
\tag{4.18}
$$

which is precisely (4.16) for $n \equiv 2$, $x_0 \equiv a$, $x_1 \equiv (a + b)/2$, and $x_2 \equiv b$.

As mentioned in Chapter 2, linear systems such as the one appearing in (4.16) become ill-conditioned for moderately large n, potentially leading to inaccurate results.[3] Therefore, the computation of quadrature weights by means of solving system (4.16) is not advisable for large n. In Section 4.4, we will see how to

---

3 In Part II we will introduce the concept of condition number of a matrix and how to measure its impact on the accuracy of the results.

compute accurate quadrature weights by imposing exactness using suitable sets of polynomials (not monomials) and nodes (not equidistant) that lead to well-conditioned linear systems.

Summary: Interpolatory Quadrature Formulas

For a given set of distinct nodes $\{x_0, x_1, \ldots, x_n\}$ in the interval $[a, b]$, interpolatory quadrature formulas approximate the value of the integral

$$I(f) = \int_a^b f(x)\,\mathrm{d}x$$

by the sum

$$I_n(f) = \int_a^b \Pi_n f(x)\,\mathrm{d}x = \sum_{j=0}^n f_j w_j,$$

where $\Pi_n f(x)$ is the interpolant of the integrand $f(x)$ obtained from the ordinates $f_j = f(x_j)$. The quantities w_j are called *quadrature weights* and they are the result of integrating the jth cardinal polynomial:

$$w_j \doteq \int_a^b \ell_j(x)\,\mathrm{d}x \quad (j = 0, 1, \ldots, n).$$

For small or moderate values of n, these weights can also be obtained by solving the linear system

$$\begin{bmatrix} 1 & 1 & \cdots & 1 \\ x_0 & x_1 & \cdots & x_n \\ x_0^2 & x_1^2 & \cdots & x_n^2 \\ \vdots & \vdots & & \vdots \\ x_0^n & x_1^n & \cdots & x_n^n \end{bmatrix} \begin{bmatrix} w_0 \\ w_1 \\ w_2 \\ \vdots \\ w_n \end{bmatrix} = \begin{bmatrix} b - a \\ (b^2 - a^2)/2 \\ (b^3 - a^3)/3 \\ \vdots \\ (b^{n+1} - a^{n+1})/(n+1) \end{bmatrix},$$

which arises when imposing the quadrature formula to be exact for any polynomial $p(x) \in \mathbb{R}_n[x]$.

4.2.1 Newton–Cotes Quadratures

Formulas (4.10)–(4.12) are particular cases of a more general family of quadrature rules usually known as *Newton–Cotes formulas*, henceforth referred to as NCFs. These quadratures are particular cases of (4.9) for equally distributed nodes within the interval $[a, b]$. If $x_0 = a$ and $x_n = b$ (such as in the trapezoidal or Simpson rules), the NCF is said to be *closed*, whereas if $a < x_0$ and $x_n < b$

(such as in the midpoint rule) then the formula is said to be *open*. In general, closed and open NCFs are prescribed as follows:

Closed and Open Newton–Cotes Formulas

NCFs are particular cases of the quadrature formula (4.9) for equally spaced nodes. In a closed ncf, the nodes are given by the distribution

$$x_k = a + kh, \quad \text{with } n \geq 1, \ h = \frac{b-a}{n} \text{and } k = 0, 1, \ldots, n, \qquad (4.19)$$

where the limits of integration belong to the set of nodes: $x_0 = a$ and $x_n = b$.
In an open NCF, the nodes are

$$x_k = a + (k+1)h, \quad \text{with } n \geq 0, \ h = \frac{b-a}{n+2} \text{and } k = 0, 1, \ldots, n, \qquad (4.20)$$

where the limits of integration are not included in the nodes: $x_0 = a + h$ and $x_n = b - h$.
In both cases, the quadrature formula is given by the expression

$$I_n(f) = h \sum_{j=0}^{n} \omega_j f(x_j), \quad \text{with} \quad \begin{cases} h = \dfrac{b-a}{n} & \text{(closed)}, \\[2mm] h = \dfrac{b-a}{n+2} & \text{(open)}. \end{cases} \qquad (4.21)$$

Table (4.1) contains the coefficients ω_j corresponding to the open and closed Newton–Cotes quadrature formulas (4.21) up to degrees 3 and 4, respectively. Notice that these coefficients *are not* the quadrature weights w_j appearing in

Table 4.1 Coefficients ω_j of open and closed Newton–Cotes quadrature formulas (4.21).

n	Closed				Open			
	1	2	3	4	0	1	2	3
ω_0	1/2	1/3	3/8	14/45	2	3/2	8/3	55/24
ω_1	1/2	4/3	9/8	64/45	—	3/2	−4/3	5/24
ω_2	—	1/3	9/8	24/45	—	—	8/3	5/24
ω_3	—	—	3/8	64/45	—	—	—	55/24
ω_4	—	—	—	14/45	—	—	—	—
Name	Trapezoidal	Simpson	3/8th	Milne	Midpoint		Steffensen	

(4.9), but proportional to them according to the relation $w_j = h\omega_j$, where h takes the value following (4.21) depending on whether the formula is open or closed. For example, let us approximate the integral

$$I(f) = \int_0^\pi e^x \sin(x) \, dx = \frac{e^\pi + 1}{2} \approx 12.07, \tag{4.22}$$

using the closed NCF for $n = 3$ (usually known as the "*3/8ths rule*"). In this case, $h = \pi/3$ and (4.21) provides the approximation

$$\frac{\pi}{3}\left[\frac{3}{8}f(0) + \frac{9}{8}f\left(\frac{\pi}{3}\right) + \frac{9}{8}f\left(\frac{2\pi}{3}\right) + \frac{3}{8}f(\pi)\right] \approx 11.19. \tag{4.23}$$

We can proceed similarly using an open NCF with $n = 3$ (which is one of what are usually known as *Steffensen rules* – rightmost column of Table 4.1). In this case, $h = \pi/5$ and the quadrature is

$$\frac{\pi}{5}\left[\frac{55}{24}f\left(\frac{\pi}{5}\right) + \frac{5}{24}f\left(\frac{2\pi}{5}\right) + \frac{5}{24}f\left(\frac{3\pi}{5}\right) + \frac{55}{24}f\left(\frac{4\pi}{5}\right)\right] \approx 13.29. \tag{4.24}$$

Clearly, the two previous quadratures are not very accurate – they lead to approximations with relative errors ranging from 7% to nearly 10%. Theoretically, we could obtain more accurate approximations of (4.22) by using Newton–Cotes formulas based on interpolants of degree $n = 5, 6, \ldots$, and higher. However, the reader should remember that this strategy will eventually fail for large n since NCF are implicitly based on interpolants obtained from equally spaced nodes, therefore leading to inherently ill-conditioned computations.

As in numerical differentiation, there are two possible strategies to improve the accuracy of interpolatory quadratures. The first option is based on the *divide and conquer* principle, where the domain of integration $[a, b]$ is partitioned in a large number of small subintervals. Since the integrand $f(x)$ is expected to have small changes in each one of those intervals, any low degree quadrature rule[4] (such as the NCF seen before) applied on these small domains should provide better accuracy. The resulting approximation is therefore the sum of these low degree quadratures, usually termed as *composite* or *compound* quadrature rules. To some extent, we proceeded similarly in Chapter 3 to approximate the derivative of functions on large domains by means of finite difference formulas. Recall that we used *local* interpolants of low or moderate degree to obtain such approximations at different locations of the domain. The second option is, as the reader may have already guessed, to use just one interpolant of high order based on suitable nodes that lead to well-conditioned computations. In the rest of this section, we will address the first option.

---

4 That is, obtained from a low degree interpolant

4.2.2 Composite Quadrature Rules

One of the most important formulas in numerical mathematics is the *composite trapezoidal rule*. In this quadrature, the domain of integration $[a, b]$ is partitioned in m subintervals $[x_k, x_{k+1}]$ of the same length $h \equiv (b - a)/m$, where the abscissas x_k are

$$x_k \equiv a + kh, \tag{4.25}$$

for $k \equiv 0, 1, \ldots, m$ (see Figure 4.2). The original integral is then decomposed as the sum of the integrals over each subinterval:

$$I(f) \equiv \int_a^b f(x) \, dx \equiv \int_{x_0}^{x_1} f(x) \, dx + \int_{x_1}^{x_2} f(x) \, dx + \cdots + \int_{x_{m-1}}^{x_m} f(x) \, dx. \tag{4.26}$$

The composite trapezoidal rule is obtained when applying the simple quadrature template (4.11) on each one of the integrals in (4.26) so that

$$\int_{x_k}^{x_{k+1}} f(x) \, dx \approx \frac{x_{k+1} - x_k}{2} [f(x_k) + f(x_{k+1})] \equiv \frac{h}{2}(f_k + f_{k+1}), \tag{4.27}$$

for $k \equiv 0, 1, \ldots, m - 1$, and where we have used the notation $f_j \equiv f(x_j)$ for brevity. As a result, we obtain the approximated integral

$$I(f) \approx \frac{h}{2}(f_0 + f_1) + \frac{h}{2}(f_1 + f_2) + \cdots + \frac{h}{2}(f_{m-1} + f_m). \tag{4.28}$$

Each one of the terms $h(f_k + f_{k+1})/2$ of the sum above is in fact the area of the hatched trapeziums of Figure 4.2. Rearranging the elements appearing in (4.28) leads to the composite trapezoidal formula, denoted by $I_{1,m}(f)$, where the subscripts indicate that we have used local interpolants of degree 1 in each one of the m subdomains:

$$I_{1,m}(f) \equiv \frac{h}{2}(f_0 + 2f_1 + 2f_2 + \cdots + 2f_{m-1} + f_m)$$
$$= h\left(\frac{1}{2}f_0 + f_1 + f_2 + \cdots + f_{m-1} + \frac{1}{2}f_m\right) \equiv h \sum_{k=0}^{m}{}' f_k. \tag{4.29}$$

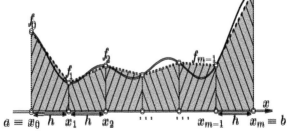

Figure 4.2 Areas associated with the composite trapezoidal rule (hatched regions below dashed lines).

The prime sign appearing in the previous sum means that the first and last terms of the sum are to be halved. Overall, the composite trapezoidal rule, frequently referred to as *trapezoidal quadrature*,[5] reads

$$I_{1,m}(f) = h \sum_{k=0}^{m}{}' f(a + kh)$$

$$= h \left(\frac{f(a)}{2} + f(a + h) + \cdots + f(b - h) + \frac{f(b)}{2} \right). \tag{4.30}$$

As we will see later, this formula provides outstanding accuracy when it is used to approximate integrals of periodic functions. For this reason, formula (4.30) is *extremely* important as it lies at the very heart of the so-called DFT (discrete Fourier transform) for the trigonometric interpolation of periodic functions that will be addressed in Part II. In the trapezoidal rule, the integrations for each one of the partitioned subintervals were approximated by means of linear interpolants. A more accurate composite quadrature can be obtained by considering local quadratic interpolants, i.e. using Simpson's quadrature rule (4.12). To obtain a *composite Simpson quadrature*, the domain of integration $[a, b]$ is partitioned in $2m$ subintervals of width $h = (b - a)/m$ (see Figure 4.3). Consider the $2m + 1$ abscissas

$$x_k = a + k\frac{h}{2}, \tag{4.31}$$

for $k = 0, 1, \ldots, 2m$. Notice that in this case the distance between adjacent abscissas is $h/2$, not h. If we split $I(f)$ as the sum of the integrals

$$I(f) = \int_a^b f(x)\, \mathrm{d}x = \int_{x_0}^{x_2} f(x)\, \mathrm{d}x + \int_{x_2}^{x_4} f(x)\, \mathrm{d}x + \cdots + \int_{x_{2m-2}}^{x_{2m}} f(x)\, \mathrm{d}x, \tag{4.32}$$

we can apply the simple Simpson's quadrature template (4.12) on each one of the subintervals of integration $[x_0, x_2]$, $[x_2, x_4]$, etc., appearing in (4.32) so that

$$\int_{x_k}^{x_{k+2}} f(x)\, \mathrm{d}x \approx h \left[\frac{1}{6}f(x_k) + \frac{2}{3}f(x_{k+1}) + \frac{1}{6}f(x_{k+2}) \right], \tag{4.33}$$

for $k = 0, 1, \ldots, 2m - 2$. Figure 4.3 shows that the parabolic upper boundaries of each one of the hatched subdomains $[x_k, x_{k+2}]$ are the dashed parabolas resulting from the quadratic interpolation obtained from the local nodes

---

5 Since the primitive trapezoidal template (4.11) is rarely used in practice.

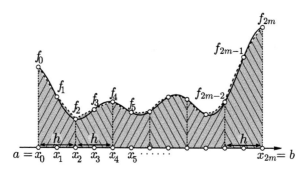

Figure 4.3 Areas associated with the composite Simpson's rule (hatched regions below dashed parabolic curves).

$\{x_k, x_{k+1}, x_{k+2}\}$. As a result, we obtain the approximated integral

$$I(f) \approx h\left[\frac{f_0}{6} + \frac{2f_1}{3} + \frac{f_2}{6}\right] + h\left[\frac{f_2}{6} + \frac{2f_3}{3} + \frac{f_4}{6}\right] + \cdots$$

$$+ h\left[\frac{f_{2m-2}}{6} + \frac{2f_{2m-1}}{3} + \frac{f_{2m}}{6}\right]$$

$$= \frac{h}{6}[f_0 + 4f_1 + 2f_2 + 4f_3 + 2f_4 + \cdots + 2f_{2m-2} + 4f_{2m-1} + f_{2m}].$$

Rearranging the sum above leads to the *composite Simpson's quadrature*:

$$I_{2,m}(f) = \frac{h}{6}\left[f_0 + 4\sum_{k=1}^{m} f_{2k-1} + 2\sum_{k=1}^{m-1} f_{2k} + f_{2m}\right]. \tag{4.34}$$

At this point, it should be emphasized that composite formulas (4.30) and (4.34) are the cumulative result of stencil quadratures (4.11) and (4.12) arising from *local* linear and quadratic interpolations, respectively. In that sense, this technique is essentially the same as the one used in finite difference formulas of Chapter 3, where FD1, BD1, or CD2 stencils were used in different parts of the partitioned domain.

Codes 6 and 7 provide simple implementations of composite quadratures (4.30) and (4.34), respectively. These codes require the external function **fun** to be evaluated on sets of abscissas (vectors), so the Matlab *element-wise* operations .*, ./, or .ˆ must be used when constructing that function:

```
%Code 6: Composite Trapezoidal Quadrature
% Input: a-b (low-up lim.); m (# intervals); fun (func. name)
% Output: I_{1,m}(f)
 function T = ctrap(a,b,m,fun)
 h = (b-a)/m;x = a + [0:m]'*h;f = feval(fun,x); N = m + 1;
 T = h*(.5*f(1)+sum(f(2:N-1))+.5*f(N));
```

(Continued)

```
%Code 7: Composite Simpson's Quadrature
% Input: a-b (low-up lim.); m (# intervals); fun (func. name)
% Output: I_{2,m}(f)
   function S = csimp(a,b,m,fun)
   h = (b-a)/m;x = a + [0:2*m]*h/2;f = feval(fun,x);N = 2*m+1;
   S = (h/6)*(f(1)+4*sum(f(2:2:N-1))+2*sum(f(3:2:N-2))+f(N));
```

Let us now explore the accuracy of these composite quadrature rules by approximating again the integral (4.22). Taking for example $m = 16$ in (4.30) provides $I_{1,16}(f) \approx 11.9929$, with a relative error smaller than 1%. In order to make a fair comparison with Simpson's composite quadrature, we take $m = 8$ (since the number of function evaluations in $I_{2,m}(f)$ is $2m + 1$). The reader may check that (4.34) provides in that case $I_{2,8}(f) \approx 12.0699$, which is nearly 200 times more accurate than $I_{1,16}(f)$. This simple example already reveals that we may achieve outstanding higher accuracy not necessarily only by increasing the number of partitions, but also by increasing (if possible) the degree of the interpolant.

4.3   Accuracy of Quadrature Formulas

In this section, we study the accuracy of quadrature formulas arising from equispaced interpolation. We start with two definitions:

Error and Degree of Exactness of a Quadrature Formula

We define the *quadrature error* of an interpolatory quadrature formula

$$I_n(f) \equiv \sum_{k=0}^{n} w_k f(x_k),$$

approximating the value of $I(f) = \int_a^b f(x)\, dx$, as the difference

$$\boxed{E_n(f) \equiv I(f) - I_n(f)},$$ (4.35)

In addition, the *degree of exactness* (also known as *order of accuracy*) of the quadrature rule $I_n(f)$ is the maximum integer $r \geq 0$ for which

$$\boxed{E_n(p) \equiv 0, \quad \forall p \in \mathbb{R}_r[x]},$$ (4.36)

that is, if the quadrature formula is exact for polynomials of degree r at most.

The method of undetermined coefficients provides a systematic procedure to generate the quadrature weights $\{w_0, w_1, \ldots, w_n\}$ by imposing the quadrature formula to be exact for all polynomials in $\mathbb{R}_n[x]$, that is, to satisfy (4.36) with $r = n$. It may occur, however, that the quadrature formula produced by this method may have a degree of exactness *higher* than n, as we will see later. Besides, the exactness of a quadrature formula does not provide a quantitative measurement of its numerical accuracy, that is, of its quadrature error (4.35). In this section, we illustrate how to obtain this quantitative measurement of the error in a simple case, namely, the midpoint quadrature rule. Before doing that, however, we first need to review a very important result from the theory of integration.

Consider two functions $u(x)$ and $v(x)$ that are continuous in $[a, b]$, and let us assume that $v(x) > 0$. Also, let A and B be the lower and upper bounds of $u(x)$ in the interval, that is,

$$A \leq u(x) \leq B, \quad \forall x \in [a, b]. \tag{4.37}$$

Therefore, according to the *intermediate value theorem*, for any intermediate value C between A and B, we can always find $\xi \in [a, b]$ such that $u(\xi) = C$. Since $v(x) > 0$, and by virtue of (4.37), we may also write

$$Av(x) \leq u(x)v(x) \leq Bv(x), \quad \forall x \in [a, b]. \tag{4.38}$$

Let us define $I \equiv \int_a^b v(x) \, dx$. If we integrate between a and b throughout the inequality (4.38), we obtain

$$A\, I \leq \int_a^b u(x)v(x) \, dx \leq B\, I, \tag{4.39}$$

or, dividing by I,

$$A \leq I^{-1} \int_a^b u(x)v(x) \, dx \leq B. \tag{4.40}$$

The previous inequality tells us that the quantity $I^{-1} \int_a^b u(x)v(x) \, dx$ appearing in the middle is an intermediate value between A and B, which are precisely the lower and upper bounds of $u(x)$. Therefore, $u(x)$ must take this intermediate value at some $x \equiv \xi \in [a, b]$, that is,

$$u(\xi) = I^{-1} \int_a^b u(x)v(x) \, dx \equiv \frac{\displaystyle\int_a^b u(x)v(x) \, dx}{\displaystyle\int_a^b v(x) \, dx}. \tag{4.41}$$

In other words, we can always find an intermediate abscissa $\xi \in [a, b]$ such that

$$\int_a^b u(x)v(x) \, dx = u(\xi) \int_a^b v(x) \, dx.$$

(4.42)

This result is usually known as the *mean value theorem for integrals*, and it can be used to provide an estimation of the quadrature error (4.35). As an illustration, let us approximate

$$I(f) = \int_a^b f(x) \, dx$$

(4.43)

using the midpoint quadrature rule (4.10),

$$I_0(f) = (b - a)f\left(\frac{a+b}{2}\right).$$

Unless stated otherwise, we will henceforth assume that our integrand $f(x)$ is a smooth function. We can approximate $f(x)$ by its Taylor expansion at the midpoint of the interval $x_0 = (a + b)/2$:

$$f(x) = f(x_0) + f'(x_0)(x - x_0) + \frac{1}{2}(x - x_0)^2 f''(\eta(x)),$$

(4.44)

where, according to Taylor's theorem, $\eta(x) \in [x_0, x]$ if $x_0 < x$, or $\eta(x) \in [x, x_0]$ if $x_0 > x$. To avoid cumbersome notation, let us define $u(x) \doteq f''(\eta(x))$. Introducing expansion (4.44) in (4.43),

$$I(f) = \int_a^b \left[f(x_0) + f'(x_0)(x - x_0) + \frac{(x - x_0)^2}{2} u(x) \right] dx.$$

After integrating each term, we obtain

$$I(f) = (b - a)f(x_0) + f'(x_0)\frac{1}{2}[(b - x_0)^2 - (a - x_0)^2]$$
$$+ \frac{1}{2} \int_a^b (x - x_0)^2 u(x) \, dx.$$

Since $x_0 = (a + b)/2$, the first term of the previous expression is in fact the midpoint rule and the second term cancels out. As a result, we may write

$$I(f) = I_0(f) + \frac{1}{2} \int_a^b (x - x_0)^2 u(x) \, dx.$$

(4.45)

In other words, the quadrature error of the midpoint rule is

$$E_0(f) = I(f) - I_0(f) = \frac{1}{2} \int_a^b (x - x_0)^2 u(x) \, dx.$$

(4.46)

In this last integral, we can identify the binomial $(x - x_0)^2$ as the positive function $v(x)$ appearing in (4.42). As a result, there exists an intermediate abscissa $\xi \in [a, b]$ such that

$$\frac{1}{2} \int_a^b (x - x_0)^2 u(x) \, dx = \frac{1}{2} u(\xi) \int_a^b (x - x_0)^2 \, dx = \frac{1}{24} u(\xi)(b - a)^3.$$

(4.47)

Finally, we conclude that the quadrature error of the midpoint rule is

$$E_0(f) = I(f) - I_0(f) = \frac{1}{24}(b - a)^3 f''(\xi),$$

(4.48)

for some $\xi \in [a, b]$. Since $E_0(f)$ is proportional to $f''(\xi)$, the midpoint quadrature is exact in $\mathbb{R}_1[x]$. Remember that when we applied the method of undetermined coefficients to obtain w_0, Eq. (4.10) implicitly implied exactness just in $\mathbb{R}_0[x]$. The quadrature error (4.48) can also be expressed in terms of $h = (b - a)/2$ as

$$E_0(f) = \frac{h^3}{3} f''(\xi).$$

(4.49)

We have obtained an explicit expression of the quadrature error for the midpoint quadrature rule. In general, the mean value theorem for integrals can be used to measure the quadrature error of higher order quadrature formulas. For example, it can be shown that the quadrature errors of NCF satisfy the following:

Quadrature Errors of Newton–Cotes Formulas

$$E_n(f) \propto \begin{cases} \dfrac{h^{n+3}}{(n+2)!} f^{(n+2)}(\xi), & n = 0, 2, 4, \ldots; \\[3mm] \dfrac{h^{n+2}}{(n+1)!} f^{(n+1)}(\xi), & n = 1, 3, 5, \ldots, \end{cases}$$

(4.50)

for some $\xi \in [a, b]$ and $h = (b - a)/n$ or $h = (b - a)/(n + 2)$ if the quadrature formula is closed or open, respectively.

For more details regarding the proportionality factors in (4.50) we refer the reader to the Complementary Reading section, at the end of the chapter. Expressions (4.50) reveal two important facts. On the one hand, NCFs have an extra degree of exactness if n is *even*, due to the $(n + 2)$th derivative of $f(x)$. On the other hand, the quadrature errors of NCF are $O(h^{n+3})$ and $O(h^{n+2})$, for even and odd n, respectively. For this reason, many numerical analysis textbooks explicitly mention that even and odd NCFs have *order of infinitesimal* h^{n+3} and h^{n+2}, respectively. Since $h = O(n^{-1})$ in closed and open formulas, NCFs have quadrature errors that are essentially $O(n^{-n})$, that is, they provide *geometrical* convergence in n. However, as mentioned in Chapter 2, increasing n with equispaced nodes is numerically ill-conditioned. This is precisely the reason for the existence of composite NCFs, since they reduce the quadrature error not by increasing n but by diminishing the order of infinitesimal after performing a partition of the domain $[a, b]$ in m subdomains of size $h = (b-a)/m$, that is, $h = O(m^{-1})$. As a result, it can be shown that the composite NCFs satisfy the following:

Quadrature Errors of Composite Newton–Cotes Formulas

The quadrature error of a composite NCF

$$E_{n,m}(f) \equiv I(f) - I_{n,m}(f) \tag{4.51}$$

based on local interpolants of degree n applied over m partitions of the domain $[a, b]$ satisfies

$$E_{n,m}(f) \propto \begin{cases} m^{-(n+2)}, & n = 0, 2, 4, \ldots; \\ m^{-(n+1)}, & n = 1, 3, 5, \ldots. \end{cases} \tag{4.52}$$

The algebraic convergence rates shown above are inherently of the same nature as those seen in (3.33) for finite difference templates obtained from the differentiation of local interpolants of degree n.

To illustrate the convergence rates (4.52), we explore the accuracy of composite trapezoidal and Simpson rules (4.30) and (4.34) in the approximation of integral (4.22) for a wide range of values of m. Table 4.2 contains the numerical approximations provided by the two composite rules. As expected, composite Simpson's rule performs much better than its trapezoidal counterpart. Figure 4.4 shows the absolute quadrature error of the trapezoidal $I_{1,m}(f)$ and Simpson's $I_{2,m}(f)$ rules for m ranging between 10 and 10^3. According to (4.52), the errors $E_{1,m}(f)$ and $E_{2,m}(f)$ should decrease as m^{-2} and m^{-4}, respectively. This is in perfect agreement with the curves shown in Figure 4.4.

Even though Simpson's rule seems to perform outstandingly well, it still requires about $m = 10^3$ partitions (i.e. nearly 2000 function evaluations) to get 12 exact decimals. This is computationally very expensive. The reader may

Table 4.2 Trapezoidal $I_{1,m}(f)$ and Simpson $I_{2,m}(f)$ composite quadrature approximations of $I(f) = \int_0^\pi e^x \sin x \, dx = (e^\pi + 1)/2 \approx 12.070\,346\,316\,389\,6$.

m	$I_{1,m}(f)$ (Trapezoidal)	$I_{2,m}(f)$ (Simpson)
10	11.872 453 333 353 8	12.070 182 061 832
17	12.001 722 492 663	12.070 3 26 724 722
28	12.045 031 791 874	12.070 343 657 626
46	12.060 964 541 243	12.070 345 951 571
77	12.066 997 721 000	12.070 346 269 931
129	12.069 153 207 122	12.070 346 310 492
215	12.069 916 791 618	12.070 346 315 625
359	12.070 192 260 530	12.070 346 316 291
599	12.070 290 979 571	12.070 346 316 376
1000	12.070 326 461 472	12.070 346 316 388

a) The converged figures are shadowed.

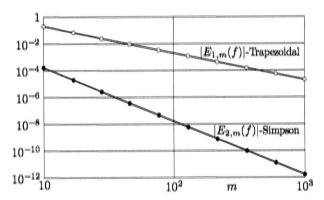

Figure 4.4 Absolute quadrature errors of composite trapezoidal and Simpson rules in the approximation of integral (4.22).

wonder whether there are quadrature rules capable of achieving that accuracy at a much reduced computational cost, i.e. with considerably smaller number of function evaluations. For sufficiently smooth integrands the answer is affirmative, and in the Section 4.4 we will study these quadratures.

Practical 4.1:   Local Versus Global Equispaced Quadratures

(a) Apply Simpson's formula (4.34) to compute the trivial integrals

$$J_k = \int_0^1 x^k \, dx = \frac{1}{k+1},$$

for $k = 7, 4, 3$. For $m = 10, 20, 30, \ldots, 100$, compute the absolute quadrature error $\varepsilon_m = |E_{2,m}|$. Plot ε_m versus m on a $\texttt{loglog}$ graph and identify the order (exponent) of algebraic convergence. What order do you observe? Is it the same in all the integrals? If not, explain.

(b) One possible way for estimating π is through the numerical approximation of the integral

$$M = \int_0^1 \frac{1}{1+x^2} \, dx = \arctan(1) = \frac{\pi}{4}, \text{ so that } \pi = 4M.$$

Using composite Simpson's rule, approximate π with an absolute error smaller than 10^{-13}. How many function evaluations does it require to achieve that accuracy?

(c) Alternatively, devise your own quadrature rule based on equidistant nodes $x_j = j/n$, $j = 0, 1, \ldots, n$. Compute the weights by imposing exactness in $\mathbb{R}_n[x]$, i.e. by solving linear system (4.16) for $a = 0$ and $b = 1$

$$\sum_{\ell=0}^n x_\ell^k w_\ell = (k+1)^{-1}, \quad k = 0, 1, \ldots, n.$$

Here you have a simple code for $n = 4$:

```
n = 4; j = [0:n]; x = j/n; V = zeros(n+1);
for k = j; V(k+1,:) = x.^k; end;w = V\(1./(1+j))';
```

Redo (b) by increasing n. Notice (but do not worry for now) the warning message displayed by Matlab. Have you managed to achieve the same accuracy as in (b)? If not, provide your best approximation, the number of function evaluations required, and explain the origin of such limited accuracy. Would you use this alternative quadrature formula in general?

4.4   Clenshaw–Curtis Quadrature

Practical 4.1 reveals that if it not were for the ill-conditioning, quadratures based on one global interpolant of high degree seemed to be much more accurate

(using the same number of function evaluations) than those based on different local low degree interpolants, such as composite Simpson. In Chapter 3 we
already saw that a global interpolant based on a modest number of Chebyshev nodes could provide much better accuracy than finite difference formulas
based on local equispaced interpolants. A natural question is whether there
exist quadrature formulas based on Chebyshev nodes providing geometric convergence. The answer is affirmative and in this section we will provide such
quadrature formula.

Let us start by considering the integral

$$I(f) = \int_{-1}^{1} f(x) \, dx. \tag{4.53}$$

Unless stated otherwise, we will assume that the integrand $f(x)$ is sufficiently
smooth within the integration domain $[-1, 1]$, with bounded derivatives of order
as high as needed. In Chapter 2 we already introduced the Chebyshev nodes
of the second kind, also known as Chebyshev's practical abscissas:

$$x_j = \cos\left(\frac{\pi}{n}j\right), \tag{4.54}$$

for $j = 0, 1, \ldots, n$. At this point, it will be useful to introduce a new basis of
polynomials in $\mathbb{R}_n[x]$ over the interval $[-1, 1]$.

Chebyshev Polynomials

Consider the function $T_n(x) = \cos(n\theta)$, where n is a nonnegative integer,
$x = \cos\theta$, and $\theta \in [0, \pi]$. For any x in $[-1, 1]$, the single-valued function

$$\boxed{T_n(x) = \cos(n \arccos x),} \tag{4.55}$$

is called the *Chebyshev polynomial of degree n*.

Using the definition above, it is obvious that $T_0(x) = 1$ and $T_1(x) = x$. For
$n = 2$, we can use the trigonometric identities to obtain $T_2(x) = \cos 2\theta = 2\cos^2\theta - 1 = 2x^2 - 1 = 2xT_1(x) - T_0(x)$. In general, Chebyshev polynomials
(4.55) [6] satisfy the recurrence

$$T_{n+1}(x) = 2xT_n(x) - T_{n-1}(x), \quad (n \geq 1). \tag{4.56}$$

Figure 4.5 shows the first four Chebyshev polynomials. This family of polynomials also satisfy the parity rule

$$T_n(-x) = (-1)^n T_n(x). \tag{4.57}$$

---

6 Also known as Chebyshev polynomials of the first kind.

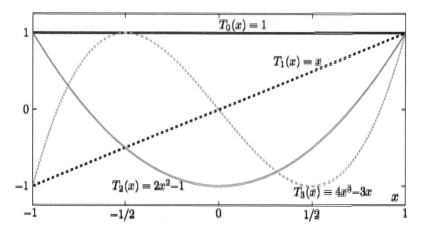

Figure 4.5 Chebyshev polynomials T_0 (solid black), T_1 (dashed black), T_2 (solid gray), and T_3 (dashed gray).

Since $x(\theta) \equiv \cos\theta$ or, equivalently, $\theta(x) \equiv \arccos x$, the derivative of $T_n(x)$ is

$$\frac{\mathrm{d}}{\mathrm{d}x}T_n(x) \equiv \frac{\mathrm{d}}{\mathrm{d}x}\cos(n\theta(x)) \equiv -n\sin(n\theta)\frac{\mathrm{d}\theta}{\mathrm{d}x} \equiv n\frac{\sin n\theta}{\sin\theta}.$$

As a consequence, for $n \geq 2$, we have

$$\frac{\mathrm{d}}{\mathrm{d}x}\left[\frac{T_{n+1}(x)}{n+1} - \frac{T_{n-1}(x)}{n-1}\right] \equiv \frac{\sin[(n+1)\theta] - \sin[(n-1)\theta]}{\sin\theta}$$
$$= 2\cos(n\theta) \equiv 2T_n(x).$$

In general, the antiderivative of the Chebyshev polynomials $T_n(x)$ is

$$\int^x T_n(z)\,\mathrm{d}z \equiv \begin{cases} x, & (n \equiv 0) \\[2ex] \dfrac{x^2}{2}, & (n \equiv 1) \\[2ex] \dfrac{1}{2}\left[\dfrac{T_{n+1}(x)}{n+1} - \dfrac{T_{n-1}(x)}{n-1}\right], & (n \geq 2), \end{cases} \qquad (4.58)$$

where we omitted the constants. So far in this book, we have only considered the *canonical* or *monomial* basis $\{1, x, x^2, \ldots, x^n\}$ of $\mathbb{R}_n[x]$. By default, Chebyshev polynomials of degree less than or equal to n are expressed as a linear combination of these monomials. Conversely, every element of the canonical basis x^k of $\mathbb{R}_n[x]$ can also be expressed as a linear combination of the Chebyshev polynomials $\{T_0, T_1, T_2, \ldots, T_n\}$:

$$1 \equiv T_0, \quad x \equiv T_1, \quad x^2 \equiv (T_0 + T_2)/2, \quad x^3 \equiv (3T_1 + T_3)/4, \ldots.$$

In other words, we may use the set $\{T_0, T_1, T_2, \ldots, T_n\}$ as an alternative basis of $\mathbb{R}_n[x]$.

We saw in Chapter 2 that the interpolants resulting from sampling functions at the Chebyshev nodes (4.54) were extremely accurate. Accordingly, it is legitimate to expect the integral $I(f)$ in (4.53) to be accurately approximated by $I(\Pi_n f)$, where $\Pi_n f(x)$ is the Chebyshev interpolant of $f(x)$. As in (4.9), let us consider the quadrature formula

$$I_n(f) \equiv I(\Pi_n f) = f_0 w_0 + f_1 w_1 + \cdots + f_n w_n \equiv \sum_{j=0}^{n} f_j w_j, \qquad (4.59)$$

where f_j are the ordinates of the function $f(x)$ at the Chebyshev nodes (4.54):

$$f_j \equiv f(x_j) = f\left(\cos\left(j\frac{\pi}{n}\right)\right), \qquad (4.60)$$

for $j = 0, 1, \ldots, n$. In order to obtain the quadrature weights w_j, we proceed by demanding maximum degree of exactness of (4.59), as we did before using the method of undetermined coefficients. In other words, we must impose

$$I_n(p) \equiv I(p), \quad \forall p \in \mathbb{R}_n[x]. \qquad (4.61)$$

Our experience in the past tells us that imposing exactness using the monomial basis $\{x^k\}$ is a bad idea, since the resulting Vandermonde matrix leads to numerical ill-conditioning (see Matlab's warning messages of Practical 4.1). This time we proceed by imposing exactness using the Chebyshev polynomial basis, that is,

$$I_n(T_\ell) \equiv I(T_\ell) \equiv \int_{-1}^{1} T_\ell(x)\, dx, \quad \ell = 0, 1, \ldots, n. \qquad (4.62)$$

By virtue of (4.57) and (4.58), the $n + 1$ exactness conditions are

$$I_n(T_\ell) \equiv \sum_{j=0}^{n} w_j T_\ell\left(\cos\left(j\frac{\pi}{n}\right)\right) \equiv \begin{cases} 0, & (\ell \quad \text{odd}) \\ \dfrac{2}{1 - \ell^2}, & (\ell \quad \text{even}), \end{cases} \qquad (4.63)$$

for $\ell = 0, 1, 2, \ldots, n$. In matrix-vector form, the exactness conditions read

$$
\begin{bmatrix}
1 & 1 & 1 & \cdots & 1 \\
1 & \cos(\pi/n) & \cos(2\pi/n) & \cdots & -1 \\
1 & \cos(2\pi/n) & \cos(4\pi/n) & \cdots & 1 \\
1 & \cos(3\pi/n) & \cos(6\pi/n) & \cdots & -1 \\
\vdots & \vdots & \vdots & & \vdots \\
1 & -1 & 1 & \cdots & (-1)^n
\end{bmatrix}
\begin{bmatrix}
w_0 \\ w_1 \\ w_2 \\ w_3 \\ \vdots \\ w_n
\end{bmatrix}
\equiv
\begin{bmatrix}
2 \\ 0 \\ -2/3 \\ 0 \\ -2/15 \\ \vdots
\end{bmatrix}, \qquad (4.64)
$$

Fortunately, the matrix appearing in (4.64) with elements $V_{\ell j} \equiv \cos(\ell j \pi/n)$ is extremely well-conditioned and the weights can be accurately computed by solving that system. The weights arising from the solution of (4.64) are usually known as the *Clenshaw–Curtis* weights. Accordingly, quadrature formula (4.59) is named *Clenshaw–Curtis quadrature*.

```
n = 6; l = [0:n]'; k = [2:n]';
w = cos(l*l'*pi/n)\[2;0;(1+(-1).^k)./(1-k.^2)];
```

The two Matlab command lines above exemplify a rather simple (but computationally not very efficient) way to compute the Clenshaw–Curtis quadrature weights for $n = 6$, where we have used the dyadic or tensor product `l*l'` to generate the matrix of system (4.64). For even n, it can be shown that the quadrature weights w_j that solve (4.64) are explicitly given by

$$
w_j = \begin{cases} \dfrac{1}{n^2 - 1}, & (j = 0, n), \\[2mm] \dfrac{4}{n} \sum_{k=0}^{n/2}{}' \dfrac{1}{1 - 4k^2} \cos\left(\dfrac{2\pi k j}{n}\right), & (1 \le j \le n - 1). \end{cases}
\tag{4.65}
$$

As usual, the prime sign means that the first and last terms of the sum are to be halved. Quadrature (4.59) based on the ordinates (4.60) of the function at the Chebyshev nodes can only be used to approximate integrals within the integration domain $[-1, 1]$. For an arbitrary integral

$$
\int_a^b f(z) \, dz,
\tag{4.66}
$$

with $f(z)$ smooth in $[a, b]$, we perform the change of variable

$$
z(x) = a + \frac{b - a}{2}(x + 1),
\tag{4.67}
$$

for $x \in [-1, 1]$, so that integral (4.66) now reads

$$
\frac{b - a}{2} \int_{-1}^1 g(x) \, dx,
\tag{4.68}
$$

where $g(x) \doteq f(z(x))$. Therefore, it is just a matter of applying Clenshaw–Curtis quadrature to approximate the integral above, that is,

$$
\int_a^b f(z) \, dz \approx \frac{b - a}{2} \sum_{j=0}^n w_j g_j,
\tag{4.69}
$$

where w_j are the Clenshaw–Curtis weights and g_j are the ordinates of $g(x)$ at the Chebyshev nodes, that is,

$$
g_j = g(x_j) = f\left(a + \frac{b - a}{2}\left[\cos\left(\frac{j\pi}{n}\right) + 1\right]\right), \quad j = 0, 1, \ldots, n.
\tag{4.70}
$$

Summary: Clenshaw–Curtis Quadrature

Clenshaw–Curtis quadrature approximation of the integral

$$I(f) = \int_{-1}^{1} f(x) \, dx,$$

is given by $I(\Pi_n f(x))$, where $\Pi_n f(x)$ is the Chebyshev interpolant of $f(x)$ based on the ordinates $f_j = f(x_j)$ evaluated at the Chebyshev nodes (*practical abscissas*) $x_j = \cos(j\pi/n)$, $j = 0, 1, \ldots, n$:

$$I_n^{\mathrm{CC}}(f) = \sum_{j=0}^{n} w_j f_j, \tag{4.71}$$

where the weights $\{w_0, w_1, \ldots, w_n\}$, usually known as *Clenshaw–Curtis weights*, are the solution of system (4.64). For even n, the w_j are

$$w_j = \begin{cases} (n^2 - 1)^{-1}, & (j = 0, n), \\ \dfrac{4}{n} \displaystyle\sum_{k=0}^{n/2}{}' \dfrac{1}{1 - 4k^2} \cos\left(\dfrac{2\pi k j}{n}\right), & (1 \leq j \leq n-1). \end{cases} \tag{4.72}$$

The approximation for arbitrary (bounded) integration domains is

$$\int_a^b f(z) \, dz \approx \frac{b - a}{2} \sum_{j=0}^{n} w_j f\left(a + \frac{b-a}{2}(x_j + 1)\right). \tag{4.73}$$

Code 8 contains Matlab's function `qclencurt`, a simple implementation of Clenshaw–Curtis quadrature for arbitrary domains (4.73) and values of n. Code 8 computes the weights by solving the linear system (4.64). In practice, efficient Clenshaw–Curtis implementations make use of trigonometric transformations that considerably reduce the computational cost. We refer the reader to the Complementary Reading section at the end of this chapter for more details.

```
%Code 8: Clenshaw-Curtis Quadrature function
% Input: a-b (low-up lim.); n (# nodes-1); fun (func. name)
% Output: I_n(f)
function Icc = qclencurt(a,b,n,fun)
l = [0:n]'; k = [2:n]'; x = cos(l*pi/n);
w = cos(l*l'*pi/n)\[2;0;(1+(-1).^k)./(1-k.^2)];
z = a+.5*(b-a)*(x+1); f = feval(fun,z);
Icc =.5*(b-a)*w'*f;
```

In order to compare the accuracy of Clenshaw–Curtis quadrature with composite trapezoidal and Simpson rules, we proceed to approximate the integral appearing in (4.22) using Code 8. The resulting approximations can be found in Table 4.3, to be compared with those appearing in Table 4.2. For $n = 15$(which involves just 16 function evaluations), Clenshaw–Curtis already provides full accuracy in double precision arithmetic (Simpson's rule required nearly 2000 evaluations to get an error of 10^{-12}). Figure 4.6 represents the Clenshaw–Curtis absolute quadrature error $|E_n^{CC}(f)| \equiv |I(f) - I_n^{CC}(f)|$, clearly showing a geometric decay as a function of n on a `loglog` graph, to be compared with the algebraic convergence rates exhibited by trapezoidal and Simpson composite rules already seen in Figure 4.4. In general, the Clenshaw–Curtis quadrature exhibits geometric convergence if the integrand is sufficiently smooth within (and near) the domain of integration. Providing rigorous bounds of $|E_n^{CC}(f)|$ is out of the scope of this introductory book. We recommend the interested reader to have a look at modern treatises on numerical quadrature recommended at the end of this book.

Table 4.3 Clenshaw–Curtis quadrature approximation of
$I(f) \equiv \int_0^\pi e^x \sin x \, dx = (e^\pi + 1)/2 \approx 12.070\ 346\ 316\ 389\ 6.$

n	$I_n(f)$
3	12.582 248 553 443 8
5	12.069 269 698 472 4
9	12.070 346 336 544 9
15	12.070 346 316 389 6

a)   The converged figures are shadowed.

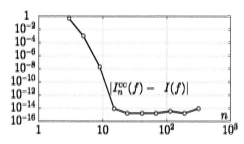

Figure 4.6   Absolute quadrature error in the Clenshaw–Curtis quadrature approximation of integral (4.22).

Practical 4.2: Gravitational Potential of a Ring

A ring of radius a and linear mass density λ lies on the plane $z = 0$, as shown in the figure on the right. The gravitational potential dV created by a ring differential of length $d\ell = a\,d\theta$ and mass dm at a point $(0, y, z)$ of the plane $x = 0$ is

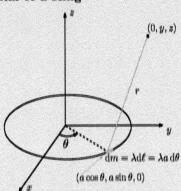

$$dV = -G\frac{\lambda a\,d\theta}{\sqrt{a^2\cos^2\theta + (y - a\sin\theta)^2 + z^2}},$$

where G is the gravitational constant.

For simplicity, take $G = 1$, $a = 1$, and $\lambda = 1$ so that the potential at an arbitrary point $(y, z) \neq (\pm 1, 0)$ is given by the integral

$$V(y, z) \equiv -\int_0^{2\pi} \frac{d\theta}{\sqrt{\cos^2\theta + (y - \sin\theta)^2 + z^2}}.$$

Using the Clenshaw–Curtis quadrature $I_n^{CC}(f)$, devise a Matlab function that approximates the integral above in order to provide $V(y, z)$ along the $x = 0$ plane for any point $(y, z) \neq (\pm 1, 0)$. Compute $V(1.5, 1)$ with an absolute error smaller than 10^{-6}. What is the minimum value of n required to achieve that accuracy?

(a) For $y = 1.5$, compute with absolute accuracy 10^{-6} the potential $V(z)$ within the range $z \in [-5, 5]$. Did you require the same n for all values of z to reach that precision? If not, plot $n(z)$ and try to explain what may be the cause for this.

(b) Compute the equipotential curves $V(y, z) = \text{const.}$ for $x = 0$ and $(y, z) \in [-1.5, 1.5] \times [-1, 1]$. Approximate the gradient $\nabla V = [\partial_y V\ \partial_z V]^T$ at (y_0, z_0) using finite differences (take $h \approx 10^{-4}$):

$$\nabla V \approx \frac{1}{h}[V(y_0 + h, z_0) - V_0\ \ V(y_0, z_0 + h) - V_0]^T,$$

and move perpendicularly to the gradient to a nearby point $[y_1\ z_1]^T \equiv [x_0\ y_0]^T + h[-\partial_z V, \partial_y V]^T$. Optionally, use Matlab's contour function.

4.5 Integration of Periodic Functions

In this section, we will study optimal strategies to integrate periodic functions. We will illustrate these techniques by solving the classical mathematical problem of calculating the circumference or length ℓ of an ellipse of semiaxes a and b:

$$\frac{x^2}{a^2} + \frac{y^2}{b^2} = 1, \tag{4.74}$$

that can be parameterized as

$$x(\theta) = a\cos\theta, \quad y(\theta) = b\sin\theta, \tag{4.75}$$

for $0 \leq \theta < 2\pi$. The length $d\ell$ of a differential elliptical arc is given by

$$d\ell = \sqrt{(dx)^2 + (dy)^2} = d\theta\sqrt{a^2\sin^2\theta + b^2\cos^2\theta}. \tag{4.76}$$

Without loss of generality, we can assume that $a > b$ so that $d\ell$ may be expressed as

$$d\ell = d\theta\sqrt{a^2 - (a^2 - b^2)\cos^2\theta} = a\,d\theta\sqrt{1 - k^2\cos^2\theta}, \tag{4.77}$$

where we have introduced the new variable $k^2 \doteq (a^2 - b^2)/a^2$. As a result, the length of the ellipse is

$$\ell = a\int_0^{2\pi}\sqrt{1 - k^2\cos^2\theta}\,d\theta. \tag{4.78}$$

The trigonometric integral appearing in (4.77) is of the *elliptic* type,[7] and must be calculated numerically. For example, according to (4.77), the length of an ellipse with semiaxes $a = 5$ and $b = 4$ (see Figure 4.7a) is

$$\ell = 5\int_0^{2\pi}\sqrt{1 - \frac{9}{25}\cos^2\theta}\,d\theta. \tag{4.79}$$

Figure 4.7b plots the integrand within the interval $[0, 2\pi]$. Owing to the periodicity of the integrand (or obvious symmetry of the ellipse), the length ℓ is twice the length of the gray dashed arc of length $\ell/2$ that can be calculated by integrating between 0 and π (gray dashed curve in Figure 4.7b), that is,

$$\ell = 10\int_0^{\pi}\sqrt{1 - \frac{9}{25}\cos^2\theta}\,d\theta. \tag{4.80}$$

We will approximate the integral appearing in (4.80) using different strategies. The first one, as the reader may already have guessed, is to use the Clenshaw–Curtis quadrature. The second one is the compound trapezoidal rule.

---

7 More precisely, an *incomplete elliptic integral of the second kind*.

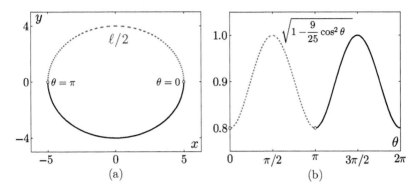

Figure 4.7 Computation of the length ℓ of the ellipse $x^2/25 + y^2/16 = 1$. (a) We compute the length $\ell/2$ of the dashed gray arc by integrating just half of the interval since the integrand is π-periodic; dashed curve in (b).

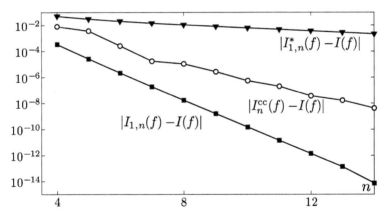

Figure 4.8 Semi-logarithmic plot of the absolute quadrature errors corresponding to the three methods used to approximate $I(f) = 10 \int_0^\pi (1 - (9/25)\cos^2\theta)^{1/2}\,d\theta$. From bottom to top: Composite trapezoidal (black squares), Clenshaw–Curtis (hollow circles), and composite trapezoidal applied with previous splitting of the domain of integration $[0, \pi/4]$ and $[\pi/4, \pi]$ (black triangles).

Since the integrand of (4.80) is regular within the integration domain, we expect Clenshaw–Curtis to provide geometrical convergence. However, the reader may at this point wonder why use the composite trapezoidal rule, since this rule is known to have just algebraic convergence of order 2. The answer is in Figure 4.8, showing the quadrature error of the two strategies. The most unexpected result is that composite trapezoidal rule outperforms by far Clenshaw–Curtis quadrature. Both quadratures exhibit geometrical convergence, although the

Table 4.4 Composite trapezoidal $I_{1,n}(f)$, Clenshaw–Curtis $I_n^{cc}(f)$, and composite trapezoidal with splitting $I_{1,n}^*(f)$ quadrature approximations of

$$I(f) = 10 \int_0^\pi (1 - (9/25)\cos^2\theta)^{1/2} \, d\theta \approx 28.361\,667\,888\,974\,5.$$

n	$I_{1,n}(f)$	$I_n^{cc}(f)$	$I_{1,n}^*(f)$
4	28.361 332 653 853 4	28.369 716 926 131 7	28.309 281 417 700 5
8	28.361 667 871 864 7	28.361 656 928 116 9	28.348 815 901 346 9
12	28.361 667 888 973 1	28.361 667 925 755 9	28.355 975 581 517 9
16	28.361 667 888 974 5	28.361 667 888 681 1	28.358 469 826 284 3
20	28.361 667 888 974 5	28.361 667 888 977 6	28.359 622 270 573

a) The converged figures are shadowed.

composite trapezoidal rule clearly requires nearly half the number of function evaluations to achieve the same accuracy as Clenshaw–Curtis. The resulting approximations of the length ℓ have been summarized in the first two columns of Table 4.4

What we have just described is a particular case of a more general property of the composite trapezoidal rule:

Geometrical Convergence of Composite Trapezoidal Rule

Consider the integral

$$I(f) = \int_a^b f(x) \, dx,$$

where $f(x)$ is a smooth function whose *odd* derivatives satisfy

$$f^{(j)}(a) = f^{(j)}(b),$$

for $j = 1, 3, 5, \ldots, 2k - 1$. Then, the composite trapezoidal rule

$$I_{1,n}(f) = h \left(\frac{1}{2} f_0 + f_1 + f_2 + \cdots + f_{n-1} + \frac{1}{2} f_n \right), \tag{4.81}$$

with $f_j = f(a + hj)$ and $h = (b - a)/n$, approximates $I(f)$ with a quadrature error satisfying

$$|E_{1,n}(f)| = |I_{1,n}(f) - I(f)| = O(n^{-(2k+1)}). \tag{4.82}$$

As a consequence, if the integrand $f(x)$ is a smooth periodic function with period $b - a$, the trapezoidal approximation $I_{1,n}(f)$ converges as a function of n *geometrically* to $I(f)$.

This result has very important implications in many other areas of numerical mathematics such as in the computation of Fourier transforms of periodic functions (covered in Part II) or in the approximation of Cauchy closed contour integrals in the complex plane, for example.

For periodic integrands, the trapezoidal rule only provides geometrical convergence if the integration limits are suitably chosen. For example, if we split the integral appearing in (4.80) in two,

$$\int_0^\pi \sqrt{1 - \frac{9}{25}\cos^2\theta}\, d\theta = \int_0^{\pi/4} \sqrt{1 - \frac{9}{25}\cos^2\theta}\, d\theta + \int_{\pi/4}^\pi \sqrt{1 - \frac{9}{25}\cos^2\theta}\, d\theta,$$

and approximate these two integrals on the right-hand side by the sum

$$I_{1,n}^*(f) \equiv I_{1,n}^{[0,\pi/4]}(f) + I_{1,n}^{[\pi/4,\pi]}(f)$$

corresponding to the trapezoidal rules applied on the intervals $[0, \pi/4]$ and $[\pi/4, \pi]$, the geometrical convergence is lost (in both of them). This phenomenon is shown in Figure 4.8 (top curve) and Table 4.4 (third column). The reason is that the integrand of the two integrals no longer satisfies the identity of its odd derivatives at the end of the intervals.

4.6 Improper Integrals

Improper integrals appear in physics with more frequency than one may expect (Debye's integral for the statistical mechanics of crystals was just one example). In this book, we will consider improper integrals of two different types:

$$I_1 \equiv \int_0^\infty f(x)\, dx, \tag{4.83}$$

usually known as **improper integrals of the first kind**, where the integrand is smooth and where the domain of integration is unbounded,

and

$$I_2 \equiv \int_{-1}^1 f(x)\, dx, \tag{4.84}$$

usually known as **improper integrals of the second kind**, where the integrand is singular at one or both ends of the integration domain.

Other improper cases can easily be reduced to I_1 or I_2 by introducing a simple change of variable, or by splitting the integration domain. Unless stated otherwise, we will always assume that improper integrals I_1 and I_2 *exist* and have a finite value.

The first rule of numerical approximation of improper integrals is to perform whenever possible suitable transformations or integration by parts that make the integral regular, and then apply Newton–Cotes or Clenshaw–Curtis quadrature formulas. However, finding such transformations is not always a trivial task. In such situations, we will describe succinctly a few techniques to approximate such integrals. Unfortunately, these techniques are not always as robust as Simpson or Clenshaw–Curtis quadrature formulas, for example. Sometimes, the integrands may exhibit a wide variety of strange pathologies, such as complicated asymptotic divergences nearby the limits of integration or highly oscillatory behavior in some regions of the unbounded domain of integration. In these particular cases, *ad hoc* methods must be devised.

4.6.1 Improper Integrals of the First Kind

In order to approximate

$$I(f) = \int_0^\infty f(x) \, dx, \tag{4.85}$$

we consider the change of variable

$$x = L\cot^2\left(\frac{t}{2}\right), \quad t \in (0, \pi), \tag{4.86}$$

where L is a suitable positive scale factor and whose precise role will be described later. Introducing (4.86) in (4.85) leads to the integral

$$I(f) = \int_0^\pi f\left\{L\cot^2\left(\frac{t}{2}\right)\right\} \frac{\sin t}{(1 - \cos t)^2} \, dt. \tag{4.87}$$

The previous integral is approximated using the quadrature rule

$$I_n(f) = \sum_{j=1}^n w_j f\left\{L\cot^2\left(\frac{t_j}{2}\right)\right\}, \tag{4.88}$$

based on the abscissas[8]

$$t_j = \frac{\pi j}{n+1}, \tag{4.89}$$

and quadrature weights

$$w_j = \frac{4L}{n+1} \frac{\sin t_j}{(1 - \cos t_j)^2} \sum_{m=1}^n \sin(m t_j) \frac{1 - \cos(m\pi)}{m}, \tag{4.90}$$

for $j = 1, 2, \ldots, n$. Transformation (4.86) maps equidistant abscissas t_j (4.89) on the interval $(0, \pi)$ to the semi-infinite domain $(0, +\infty)$; see Figure 4.9. The

---

8 Notice that $t_1 = \pi/(n+1) > 0$ and $t_n = \pi n/(n+1) < \pi$.

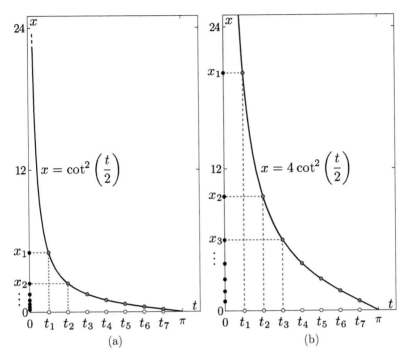

Figure 4.9 Cotangent transformation $x = L\cot^2(t/2)$ applied on the abscissas t_j (4.89) for $n = 7$ and for scale factors (a) $L = 1$ and (b) $L = 4$.

scaling factor L must be chosen according to the observed convergence as n is increased. In general, quadrature formula (4.88) provides geometrical convergence if the integrand decays exponentially for $x \to +\infty$. However, this fast convergence usually requires a previous tuning of the scale factor L.

```
%Code 9: Cot-Map for 1st kind improper integrals [0,+inf)
% Input: n: #abscissas; L: scaling factor
% Output: I_n(f)
function In = qcot(n,L,fun)
m = [1:n]; t = m'*pi/(n+1); x = L*(cot(t/2)).^2; w = 0*x.';
f = feval(fun,x);
for ii=1:n
    c1 = (4*L)*sin(t(ii))/((n+1)*(1-cos(t(ii)))^2);
    w(ii) = c1*sin(m*t(ii))*((1-cos(m'*pi))./m');
end
In = w*f;
```

Let us approximate the integral

$$I(f) \equiv \int_0^{+\infty} \frac{x}{e^x + 1}\, dx = \frac{\pi^2}{12} \approx 0.822\ 467\ 033\ 424\ 113, \tag{4.91}$$

using quadrature formula (4.88) implemented in Code 9. The numerical results are outlined in Table 4.5 for three different values of the scaling factor L. Figure 4.10a shows the absolute quadrature error $|E_n(f)| \equiv |I_n(f) - I(f)|$ for $10 \le n \le 40$ and for the same scaling factors L reported in Table 4.5. We can easily notice that not all scaling factors provide the same convergence rate. For this particular integral, the best results are obtained for $L \approx 9$, as shown in Figure 4.10b, where we have represented E_{40} for a wide range of values

Table 4.5 Cotangent quadrature rule approximation $I_n(f)$ (4.88) of integral (4.91) for $L = 1$, 9, and 20.

n	$L = 1$	$L = 9$	$L = 20$
10	0.824 847 189 762 059	0.823 225 172 456 285	0.816 835 689 543 49
20	0.822 438 795 855 724	0.822 466 782 440 903	0.822 458 402 657 222
30	0.822 467 185 569 772	0.822 467 033 399 751	0.822 467 035 408 823
40	0.822 467 168 684 533	0.822 467 033 424 109	0.822 467 033 552 936

a) The converged figures are shadowed.

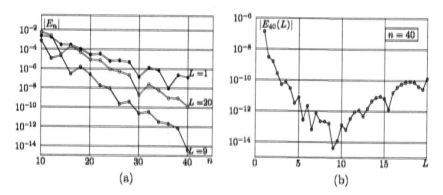

(a) (b)

Figure 4.10 Absolute quadrature error $|E_n(f)| \equiv |I_n(f) - I(f)|$ of cotangent quadrature formula (4.88) in the approximation of integral (4.91). (a) Error as a function of n for three different scaling factors L. (b) Error for $n = 40$ and for a wide range of scaling factors L showing a minimum at $L \approx 9$.

of L. In general, different integrands will require a different scaling factor L for an optimal convergence rate of quadrature (4.88). As shown in Figure 4.9, increasing L moves the sampling abscissas x_j to higher positions, thus capturing the behavior of the function for larger values of x. As a rule, integrals involving rapidly (exponentially) decaying integrands tend to be easier to approximate with moderate values of L, like the integral in (4.91). However, integrands exhibiting remarkable variations for large x (such as rapid oscillations or change of sign) require special methods that are beyond the scope of this basic book.

4.6.2 Improper Integrals of the Second Kind

In this case, we aim to approximate the integral

$$I(f) \equiv \int_{-1}^{1} f(x) \, dx, \tag{4.92}$$

where $f(x)$ is smooth in $(-1, 1)$ but may be singular at one of the integration limits $x \equiv -1$, $x \equiv +1$, or at both. There are many methodologies to solve this problem, depending on the type of singularity. In this book, we will only study two of them.

A first approximation can be obtained by applying an open version of the Clenshaw–Curtis quadrature. This can be accomplished by considering interpolatory integrands based on the *interior* nodes

$$x_k \equiv \cos\left(\frac{2k-1}{2n}\pi\right), \quad k \equiv 1, 2, \ldots, n, \tag{4.93}$$

usually known as *Chebyshev nodes of the first kind*, or *Chebyshev classical abscissas*[9] or *Chebyshev zeroes*,[10] simply because the x_k appearing in (4.93) satisfy $T_n(x_k) \equiv 0$, for $n \geq 1$. Interpolation based on the nodes (4.93) is extremely well-conditioned and with a very small Lebesgue constant. These nodes do not contain the integration limits $x \equiv \pm 1$, since $x_1 \equiv \cos(\pi/2n) < 1$ and $x_n \equiv \cos[(2n-1)\pi/2n] > -1$. We can proceed in a similar way as we did in the Clenshaw–Curtis quadrature to obtain the weights that make the formula

$$I_n(f) \equiv \sum_{k=1}^{n} w_k f\left\{\cos\left(\frac{2k-1}{2n}\pi\right)\right\} \tag{4.94}$$

---

9 The nodes $x_k \equiv \cos(\pi k/n)$, for $k \equiv 0, 1, \ldots, n$ are called *practical* abscissas.
10 The practical abscissas are *extrema* of $T_n(x)$.

exact for any $f(x) \in \mathbb{R}_{n-1}[x]$. For any even $n \geq 2$, these weights are

$$w_k = \frac{2}{n}\left\{1 - 2\sum_{m=1}^{n/2}\frac{1}{4m^2 - 1}\cos\left(\frac{2k-1}{n}m\pi\right)\right\}, \quad k = 1, 2, \ldots, n.$$

(4.95)

Quadrature formula (4.94) based on the nodes (4.93) and weights (4.95) is known as *Fejér's first quadrature formula*, or simply *Fejér's quadrature rule*.[11] Fejér's quadrature can be applied to approximate regular integrals, providing qualitative geometrical convergence rates similar to Clenshaw–Curtis. When used to approximate improper integrals of the second kind, Fejér's rule *does not* provide geometrical convergence. However, it is a very robust formula and usually provides reliable orientative approximations that can be very useful to guide other quadratures that are more accurate but less stable.

A much more accurate (but less robust) quadrature formula can be devised to approximate the improper integral (4.92) by using the transformation

$$x = \tanh u, \quad u \in (-\infty, +\infty).$$

(4.96)

With this change of variable, the integral (4.92) now reads

$$I(f) = \int_{-\infty}^{+\infty} f\{\tanh u\}\,\operatorname{sech}^2 u\,du,$$

(4.97)

from which it initially seems that we have just changed one problem for another, since the resulting integral is improper of the first kind. However, a suitable discretization of the new variable u allows for a quadrature formula that frequently provides fast convergence rates. This rule is given by

$$I_n(f) = \frac{h}{2}\sum_{j=-n}^{n} f\left\{\tanh\left(\frac{jh}{2}\right)\right\}\cosh^{-2}\left(\frac{jh}{2}\right),$$

(4.98)

where the integrand $f(x)$ is indirectly evaluated at the transformed abscissas $x_j = \tanh u_j$, with $u_j = jh/2$, for $j = -n, \ldots, n$ (see Figure 4.11), with corresponding quadrature weights $w_j = \cosh^{-2} u_j$. Quadrature formula (4.98) requires some tuning to achieve fast accuracy. The distance h between adjacent

---

11 Lipót Fejér (1880–1959), Hungarian mathematician who made important contributions to the field of harmonic analysis, doctoral advisor of famous mathematicians such as John von Neumann, Paul Erdös, or Pál Turán.

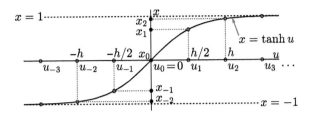

Figure 4.11 Hyperbolic tangent transformation $x = \tanh u$ applied on the abscissas $u_j = jh/2,\ (j = -n, \ldots, n)$.

abscissas u_j must be suitably chosen. As rule of thumb, this distance adopts the value

$$h = \frac{c}{\sqrt{n}}, \tag{4.99}$$

where c is a constant that typically takes values within the interval $[1, 10]$. This constant is similar to the scaling parameter L appearing in (4.86), and it must be also selected according to the convergence rate observed.

In general, quadrature rule (4.98) can be applied to arbitrary improper integrals of the form

$$\int_a^b f(z)\,\mathrm{d}z, \tag{4.100}$$

where $f(z)$ is either singular at $z = a$, $z = b$, or at both integration limits, by introducing again the linear change of variable (4.67). Code 10 is a simple implementation of quadrature formula (4.98) that performs the previous linear change:

```
%Code 10: tanh-rule for 2nd kind improper integrals (-1,+1)
% Input: n: #abscissas; (a,b): integration domain
%         c: tanh-scaling factor
% Output: I_n(f)
function In = qtanh(a,b,n,c,fun)
  h = c/sqrt(n); u = [-n:n]*h/2;
  f = feval(fun,(b-a)*.5*(tanh(u)+1)+a);
  In = (b-a)*(.25*h)*sum(f./(cosh(u).^2));
```

In order to illustrate the performance of quadrature rules (4.94) and (4.98), we approximate the singular integral

$$I(f) = \int_0^1 \frac{dx}{x^{1/2} + x^{1/3}} = 5 - 6\log 2 \approx 0.841\ 116\ 916\ 640\ 329. \quad (4.101)$$

From a numerical point of view, this integral is particularly challenging, as reflected in the first column of Table 4.6, where we can observe the poor accuracy exhibited by Fejér's rule, as well as a very slow convergence rate, as shown in Figure 4.12a. In this particular case, hyperbolic tangent rule provides much faster convergence rates, as reflected in Figure 4.12b, although this requires a suitable adjustment of the scaling factor c, as seen in Table 4.6, where the optimal results are obtained for $c = 5$.

Table 4.6 Approximation values of integral (4.101) using Fejér's formula (4.94) and hyperbolic tangent rule (4.98) for $c = 3$, 4, 5, and 6.

n	Fejér	$c = 3$	$c = 4$	$c = 5$	$c = 6$
10	0.837	0.839 418	0.840 927 0	0.841 143 815 1	0.841 464 64
20	0.839	0.840 972	0.841 109 9	0.841 116 914 2	0.841 122 58
30	0.840	0.841 096	0.841 116 3	0.841 116 910 9	0.841 117 14
40	0.840	0.841 112	0.841 116 8	0.841 116 916 1	0.841 116 93

a) The converged figures are shadowed.

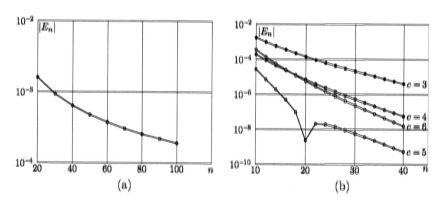

(a)   (b)

Figure 4.12 Absolute quadrature errors $|E_n(f)|$ of the approximation of integral (4.101) using two different methods. (a) Fejér's quadrature formula (4.94). (b) Hyperbolic tangent rule (4.98) for four different values of the scaling factors c.

Practical 4.3: Beads Moving Along Wires

A small bead can move along a thin frictionless wire that lies on a vertical plane (see the figure below) and whose shape is given by the curve $y = f(x)$.

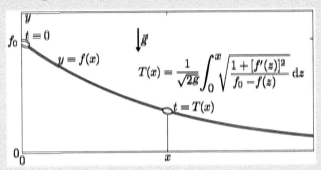

Assume that $f_0 = f(0) \geq f(x)$, for all $x > 0$. At $t = 0$, the bead (initially at rest) is released from the point of coordinates $x = 0$ and $y = f_0 = f(0)$. Afterwards, the bead starts sliding down. We want to compute the time $T(x)$ required by the bead to reach an arbitrary point $(x, f(x))$ of the wire by the action of gravity ($g = 9.81 \text{ ms}^{-2}$).

(a) Show that at the point $(x, f(x))$ the bead has an absolute speed $v = (2g)^{1/2}(f_0 - f(x))^{1/2}$ that is related to the horizontal velocity $\dot{x}$ by the equation $v = \dot{x}\sqrt{1 + [f'(x)]^2}$. Integrate to obtain the explicit expression of $T(x)$ included in the figure above.

(b) Show that the integral appearing in $T(x)$ is improper of the second kind. What local conditions on $f(x)$ near $x = 0$ are required for the integral to be convergent? **Hint:** Taylor series of $[f_0 - f(z)]^{-1/2}$.

(c) Find $T(a)$ (analytically) in the particular case of a straight wire with shape $f(x) = h(1 - x/a)$, for $h > 0$ and $a > 0$ (equivalent to an inclined plane of height h and horizontal length a).

(d) If $f(x) = he^{-x/a}$, with $h = a = 1$ m, compute numerically $T(2 \text{ m})$. Try to regularize the integrand by a suitable change of variable and make use of Clenshaw–Curtis quadrature. If you cannot find such a transformation, make use of a suitable quadrature rule for that singular integral.

Complementary Reading

For over three decades, the classical reference on numerical integration has been the monograph by Davis and Rabinowitz, *Methods of Numerical Integration*. This text is a comprehensive treatise on the topic, intended for an audience with a slightly higher mathematical background. In this reference, the reader will find detailed mathematical descriptions for many of the quadrature techniques covered in this chapter, along with many other methodologies not seen here. Such is the case of *Gaussian quadratures*, whose degree of exactness doubles the one corresponding to Clenshaw–Curtis or Fejér formulas (for the same number of function evaluations), but whose accuracy is in practice similar to these two quadratures seen here. While Clenshaw–Curtis or Fejér quadrature nodes and weights are relatively easy to obtain, the efficient computation of Gaussian nodes and weights requires numerical linear algebra techniques involving eigenvalue and eigenvector computations that are well beyond the scope of this fundamental book.

Computational integration is a very active research field and updated monographs appear from time to time, such as the highly practical Kythe and Schäferkotter's *Handbook of Computational Methods for Integration*, or the more theoretically oriented by Brass and Petras *Quadrature Theory: The Theory of Numerical Integration on a Compact Interval*, where the mathematically trained reader will find detailed error bounds for the Clenshaw–Curtis quadrature, for example. Another monograph (probably for a more specialized audience) covering details on software packages such as QUADPACK is the one by Krommer and Ueberhuber's *Computational Integration*. Detailed explanations of the exponential convergence of the composite trapezoidal rule based on Euler–Maclaurin's formula can be found in all the monographs above.

Problems and Exercises

The symbol (A) means that the problem has to be solved analytically, whereas (N) means to use numerical Matlab codes. Problems with the symbol * are slightly more difficult.

1. (N) Use the small code shown in Practical 4.1 to produce the coefficients of the Newton–Cotes formulas of Table 4.1. Recall that the definition of h depends on whether the quadrature formula is open or closed. Set Matlab in `format rat`.

2. (A) Using (4.58) and (4.57), obtain the exactness conditions (4.63) for the Clenshaw–Curtis quadrature formula.

3. (A) Find the weights of the Clenshaw–Curtis quadrature formula for $n = 3$. Use the method of undetermined coefficients to maximize exactness using monomials x^k. Repeat the exercise using Chebyshev polynomials.

4. (N) Devise your own routine to compute the Clenshaw–Curtis quadrature using the explicit formulas (4.72) for the weights. Do the weights satisfy any symmetry? If yes, can this be exploited to reduce the computational cost?

5. (N) A chain of length $\ell = 1$ u. l. (units of length) lies partially on an horizontal surface ($x < 0$ in the figure on the right). The rest of the chain lies on a surface of Gaussian shape whose height is given by $y = e^{-x^2}$. Find the abscissa x_0 corresponding to the right end of the chain.

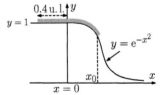

6. (N) A particle moves in the xy-plane along a closed curve (see figure on the right) whose parametric equations are

$$x(t) = e^{\sin t}, \quad y(t) = e^{\cos t},$$

for $t \in [0, 2\pi]$.

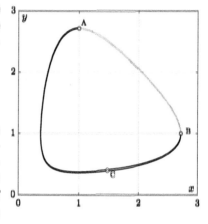

(a) Approximate the length of the closed curve using Clenshaw–Curtis and composite trapezoidal rules with at most 49 function evaluations. Which of the two rules provides more accuracy?

(b) Using the same quadrature rules and number of function evaluations, approximate the length of the arc A–B (plotted in gray in the figure on the right). In this case, which of the two quadratures is better? Why?

7. Redo Practical 4.2 using the composite trapezoidal rule.

(a) Far away from the ring, do you observe exponential convergence of the quadrature? If yes, explain why.

(b) Do you also observe poor convergence of the quadrature when computing the potential near the ring, for example, near the point $(y, z) = (1, 0)$)? If yes, explain why.

(c) If the sector of ring $\theta \in [\pi/4, 3\pi/4]$ is missing, i.e. there is no singularity at $(y, z) = (1, 0)$, what kind of convergence do you observe now near that point? Why?

8. (N*) Consider the graphs corresponding to the functions $f(x) = \cos x$ and $g(x) = \gamma x$, where γ is a positive constant. As shown in the figure on the right, the two graphs intersect for some x within the interval $(0, \pi/2)$, leading to two regions of areas S_I and S_{II} (filled in

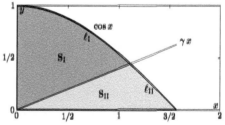

dark and light gray, respectively), and two arcs of lengths ℓ_I and ℓ_{II}.

(a) If $\gamma = 1$, find the length ℓ_I.

(b) Find the value of γ for which the two arcs have the same length.

(c) Find the value of γ for which the light and dark gray regions have the same area.
Provide at least six exact digits in your results.

9. (A-N*) At $t = 0$ s, a particle is released from rest at the point A of coordinates $(x, y) = (-a, 0)$; see the figure on the right. The particle starts sliding down on a frictionless semispherical bowl of radius $a = 1$ m, passing by the point B and later C at $t \equiv t_B$ and $t \equiv t_C$, respectively (both points are at the same height and their position vectors form an angle of $\pi/4$ with the vertical direction).

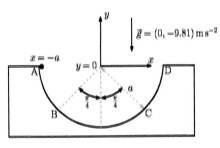

(a) Compute the lapse of time $\Delta T_{BC} \equiv t_C - t_B$.
(b) Show that the instant of time t_D at which the particle arrives at point D with coordinates $(a, 0)$ is

$$t_D = \left(\frac{2}{g}\right)^{1/2} I(1), \quad \text{with } I(1) = \int_0^1 (1 - x^2)^{-3/4} \, dx.$$

Try to approximate the last improper integral $I(1)$ with suitable quadrature formulas. If possible, find a change of variable to remove the singularity and provide at least six exact digits in your result.
(c) What is the period T of the resulting motion?

10. (A-N) A block of mass $m = 1$ kg is initially at rest located at the abscissa $x_0 = 1$ m. The block is connected to a nonlinear spring that provides a cubic force

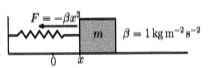

$F(x) \equiv -\beta x^3$ (x is the displacement of the block from the origin of coordinates; see the figure on the right). Neglecting frictional forces, show that the period of oscillations is given by

$$T = 4\sqrt{2} \int_0^1 (1 - x^4)^{-1/2} \, dx \approx 7.416\,298\,5 \text{ s.}$$

11. (A-N) We need to approximate improper integrals of the form

$$I(f) = \int_0^1 \log x f(x) \, dx,$$

where $f(x)$ is a smooth function in $(0, 1)$. We devise a quadrature formula of the form

$$I_n(f) = \sum_{j=0}^{n} w_j f(x_j), \quad \text{with } x_j = \frac{j}{n}, \quad j = 0, 1, 2, \ldots, n,$$

where $\{w_0, w_1, \ldots, w_n\}$ are the weights that make the formula $I_n(f)$ with the maximum exactness.

(a) For $n = 2$, obtain the exact weights (analytically).

(b) For arbitrary n, write down the system of equations whose solution would lead to the sought weights.

(c) For $n = 6$, apply the quadrature to approximate the integrals.

$$I_1 = \int_0^1 \frac{\log x}{1 + x}\, dx, \quad I_2 = \int_0^1 \log(x) \log(1 + x)\, dx$$

12. Using suitable quadrature rules, approximate the following integrals

$$I_F = \int_0^{\pi/2} \frac{\cos x}{\sqrt{x}}\, dx \quad \text{(Fresnel)} \quad \text{and} \quad I_D = \int_0^1 \frac{x}{e^x - 1}\, dx \quad \text{(Debye)},$$

with at least five exact digits.

Part II

5

Numerical Linear Algebra

5.1 Introduction

Linear algebra is one of the most powerful branches of mathematics. Amongst its many applications, two are of utmost importance in physics and engineering: solving systems of linear equations and computing eigenvalues and eigenvectors of matrices or linear operators. For example, determining the electrical current intensities of a complex circuit reduces to solving the system of linear equations arising from Kirchhoff's laws. Similarly, the spectrum of frequencies or *normal modes* exhibited by a mechanical system performing small amplitude oscillations can be predicted in advance by the computation of the eigenvalues of a certain matrix.

In this chapter, we will assume that the reader is already familiar with the classical concepts and analytical techniques involved in the solution of linear systems such as *Gaussian elimination* or in the *diagonalization* of matrices. In general, performing these techniques with a pencil and a paper is only feasible when the dimension of the matrix is small. As physical problems become complex, involving more unknowns or degrees of freedom, matrices become very large, and calculation by hand becomes impractical so that a computer is required to perform such techniques.

In Part I, we used Matlab's backslash command \ to solve linear systems involving dozens or nearly a hundred equations to determine interpolating polynomials or quadrature weights. In that part, we never mentioned what kind of algorithmic techniques lie behind this magical command, only asking the reader to use it and to rely on the results produced. In this chapter, we will address the underlying numerical algorithms behind \ that are used to solve linear systems of equations. We will study how to measure the condition number of an algebraic problem in order to have a precise idea of how accurate our computations are and, among other things, to explain the origin of those

Fundamentals of Numerical Mathematics for Physicists and Engineers, First Edition. Alvaro Meseguer.
© 2020 John Wiley & Sons, Inc. Published 2020 by John Wiley & Sons, Inc.

irritating warning messages appearing in Matlab prompt when using \ in the solution of Vandermonde systems, for example.

Throughout Part II, we will sometimes need to compute the eigenvalues and eigenvectors of matrices. Whenever this is required, we will make use of Matlab's `eig` command to solve eigenvalue problems (exactly with the same high-level view as when we used \ in Part I). Unfortunately, the study of numerical algorithms to solve eigenvalue problems is far beyond the scope of this textbook. Many numerical linear algebra monographs devote at least one or two chapters to explain these algorithms. We will refer the reader to suitable bibliography in the Complementary Reading section, at the end of this chapter.

5.2   Direct Linear Solvers

In this chapter, we will study different algorithms to solve linear systems of equations of the form

$$
\begin{bmatrix}
a_{11} & a_{12} & \cdots & a_{1n} \\
a_{21} & a_{22} & \cdots & a_{2n} \\
\vdots & \vdots & \ddots & \vdots \\
a_{n1} & a_{n2} & \cdots & a_{nn}
\end{bmatrix}
\begin{bmatrix}
x_1 \\ x_2 \\ \vdots \\ x_n
\end{bmatrix}
=
\begin{bmatrix}
b_1 \\ b_2 \\ \vdots \\ b_n
\end{bmatrix},
\tag{5.1}
$$

or simply

$$
\mathbf{A}\mathbf{x} = \mathbf{b},
\tag{5.2}
$$

where[1] $\mathbf{A} = [a_{ij}] \in \mathbb{M}_n(\mathbb{R})$, $\mathbf{x} = [x_i] \in \mathbb{R}^n$ and $\mathbf{b} = [b_i] \in \mathbb{R}^n$. In Eq. (5.2), the matrix $\mathbf{A}$ and the right-hand side vector $\mathbf{b}$ are supposed to be known and $\mathbf{x}$ is the unknown vector or *sought solution*. Unless stated otherwise, we will always assume that $\mathbf{A}$ is a *non-singular* matrix, i.e. $\det \mathbf{A} \neq 0$.

There are many different methods to solve system (5.2). In general, these methods or linear *solvers* are classified in two different types: *direct* and *iterative*. A direct solver is a method that provides the solution $\mathbf{x}$ of (5.2) in a *finite* number of operations. A classical example of direct solver is (as the reader may have probably guessed) Gaussian elimination, where the matrix is reduced to triangular form after a finite number of elementary row operations and a finite number of substitutions. By contrast, an iterative solver provides a sequence of approximations $\mathbf{x}^{(k)}$ that (hopefully) converges to the sought solution $\mathbf{x}$ of (5.2). The number of operations required by the iterative method will depend on the requested accuracy and properties of $\mathbf{A}$. We will address iterative solvers at the end of this chapter. We start this chapter by analyzing the most fundamental direct methods to solve (5.2).

---

1 $\mathbb{M}_n(\mathbb{R})$ is the linear space of $n \times n$ real matrices.

5.2.1 Diagonal and Triangular Systems

If the matrix appearing in (5.1) is *diagonal* (that is, $a_{ij} = 0$ if $i \neq j$), the solution is straightforward:

$$\begin{bmatrix} a_{11} & 0 & \cdots & 0 \\ 0 & a_{22} & \cdots & 0 \\ \vdots & \vdots & \ddots & \vdots \\ 0 & 0 & \cdots & a_{nn} \end{bmatrix} \begin{bmatrix} x_1 \\ x_2 \\ \vdots \\ x_n \end{bmatrix} = \begin{bmatrix} b_1 \\ b_2 \\ \vdots \\ b_n \end{bmatrix}, \qquad x_i = \frac{b_i}{a_{ii}}, \tag{5.3}$$

for $i = 1, 2, \ldots, n$. Note that the diagonal elements a_{ii} are nonzero since $\mathbf{A}$ is non-singular. To obtain the solution we just need to perform n quotients. For this reason, it is said that the computational cost of solving diagonal systems is $O(n)$.

Suppose now that the matrix $\mathbf{A}$ appearing in (5.1) is *lower triangular*[2] so that the linear system now reads

$$\begin{bmatrix} \ell_{11} & 0 & 0 & \cdots & 0 \\ \ell_{21} & \ell_{22} & 0 & \cdots & 0 \\ \ell_{31} & \ell_{32} & \ell_{33} & \cdots & 0 \\ \vdots & \vdots & \vdots & \ddots & \vdots \\ \ell_{n1} & \ell_{n2} & \ell_{n3} & \cdots & \ell_{nn} \end{bmatrix} \begin{bmatrix} x_1 \\ x_2 \\ x_3 \\ \vdots \\ x_n \end{bmatrix} = \begin{bmatrix} b_1 \\ b_2 \\ b_3 \\ \vdots \\ b_n \end{bmatrix} . \tag{5.4}$$

Since the matrix is non-singular and $\det \mathbf{A} = \ell_{11}\ell_{22}\cdots\ell_{nn} \neq 0$, all the diagonal elements ℓ_{ii} must be nonzero. The equation corresponding to the first row of system (5.4) is independent of the other ones, and therefore we can easily obtain $x_1 = b_1/\ell_{11}$. The second equation involves x_1 and x_2 but since x_1 has just been calculated we have

$$x_2 = (b_2 - \ell_{21}x_1)/\ell_{22}. \tag{5.5}$$

Similarly, the third equation only involves x_1, x_2, and x_3, so that

$$x_3 = (b_3 - \ell_{31}x_1 - \ell_{32}x_2)/\ell_{33}. \tag{5.6}$$

In general, x_i is given by

$$x_i = \begin{cases} \dfrac{b_1}{\ell_{11}}, & (i = 1) \\[2ex] \dfrac{1}{\ell_{ii}} \left(b_i - \displaystyle\sum_{j=1}^{i-1} \ell_{ij}x_j \right), & (2 \leq i \leq n). \end{cases} \tag{5.7}$$

---

2 A matrix with zero elements above its diagonal.

The algorithm just described is usually termed as *forward substitution* (henceforth referred to as FS), since it substitutes the previously calculated unknowns $\{x_1, x_2, x_3, \ldots, x_{i-1}\}$ in order to obtain the *next* one x_i. The reader may check (see Exercise 3) that the computational cost of FS is $O(n^2)$.

Now suppose that $\mathbf{A}$ is *upper triangular*, with zero elements below the diagonal, so that (5.1) is

$$
\begin{bmatrix}
u_{11} & u_{12} & u_{13} & \cdots & u_{1n} \\
0 & u_{22} & u_{23} & \cdots & u_{2n} \\
0 & 0 & u_{33} & \cdots & u_{3n} \\
\vdots & \vdots & \vdots & \ddots & \vdots \\
0 & 0 & 0 & \cdots & u_{nn}
\end{bmatrix}
\begin{bmatrix}
x_1 \\ x_2 \\ x_3 \\ \vdots \\ x_n
\end{bmatrix}
=
\begin{bmatrix}
b_1 \\ b_2 \\ b_3 \\ \vdots \\ b_n
\end{bmatrix}.
\tag{5.8}
$$

Since $u_{ii} \neq 0$ for $0 \leq i \leq n$, we can proceed in a similar way as we did in FS. In this case, the last equation only involves x_n and therefore $x_n \equiv b_n/u_{nn}$. We can obtain x_{n-1} by substitution of x_n in the $(n-1)$th equation:

$$
x_{n-1} \equiv (b_{n-1} - u_{n-1,n}x_n)/u_{n-1,n-1},
\tag{5.9}
$$

and similarly with x_{n-2}

$$
x_{n-2} \equiv (b_{n-2} - u_{n-2,n-1}x_{n-1} - u_{n-2,n}x_n)/u_{n-2,n-2}.
\tag{5.10}
$$

In this case, the general solution is given by

$$
x_i \equiv
\begin{cases}
\dfrac{b_n}{u_{nn}}, & (i \equiv n) \\[2ex]
x_i \equiv \dfrac{1}{u_{ii}}\left(b_i - \displaystyle\sum_{j=i+1}^{n} u_{ij}x_j\right), & (1 \leq i \leq n-1).
\end{cases}
\tag{5.11}
$$

Since in this case the substitutions are performed in reverse order (to obtain x_i we need $x_{i+1}, x_{i+2}, \ldots, x_n$), this algorithm is usually called *backward substitution* or BS. The computational cost of BS is also $O(n^2)$.

In formulas (5.7) and (5.11), the matrix elements ℓ_{ij} and u_{ij} are accessed *by row*, respectively. For example, in order to obtain x_i, formula (5.7) has access to the elements[3]

$$
\ell_{i,1:i-1} \equiv \boxed{\ell_{i,1}\ \ell_{i,2}\ \ell_{i,3}\ \cdots\ \ell_{i,i-1}},
\tag{5.12}
$$

that is, the first $i-1$ elements of the ith row of matrix ℓ_{ij}. The same can be said about BS for the access to elements $u_{i,i+1:n}$ in (5.11). For this rea-

---

3 To simplify notation, we will often make use of Matlab's row–column range indexing.

son, algorithms corresponding to formulas (5.7) and (5.11) are usually termed as *row oriented*. Codes 11 and 12 are row oriented versions of FS and BS, respectively:

```
% Code 11: Forward Substitution for Lower Triangular Systems
% Input    L: Low Triangular non-singular square matrix
%          b: column right-hand side
% Output   x: solution of Lx = b
function x = fs(L,b)
x = 0*b; n = length(b); x(1) = b(1)/L(1,1);
for ii = 2:n
    x(ii) = (b(ii)-L(ii,1:ii-1)*x(1:ii-1))/L(ii,ii);
end
```

```
% Code 12: Backward Substitution for Upper Triangular Systems
% Input    U: Upp. Triangular non-singular square matrix
%          b: column right-hand side
% Output   x: solution of Ux = b
function x = bs(U,b)
x = 0*b; n = length(b); x(n) = b(n)/U(n,n);
for ii = n-1:-1:1
    x(ii) = (b(ii)-U(ii,ii+1:n)*x(ii+1:n))/U(ii,ii);
end
```

Equivalent *column oriented* algorithms for triangular systems can be found in the bibliography included in the Complementary Reading section, at the end of the chapter. In general, choosing row or column orientation may depend on many factors, such as how matrices are stored in a given programming language.

5.2.2 The Gaussian Elimination Method

In what follows, we will assume that the linear system

$$
\begin{bmatrix}
a_{11} & a_{12} & \cdots & a_{1n} \\
a_{21} & a_{22} & \cdots & a_{2n} \\
\vdots & \vdots & \ddots & \vdots \\
a_{n1} & a_{n2} & \cdots & a_{nn}
\end{bmatrix}
\begin{bmatrix}
x_1 \\
x_2 \\
\vdots \\
x_n
\end{bmatrix}
\equiv
\begin{bmatrix}
b_1 \\
b_2 \\
\vdots \\
b_n
\end{bmatrix}
\tag{5.13}
$$

is not diagonal or triangular. One way to solve system (5.13) is to triangularize it by means of *elementary transformations*[4] and use BS afterwards. Let us write system (5.13) as

$$
\underbrace{\begin{bmatrix}
a_{11}^{(1)} & a_{12}^{(1)} & a_{13}^{(1)} & \cdots & a_{1n}^{(1)} \\
a_{21}^{(1)} & a_{22}^{(1)} & a_{23}^{(1)} & \cdots & a_{2n}^{(1)} \\
a_{31}^{(1)} & a_{32}^{(1)} & a_{33}^{(1)} & \cdots & a_{3n}^{(1)} \\
\vdots & \vdots & \vdots & \ddots & \vdots \\
a_{n1}^{(1)} & a_{n2}^{(1)} & a_{n3}^{(1)} & \cdots & a_{nn}^{(1)}
\end{bmatrix}}_{\mathbf{A}^{(1)}}
\underbrace{\begin{bmatrix} x_1 \\ x_2 \\ x_3 \\ \vdots \\ x_n \end{bmatrix}}_{\mathbf{x}}
=
\underbrace{\begin{bmatrix} b_1^{(1)} \\ b_2^{(1)} \\ b_3^{(1)} \\ \vdots \\ b_n^{(1)} \end{bmatrix}}_{\mathbf{b}^{(1)}},
\tag{5.14}
$$

or more briefly

$$
\mathbf{A}^{(1)}\mathbf{x} = \mathbf{b}^{(1)},
\tag{5.15}
$$

where the superscript (1) means that the system is at the initial stage of the algorithm that we are about to start. Assuming that $a_{11} \neq 0$, we can define the $n-1$ quantities

$$
m_{i1} \equiv \frac{a_{i1}^{(1)}}{a_{11}^{(1)}} \quad (i = 2, 3, \ldots, n).
\tag{5.16}
$$

The triangularization of the matrix starts by canceling the elements corresponding to the first column below the diagonal. This is accomplished by subtracting from rows $\{r_2, r_3, \ldots, r_n\}$ suitable multiples of the first row r_1. The reader may easily check that these suitable multiples are the factors defined in (5.16), and this is precisely the reason why these quotients are known as *multipliers*. After these elementary transformations, the resulting system reads

$$
\begin{matrix}
r_1 \\
r_2 - m_{21}r_1 \\
r_3 - m_{31}r_1 \\
\vdots \\
r_n - m_{n1}r_1
\end{matrix}
\underbrace{\begin{bmatrix}
a_{11}^{(1)} & a_{12}^{(1)} & a_{13}^{(1)} & \cdots & a_{1n}^{(1)} \\
0 & a_{22}^{(2)} & a_{23}^{(2)} & \cdots & a_{2n}^{(2)} \\
0 & a_{32}^{(2)} & a_{33}^{(2)} & \cdots & a_{3n}^{(2)} \\
\vdots & \vdots & \vdots & \ddots & \vdots \\
0 & a_{n2}^{(2)} & a_{n3}^{(2)} & \cdots & a_{nn}^{(2)}
\end{bmatrix}}_{\mathbf{A}^{(2)}}
\begin{bmatrix} x_1 \\ x_2 \\ x_3 \\ \vdots \\ x_n \end{bmatrix}
=
\underbrace{\begin{bmatrix} b_1^{(1)} \\ b_2^{(2)} \\ b_3^{(2)} \\ \vdots \\ b_n^{(2)} \end{bmatrix}}_{\mathbf{b}^{(2)}},
\tag{5.17}
$$

element. transf.

or

$$
\mathbf{A}^{(2)}\mathbf{x} = \mathbf{b}^{(2)},
\tag{5.18}
$$

where elements $a_{2:n,1}^{(1)}$ have been directly substituted by zeros so that x_1 is only involved in the first equation. Unfortunately, these elementary transformations

---

4 For example, subtract from any row a linear combination of other rows.

have modified the rest of the system by transforming $a_{ij}^{(1)}$ and $b_i^{(1)}$ into the new quantities $a_{ij}^{(2)}$ and $b_i^{(2)}$, respectively, which are given by

$$a_{ij}^{(2)} = a_{ij}^{(1)} - m_{i1}a_{1j}^{(1)},$$
$$b_i^{(2)} = b_i^{(1)} - m_{i1}b_1^{(1)},$$

(5.19)

for $i, j = 2, \ldots, n$. Equations (5.19) are frequently known as *Gauss transformations*. This transformation renders a new matrix $\mathbf{A}^{(2)}$ and right-hand side vector $\mathbf{b}^{(2)}$ whose first rows are still those of $\mathbf{A}^{(1)}$ and $\mathbf{b}^{(1)}$, respectively.

Assuming that $a_{22}^{(2)} \neq 0$ in (5.17), we can cancel the elements $a_{3:n,2}^{(2)}$ of $\mathbf{A}^{(2)}$ by introducing the $n - 2$ new multipliers

$$m_{i2} \equiv \frac{a_{i2}^{(2)}}{a_{22}^{(2)}} \quad (i = 3, \ldots, n),$$

(5.20)

and performing the Gauss transformation

$$a_{ij}^{(3)} = a_{ij}^{(2)} - m_{i2}a_{2j}^{(2)},$$
$$b_i^{(3)} = b_i^{(2)} - m_{i2}b_2^{(2)},$$

(5.21)

for $i, j = 3, \ldots, n$, leading to the transformed system

$$
\underbrace{
\begin{bmatrix}
a_{11}^{(1)} & a_{12}^{(1)} & a_{13}^{(1)} & \cdots & a_{1n}^{(1)} \\
0 & a_{22}^{(2)} & a_{23}^{(2)} & \cdots & a_{2n}^{(2)} \\
0 & 0 & a_{33}^{(3)} & \cdots & a_{3n}^{(3)} \\
\vdots & \vdots & \vdots & \ddots & \vdots \\
0 & 0 & a_{n3}^{(3)} & \cdots & a_{nn}^{(3)}
\end{bmatrix}
}_{\mathbf{A}^{(3)}}
\begin{bmatrix}
x_1 \\ x_2 \\ x_3 \\ \vdots \\ x_n
\end{bmatrix}
=
\underbrace{
\begin{bmatrix}
b_1^{(1)} \\ b_2^{(2)} \\ b_3^{(3)} \\ \vdots \\ b_n^{(3)}
\end{bmatrix}
}_{\mathbf{b}^{(3)}},
$$

(5.22)

or

$$\mathbf{A}^{(3)}\mathbf{x} = \mathbf{b}^{(3)}.$$

(5.23)

In general, at stage k, the transformed linear system is

$$\mathbf{A}^{(k)}\mathbf{x} = \mathbf{b}^{(k)},$$

(5.24)

where the matrix $\mathbf{A}^{(k)}$, along with the right-hand side vector $\mathbf{b}^{(k)}$, has the structure

$$
[\mathbf{A}^{(k)}|\mathbf{b}^{(k)}] \doteq
\left[
\begin{array}{cccccc|c}
a_{11}^{(1)} & a_{12}^{(1)} & \cdots & a_{1k}^{(1)} & \cdots & a_{1n}^{(1)} & b_1^{(1)} \\
0 & a_{22}^{(2)} & \cdots & a_{2k}^{(2)} & \cdots & a_{2n}^{(2)} & b_2^{(2)} \\
\vdots & & 0 & \ddots & & \vdots & \vdots \\
\vdots & \vdots & 0 & a_{kk}^{(k)} & \cdots & a_{kn}^{(k)} & b_k^{(k)} \\
\vdots & \vdots & \vdots & \vdots & \ddots & \vdots & \vdots \\
0 & \cdots & 0 & a_{nk}^{(k)} & \cdots & a_{nn}^{(k)} & b_n^{(k)}
\end{array}
\right].
$$

(5.25)

Assuming $a_{kk}^{(k)} \neq 0$, the next stage $\mathbf{A}^{(k+1)}\mathbf{x} \equiv \mathbf{b}^{(k+1)}$ is obtained by calculating the corresponding multipliers

$$m_{ik} \equiv \frac{a_{ik}^{(k)}}{a_{kk}^{(k)}}, \tag{5.26}$$

for $i \equiv k+1, \ldots, n$, and afterwards performing the Gauss transformations

$$\begin{aligned} a_{ij}^{(k+1)} &\equiv a_{ij}^{(k)} - m_{ik}a_{kj}^{(k)}, \\ b_i^{(k+1)} &\equiv b_i^{(k)} - m_{ik}b_k^{(k)}, \end{aligned} \tag{5.27}$$

for $i, j \equiv k+1, \ldots, n$. Overall, assuming that none of the elements $a_{11}^{(1)}$, $a_{22}^{(2)}, \ldots, a_{nn}^{(n)}$ (usually known as *pivots*) is zero, the $n-1$ Gauss transformations (5.27) finally lead to the upper triangular system

$$\underbrace{\begin{bmatrix} a_{11}^{(1)} & a_{12}^{(1)} & a_{13}^{(1)} & \ldots & a_{1n}^{(1)} \\ 0 & a_{22}^{(2)} & a_{23}^{(2)} & \ldots & a_{2n}^{(2)} \\ 0 & 0 & a_{33}^{(3)} & \ldots & a_{3n}^{(3)} \\ \vdots & \vdots & \vdots & \ddots & \vdots \\ 0 & 0 & 0 & \ldots & a_{nn}^{(n)} \end{bmatrix}}_{\mathbf{A}^{(n)}} \underbrace{\begin{bmatrix} x_1 \\ x_2 \\ x_3 \\ \vdots \\ x_n \end{bmatrix}}_{\mathbf{x}} = \underbrace{\begin{bmatrix} b_1^{(1)} \\ b_2^{(2)} \\ b_3^{(3)} \\ \vdots \\ b_n^{(n)} \end{bmatrix}}_{\mathbf{b}^{(n)}}, \tag{5.28}$$

or

$$\mathbf{A}^{(n)}\mathbf{x} \equiv \mathbf{b}^{(n)}, \tag{5.29}$$

which can be solved using BS. The algorithm just described is the *Gaussian elimination method* [5] (GEM for short) and its computational cost is $2n^3/3$ (see Exercise 4). In practical applications, GEM, in its primitive form given by the iteration (5.27), is seldom used, and it *should not* be used (even when we can guarantee in advance that zero pivots will not appear) as will be justified later.

As an example, let us apply GEM to the linear system

$$\mathbf{A}^{(1)}\mathbf{x} \equiv \mathbf{b}^{(1)}: \quad \begin{cases} 3x_1 - 2x_2 + 3x_3 = 8 \\ 2x_1 + x_2 - x_3 = 1 \\ x_1 - x_2 - x_3 = -4 \end{cases}. \tag{5.30}$$

For $k \equiv 1$, the multipliers are $m_{21} \equiv 2/3$ and $m_{31} \equiv 1/3$, so that subtracting from the second and third equations the first row multiplied by m_{21} and m_{31}, respectively, gives

$$\mathbf{A}^{(2)}\mathbf{x} \equiv \mathbf{b}^{(2)}: \quad \begin{cases} 3x_1 - 2x_2 + 3x_3 = 8 \\ \frac{7}{3}x_2 - 3x_3 = -\frac{13}{3} \\ -\frac{1}{3}x_2 - 2x_3 = -\frac{20}{3} \end{cases}. \tag{5.31}$$

---

5 Originally formulated by the German mathematician and physicist Johann Carl Friedrich Gauss (1777–1855).

Finally, subtracting from the third equation the second row multiplied by $m_{32} = -1/7$ leads to the sought upper triangular system

$$\mathbf{A}^{(3)}\mathbf{x} = \mathbf{b}^{(3)} : \quad \begin{cases} 3x_1 - 2x_2 + 3x_3 = 8 \\ \frac{7}{3}x_2 - 3x_3 = -\frac{13}{3} , \\ -\frac{17}{7}x_3 = -\frac{51}{7} \end{cases} \quad (5.32)$$

whose solution (as the reader can check using BS) is $\mathbf{x} = [x_1\ x_2\ x_3]^{\mathrm{T}} = [1\ 2\ 3]^{\mathrm{T}}$.

Summary: Gaussian Elimination Method (GEM)

The linear system of equations

$$\mathbf{A}^{(1)}\mathbf{x} = \mathbf{b}^{(1)} : \quad \begin{bmatrix} a_{11}^{(1)} & a_{12}^{(1)} & a_{13}^{(1)} & \dots & a_{1n}^{(1)} \\ a_{21}^{(1)} & a_{22}^{(1)} & a_{23}^{(1)} & \dots & a_{2n}^{(1)} \\ a_{31}^{(1)} & a_{32}^{(1)} & a_{33}^{(1)} & \dots & a_{3n}^{(1)} \\ \vdots & \vdots & \vdots & \ddots & \vdots \\ a_{n1}^{(1)} & a_{n2}^{(1)} & a_{n3}^{(1)} & \dots & a_{nn}^{(1)} \end{bmatrix} \begin{bmatrix} x_1 \\ x_2 \\ x_3 \\ \vdots \\ x_n \end{bmatrix} = \begin{bmatrix} b_1^{(1)} \\ b_2^{(1)} \\ b_3^{(1)} \\ \vdots \\ b_n^{(1)} \end{bmatrix} , \quad (5.33)$$

can be transformed to the upper triangular system

$$\mathbf{A}^{(n)}\mathbf{x} = \mathbf{b}^{(n)} : \quad \begin{bmatrix} a_{11}^{(1)} & a_{12}^{(1)} & a_{13}^{(1)} & \dots & a_{1n}^{(1)} \\ 0 & a_{22}^{(2)} & a_{23}^{(2)} & \dots & a_{2n}^{(2)} \\ 0 & 0 & a_{33}^{(3)} & \dots & a_{3n}^{(3)} \\ \vdots & \vdots & \vdots & \ddots & \vdots \\ 0 & 0 & 0 & \dots & a_{nn}^{(n)} \end{bmatrix} \begin{bmatrix} x_1 \\ x_2 \\ x_3 \\ \vdots \\ x_n \end{bmatrix} = \begin{bmatrix} b_1^{(1)} \\ b_2^{(2)} \\ b_3^{(3)} \\ \vdots \\ b_n^{(n)} \end{bmatrix} , \quad (5.34)$$

by iteratively performing $n - 1$ Gauss transformations:

$$a_{ij}^{(k+1)} = a_{ij}^{(k)} - m_{ik}a_{kj}^{(k)} \quad (i, j = k+1, \dots, n),$$
$$b_i^{(k+1)} = b_i^{(k)} - m_{ik}b_k^{(k)} \quad (i = k+1, \dots, n), \quad (5.35)$$

for $k = 1, 2, \dots, n - 1$, where the multipliers appearing in (5.35) are

$$m_{k+1:n,\ k} \equiv \frac{1}{a_{kk}^{(k)}} a_{k+1:n,\ k}^{(k)} , \quad (5.36)$$

and where it is assumed that the pivots $a_{jj}^{(j)} \neq 0$ for $j = 1, \dots, n$. The computational cost of GEM is $2n^3/3$. The solution of system (5.34) is also the solution of (5.33), and is obtained using BS.

5.3 LU Factorization of a Matrix

Although GEM is never used as a direct solver, it is conceptually very important as it implicitly provides a very convenient way of factorizing a matrix $\mathbf{A}$ as the product

$$\mathbf{A} = \mathbf{LU}, \tag{5.37}$$

where $\mathbf{L}$ and $\mathbf{U}$ are lower and upper triangular matrices, respectively. As we will see soon, factorizing a matrix $\mathbf{A}$ in the form shown in (5.37) is computationally very efficient, in order to solve linear systems of the form $\mathbf{Ax} = \mathbf{b}$.

To see how to factor a matrix as in (5.37), let us revisit the example seen in Section 5.2:

$$\mathbf{A}^{(1)}\mathbf{x} = \mathbf{b}^{(1)} : \quad \begin{cases} 3x_1 - 2x_2 + 3x_3 = 8 \\ 2x_1 + x_2 - x_3 = 1 \\ x_1 - x_2 - x_3 = -4 \end{cases}.$$

Take, for example, the multipliers $m_{21} = 2/3$ and $m_{31} = 1/3$ for $k = 1$ and consider the matrix

$$\mathbf{M}_1 = \begin{bmatrix} 1 & 0 & 0 \\ -m_{21} & 1 & 0 \\ -m_{31} & 0 & 1 \end{bmatrix} = \begin{bmatrix} 1 & 0 & 0 \\ -\frac{2}{3} & 1 & 0 \\ -\frac{1}{3} & 0 & 1 \end{bmatrix}.$$

A simple calculation shows that matrix $\mathbf{A}^{(2)}$ can be obtained by performing the matrix–matrix product $\mathbf{M}_1\mathbf{A}^{(1)}$:

$$\mathbf{M}_1\mathbf{A}^{(1)} = \begin{bmatrix} 1 & 0 & 0 \\ -\frac{2}{3} & 1 & 0 \\ -\frac{1}{3} & 0 & 1 \end{bmatrix} \begin{bmatrix} 3 & -2 & 3 \\ 2 & 1 & -1 \\ 1 & -1 & -1 \end{bmatrix} = \begin{bmatrix} 3 & -2 & 3 \\ 0 & \frac{7}{3} & -3 \\ 0 & -\frac{1}{3} & -2 \end{bmatrix} = \mathbf{A}^{(2)}.$$

Similarly, to obtain $\mathbf{A}^{(3)}$, consider the multiplier $m_{32} = -1/7$ and the matrix

$$\mathbf{M}_2 = \begin{bmatrix} 1 & 0 & 0 \\ 0 & 1 & 0 \\ 0 & -m_{32} & 1 \end{bmatrix} = \begin{bmatrix} 1 & 0 & 0 \\ 0 & 1 & 0 \\ 0 & \frac{1}{7} & 1 \end{bmatrix}.$$

For $k = 2$, matrix $\mathbf{A}^{(3)}$ can be obtained by carrying out the product

$$\mathbf{M}_2\mathbf{A}^{(2)} = \begin{bmatrix} 1 & 0 & 0 \\ 0 & 1 & 0 \\ 0 & \frac{1}{7} & 1 \end{bmatrix} \begin{bmatrix} 3 & -2 & 3 \\ 0 & \frac{7}{3} & -3 \\ 0 & -\frac{1}{3} & -2 \end{bmatrix} = \begin{bmatrix} 3 & -2 & 3 \\ 0 & \frac{7}{3} & -3 \\ 0 & 0 & -\frac{17}{7} \end{bmatrix} = \mathbf{A}^{(3)}.$$

In other words, the upper triangular matrix $\mathbf{U} = \mathbf{A}^{(3)}$ that we obtained in (5.32) using GEM can also be obtained by sequential left multiplication of the original matrix $\mathbf{A}^{(1)} = \mathbf{A}$ by $\mathbf{M}_1$ and by $\mathbf{M}_2$ afterwards, that is,

$$\mathbf{U} = \mathbf{M}_2\mathbf{A}^{(2)} = \mathbf{M}_2\mathbf{M}_1\mathbf{A}^{(1)} = \mathbf{M}_2\mathbf{M}_1\mathbf{A}. \tag{5.38}$$

Before going further, we need a couple of properties (see Exercises 5 and 6):

Property 1: Product of Lower Triangular Matrices

Let $\mathbf{L}_1$, $\mathbf{L}_2 \in M_n(\mathbb{R})$ be two arbitrary lower triangular matrices. Then, the product $\mathbf{L}_1\mathbf{L}_2$ is a lower triangular matrix.

Property 2: Inverse of a Non-singular Lower Triangular Matrix

Let $\mathbf{L} \in M_n(\mathbb{R})$ be a non-singular lower triangular matrix. Then, the inverse $\mathbf{L}^{-1}$ is a lower triangular matrix.

Since $\det \mathbf{M}_1 = \det \mathbf{M}_2 = 1$, the matrices $\mathbf{M}_1$ and $\mathbf{M}_2$ are invertible. Left multiplication of (5.38) by $\mathbf{M}_2^{-1}$ and afterwards by $\mathbf{M}_1^{-1}$ gives

$$\mathbf{M}_1^{-1}\mathbf{M}_2^{-1}\mathbf{U} = \mathbf{A}. \tag{5.39}$$

By virtue of the two previous properties, since $\mathbf{M}_1$ and $\mathbf{M}_2$ are lower triangular, $\mathbf{M}_1^{-1}$, $\mathbf{M}_2^{-1}$, as well as their product $\mathbf{M}_1^{-1}\mathbf{M}_2^{-1}$ are also lower triangular matrices. So we may write

$$\mathbf{M}_1^{-1}\mathbf{M}_2^{-1}\mathbf{U} = \mathbf{LU}, \tag{5.40}$$

where $\mathbf{L} \equiv \mathbf{M}_1^{-1}\mathbf{M}_2^{-1}$ is a lower triangular matrix, so that (5.38) reads

$$\mathbf{A} = \mathbf{LU}. \tag{5.41}$$

In other words, the original matrix $\mathbf{A}$ can be factorized as the product of a lower triangular matrix $\mathbf{L}$ and an upper triangular matrix $\mathbf{U} = \mathbf{A}^{(3)}$. This is just a particular case of what is generally known as the LU-*factorization* or LU-*decomposition* of a matrix $\mathbf{A}$. At this point, it is logical to wonder whether all matrices admit this type of factorization. To answer these and many other questions we first need to introduce new concepts and a more general framework.

Definition (Dyadic Product of Two Vectors)

Let $\mathbf{u} = [u_1 \, u_2 \, \cdots \, u_n]^{\mathrm{T}}$ and $\mathbf{v} = [v_1 \, v_2 \, \cdots \, v_n]^{\mathrm{T}}$ be two arbitrary vectors in $\mathbb{R}^n$. The *dyadic* product $\mathbf{uv}^{\mathrm{T}}$ is a matrix whose elements are given by $[\mathbf{uv}^{\mathrm{T}}]_{ij} = u_i v_j$:

$$\mathbf{uv}^{\mathrm{T}} = \begin{bmatrix} u_1 \\ u_2 \\ \vdots \\ u_n \end{bmatrix} \begin{bmatrix} v_1 & v_2 & \cdots & v_n \end{bmatrix} = \begin{bmatrix} u_1v_1 & u_1v_2 & \cdots & u_1v_n \\ u_2v_1 & u_2v_2 & \cdots & u_2v_n \\ \vdots & \vdots & & \vdots \\ u_nv_1 & u_nv_2 & \cdots & u_nv_n \end{bmatrix}. \tag{5.42}$$

Consider the matrix $\mathbf{A}_k$ appearing at k-th stage of the GEM algorithm,

$$
\mathbf{A}^{(k)} \equiv
\begin{bmatrix}
a_{11}^{(1)} & a_{12}^{(1)} & \cdots & a_{1k}^{(1)} & \cdots & a_{1n}^{(1)} \\
0 & a_{22}^{(2)} & \cdots & a_{2k}^{(2)} & \cdots & a_{2n}^{(2)} \\
\vdots & 0 & \ddots & \vdots & & \vdots \\
\vdots & \vdots & 0 & a_{kk}^{(k)} & \cdots & a_{kn}^{(k)} \\
\vdots & \vdots & \vdots & \vdots & \ddots & \vdots \\
0 & \cdots & 0 & a_{nk}^{(k)} & \cdots & a_{nn}^{(k)}
\end{bmatrix},
\tag{5.43}
$$

and the multipliers

$$
m_{ik} \equiv \frac{a_{ik}^{(k)}}{a_{kk}^{(k)}} \quad (i \equiv k+1,\dots,n).
\tag{5.44}
$$

Define the vector

$$
\mathbf{m}_k \triangleq [\,\overset{(1)}{0}\ \overset{(2)}{0}\ \cdots\ \overset{(k)}{0}\ \overset{(k+1)}{m_{k+1,k}}\ \overset{(k+2)}{m_{k+2,k}}\ \cdots\ \overset{(n)}{m_{nk}}\,]^{\mathrm{T}},
\tag{5.45}
$$

and consider the kth vector of the canonical basis of $\mathbb{R}^n$

$$
\mathbf{e}_k \equiv [\,0\ 0\ \cdots\ 0\ \overset{(k)}{1}\ 0\ \cdots\ 0\,]^{\mathrm{T}}.
\tag{5.46}
$$

Let $\mathbf{I}_n$ be the identity matrix in $M_n(\mathbb{R})$. Then the matrix

$$
\mathbf{M}_k \equiv \mathbf{I}_n - \mathbf{m}_k \mathbf{e}_k^{\mathrm{T}} \equiv
\begin{matrix}
& & & & (k) & & & & \\
\end{matrix}
\begin{bmatrix}
1 & 0 & \cdots & & 0 & & \cdots & & 0 \\
0 & 1 & \ddots & & \vdots & & & & \vdots \\
\vdots & 0 & \ddots & & 0 & & & & \\
\vdots & & & 1 & & 0 & & & \\
& & & -m_{k+1,k} & & 1 & 0 & & \\
& & & -m_{k+2,k} & & \vdots & & & \vdots \\
\vdots & \vdots & & \vdots & & & & \ddots & 0 \\
0 & 0 & \cdots & -m_{n,k} & & 0 & \cdots & & 1
\end{bmatrix}
\begin{matrix} \\ \\ \\ \\ (k+1) \\ \\ \\ \\ \end{matrix}
\tag{5.47}
$$

satisfies

$$
\mathbf{A}^{(k+1)} \equiv \mathbf{M}_k\, \mathbf{A}^{(k)},
\tag{5.48}
$$

where the elements $a_{ij}^{(k+1)}$ of $\mathbf{A}^{(k+1)}$ are given by the Gauss transformation

$$a_{ij}^{(k+1)} \equiv a_{ij}^{(k)} - m_{ik}a_{kj}^{(k)} \quad (i,j = k+1,\ldots,n). \tag{5.49}$$

To prove (5.48), let us explicitly write (5.47) in terms of Kronecker's delta:

$$[\mathbf{M}_k]_{i\ell} \equiv \delta_{i\ell} - m_{ik}\,\delta_{k\ell} \quad (i,\ell = 1,2,\ldots,n). \tag{5.50}$$

The Gauss transformation (5.49) then reads

$$a_{ij}^{(k+1)} \equiv \sum_{\ell=1}^{n} \delta_{i\ell}a_{\ell j}^{(k)} - m_{ik}\delta_{k\ell}a_{\ell j}^{(k)} \equiv \sum_{\ell=1}^{n}(\delta_{i\ell} - m_{ik}\delta_{k\ell})\,a_{\ell j}^{(k)} \equiv \sum_{\ell=1}^{n}[\mathbf{M}_k]_{i\ell}\,a_{\ell j}^{(k)}, \tag{5.51}$$

or $\mathbf{A}^{(k+1)} \equiv \mathbf{M}_k\,\mathbf{A}^{(k)}$. Matrix $\mathbf{M}_k$ defined in (5.47) is usually termed as the *Gauss transformation matrix* corresponding to the kth GEM transformation (5.49). In practice, the matrices $\mathbf{M}_k$ are *never* constructed, but they will be conceptually crucial to understand the lower triangular nature of the factorization they lead to.

Since $\det \mathbf{M}_k \equiv 1$, the Gauss transformation matrices are invertible. A simple calculation shows that

$$\mathbf{M}_k^{-1} \equiv \mathbf{I}_n + \mathbf{m}_k \mathbf{e}_k^{\mathrm{T}}. \tag{5.52}$$

This is easily shown by performing the matrix product

$$\begin{aligned}
(\mathbf{I}_n + \mathbf{m}_k\mathbf{e}_k^{\mathrm{T}})\mathbf{M}_k &\equiv (\mathbf{I}_n + \mathbf{m}_k\mathbf{e}_k^{\mathrm{T}})(\mathbf{I}_n - \mathbf{m}_k\mathbf{e}_k^{\mathrm{T}}) \\
&\equiv \mathbf{I}_n - \mathbf{m}_k\mathbf{e}_k^{\mathrm{T}} + \mathbf{m}_k\mathbf{e}_k^{\mathrm{T}} - \mathbf{m}_k\mathbf{e}_k^{\mathrm{T}}\mathbf{m}_k\mathbf{e}_k^{\mathrm{T}} \\
&\equiv \mathbf{I}_n, \tag{5.53}
\end{aligned}$$

since the scalar product $\mathbf{e}_k^{\mathrm{T}}\mathbf{m}_k$ within the term $\mathbf{m}_k\mathbf{e}_k^{\mathrm{T}}\mathbf{m}_k\mathbf{e}_k^{\mathrm{T}}$ is zero:

$$\mathbf{e}_k^{\mathrm{T}}\mathbf{m}_k \equiv \begin{matrix} (1) \cdots \quad (k) \quad \cdots \ (n) \\ [0 \cdots 0 \quad 1 \quad 0 \cdots 0] \end{matrix} \begin{bmatrix} 0 \\ 0 \\ \vdots \\ 0 \\ 0 \\ m_{k+1,k} \\ m_{k+2,k} \\ \vdots \\ m_{n,k} \end{bmatrix} \begin{matrix} (1) \\ \\ \vdots \\ \\ (k) \\ (k+1) \\ \\ \vdots \\ (n) \end{matrix} \equiv 0. \tag{5.54}$$

In general, (5.54) implies that

$$\mathbf{e}_k^{\mathrm{T}}\mathbf{m}_\ell \equiv 0 \quad (\ell \geq k). \tag{5.55}$$

The previous formulation provides a very elegant way to reduce a given $n \times n$ matrix[6] $\mathbf{A}$ to upper triangular form by sequential left matrix multiplication by the Gaussian matrices $\mathbf{M}_k$, that is,

$$\mathbf{U} = \mathbf{M}_{n-1}\mathbf{M}_{n-2} \cdots \mathbf{M}_2\mathbf{M}_1\mathbf{A}. \tag{5.56}$$

Left multiplying both sides of the last expression first by $\mathbf{M}_{n-1}^{-1}$, then by $\mathbf{M}_{n-2}^{-1}$, and so on, we have

$$\mathbf{M}_1^{-1}\cdots\mathbf{M}_{n-2}^{-1}\mathbf{M}_{n-1}^{-1}\mathbf{U} = \mathbf{M}_1^{-1}\cdots\overbrace{\mathbf{M}_{n-2}^{-1}\underbrace{\mathbf{M}_{n-1}^{-1}\mathbf{M}_{n-1}}_{\mathbf{I}_n}\mathbf{M}_{n-2}}^{\mathbf{I}_n}\cdots\mathbf{M}_1\mathbf{A}$$

$$= \mathbf{A}. \tag{5.57}$$

To simplify the product of the inverse matrices $\mathbf{M}_1^{-1}\cdots\mathbf{M}_{n-1}^{-1}$ appearing on the left-hand side of the last equation, we write them in terms of the dyadic factors following (5.52):

$$\mathbf{M}_1^{-1}\mathbf{M}_2^{-1}\cdots\mathbf{M}_{n-1}^{-1} = (\mathbf{I}_n + \mathbf{m}_1\mathbf{e}_1^{\mathrm{T}})(\mathbf{I}_n + \mathbf{m}_2\mathbf{e}_2^{\mathrm{T}})\cdots(\mathbf{I}_n + \mathbf{m}_{n-1}\mathbf{e}_{n-1}^{\mathrm{T}}). \tag{5.58}$$

In the last expression, the products containing two or more dyadic factors have the structure

$$\cdots\mathbf{m}_k\mathbf{e}_k^{\mathrm{T}}\mathbf{m}_{k+1}\mathbf{e}_{k+1}^{\mathrm{T}}\cdots ,$$

and, by virtue of (5.55) for $\ell = k + 1$, they all vanish. Therefore,

$$\mathbf{M}_1^{-1}\mathbf{M}_2^{-1}\cdots\mathbf{M}_{n-1}^{-1} = \mathbf{I}_n + \mathbf{m}_1\mathbf{e}_1^{\mathrm{T}} + \mathbf{m}_2\mathbf{e}_2^{\mathrm{T}} + \cdots + \mathbf{m}_{n-1}\mathbf{e}_{n-1}^{\mathrm{T}}, \tag{5.59}$$

or

$$\mathbf{M}_1^{-1}\mathbf{M}_2^{-1}\cdots\mathbf{M}_{n-1}^{-1} = \mathbf{L} = \begin{bmatrix} 1 & 0 & 0 & \cdots & 0 \\ m_{21} & 1 & 0 & \cdots & 0 \\ m_{31} & m_{32} & 1 & & \vdots \\ \vdots & \vdots & & \ddots & 0 \\ m_{n1} & m_{n2} & \cdots & m_{n,n-1} & 1 \end{bmatrix}, \tag{5.60}$$

so that (5.57) reads

$$\mathbf{L}\mathbf{U} = \mathbf{A}. \tag{5.61}$$

The factorization shown in (5.61) of an arbitrary matrix $\mathbf{A}$ as the product of a lower triangular matrix $\mathbf{L}$ and an upper triangular matrix $\mathbf{U}$ is known as an LU-factorization of $\mathbf{A}$.

---

6 Assuming, as mentioned earlier, that the pivots $a_{kk}^{(k)}$ are nonzero.

Property (Existence and Uniqueness of LU-Factorization)

Any square matrix $\mathbf{A} \in \mathbb{M}_n(\mathbb{R})$

$$\mathbf{A} = \begin{bmatrix} a_{11} & a_{12} & \cdots & a_{1n} \\ a_{21} & a_{22} & \cdots & a_{2n} \\ \vdots & \vdots & \ddots & \vdots \\ a_{n1} & a_{n2} & \cdots & a_{nn} \end{bmatrix},$$

with nonzero associated *principal minors*, that is,

$$|A_j| \doteq \begin{vmatrix} a_{11} & a_{12} & \cdots & a_{1j} \\ a_{21} & a_{22} & \cdots & a_{2j} \\ \vdots & \vdots & \ddots & \vdots \\ a_{j1} & a_{j2} & \cdots & a_{jj} \end{vmatrix} \neq 0, \tag{5.62}$$

for $j = 1, 2, \ldots, n$, admits a unique LU-factorization of the form $\mathbf{A} = \mathbf{LU}$ such that

$$a_{ij} = \sum_{k=1}^{n} \ell_{ik}\, u_{kj}, \tag{5.63}$$

where ℓ_{ij} and u_{ij} are the elements of the lower and upper triangular matrices $\mathbf{L}$ and $\mathbf{U}$, respectively, and where $\ell_{11} = \ell_{22} = \cdots = \ell_{nn} = 1$.

Proof of the previous statement can be found in many standard references included in the Complementary Reading Section, at the end of the chapter. In practical applications, however, this result has only a minor impact. This statement requires one to check that the determinants $|A_1|$, $|A_2|$, $\ldots, |A_n|$ do not vanish (which is unpractical from a computational point of view). We will show a more robust variation in Section 5.4.

5.3.1   Solving Systems with LU

In what follows we will show why the LU-factorization is so important in numerical linear algebra. To illustrate that, let us revisit the problem of solving an arbitrary linear system of equations of the form

$$\mathbf{Ax} = \mathbf{b}, \tag{5.64}$$

where $\mathbf{b} \in \mathbb{R}^n$ and $\mathbf{A} \in \mathbb{M}_n(\mathbb{R})$ is assumed to have all its principal minors nonzero so that it admits the factorization $\mathbf{A} = \mathbf{LU}$ and system (5.64) can therefore be expressed as

$$\mathbf{LUx} = \mathbf{b}. \tag{5.65}$$

The solution of (5.65) is performed in two stages. In the first stage we define the unknown vector $\mathbf{y} \doteq \mathbf{U}\mathbf{x}$, so that (5.65) reads

$$\mathbf{L}\mathbf{y} = \mathbf{b}. \tag{5.66}$$

System (5.66) is lower triangular, so it can be solved using FS at a cost of $O(n^2)$ operations. In a second stage, the found solution $\mathbf{y}$ of system (5.66) is then used as the right-hand side of the system

$$\mathbf{U}\mathbf{x} = \mathbf{y}. \tag{5.67}$$

Since system (5.67) is upper triangular, $\mathbf{x}$ can be easily found using BS at a similar cost $O(n^2)$.

For example, in Section 5.2.2 we used GEM to transform the linear system

$$\begin{array}{rrrrrr} 3x_1 & - & 2x_2 & + & 3x_3 & = & 8 \\ 2x_1 & + & x_2 & - & x_3 & = & 1 \,, \\ x_1 & - & x_2 & - & x_3 & = & -4 \end{array} \tag{5.68}$$

to upper triangular form

$$\begin{array}{rrrrr} 3x_1 & - & 2x_2 & + & 3x_3 & = & 8 \\ & & \frac{7}{3}x_2 & - & 3x_3 & = & -\frac{13}{3} \,, \\ & & & - & \frac{17}{7}x_3 & = & -\frac{51}{7} \end{array} \tag{5.69}$$

so that after using BS, the found solution was $\mathbf{x} = [x_1\ x_2\ x_3]^{\mathrm{T}} = [1\ 2\ 3]^{\mathrm{T}}$. Later, in Section 5.3, we saw how to obtain the LU-factorization of the matrix $\mathbf{A}$ of system (5.68) using precisely the multipliers previously originating from GEM. The reader can easily check that, according to (5.60) and to the left-hand side of (5.69), the LU-factorization of matrix $\mathbf{A}$ in system (5.68) is

$$\mathbf{A} \equiv \begin{bmatrix} 3 & -2 & 3 \\ 2 & 1 & -1 \\ 1 & -1 & -1 \end{bmatrix} = \begin{bmatrix} 1 & 0 & 0 \\ \frac{2}{3} & 1 & 0 \\ \frac{1}{3} & -\frac{1}{7} & 1 \end{bmatrix} \begin{bmatrix} 3 & -2 & 3 \\ 0 & \frac{7}{3} & -3 \\ 0 & 0 & -\frac{17}{7} \end{bmatrix} = \mathbf{L}\mathbf{U}. \tag{5.70}$$

Therefore, to solve

$$\begin{bmatrix} 3 & -2 & 3 \\ 2 & 1 & -1 \\ 1 & -1 & -1 \end{bmatrix} \begin{bmatrix} x_1 \\ x_2 \\ x_3 \end{bmatrix} = \begin{bmatrix} 8 \\ 1 \\ -4 \end{bmatrix} \quad \text{or} \quad \begin{bmatrix} 1 & 0 & 0 \\ \frac{2}{3} & 1 & 0 \\ \frac{1}{3} & -\frac{1}{7} & 1 \end{bmatrix} \underbrace{\begin{bmatrix} 3 & -2 & 3 \\ 0 & \frac{7}{3} & -3 \\ 0 & 0 & -\frac{17}{7} \end{bmatrix} \begin{bmatrix} x_1 \\ x_2 \\ x_3 \end{bmatrix}}_{\mathbf{U}\mathbf{x}} = \begin{bmatrix} 8 \\ 1 \\ -4 \end{bmatrix} ,$$

first define

$$\mathbf{y} \equiv \begin{bmatrix} y_1 \\ y_2 \\ y_3 \end{bmatrix} \equiv \mathbf{U}\mathbf{x} = \begin{bmatrix} 3 & -2 & 3 \\ 0 & \frac{7}{3} & -3 \\ 0 & 0 & -\frac{17}{7} \end{bmatrix} \begin{bmatrix} x_1 \\ x_2 \\ x_3 \end{bmatrix} ,$$

then use FS to solve

$$
\underbrace{\begin{bmatrix} 1 & 0 & 0 \\ \frac{2}{3} & 1 & 0 \\ \frac{1}{3} & -\frac{1}{7} & 1 \end{bmatrix} \begin{bmatrix} y_1 \\ y_2 \\ y_3 \end{bmatrix}}_{\mathbf{Ly}} = \underbrace{\begin{bmatrix} 8 \\ 1 \\ -4 \end{bmatrix}}_{\mathbf{b}},
$$

to obtain the solution $\mathbf{y} = \begin{bmatrix} 8 & -\frac{13}{3} & -\frac{51}{7} \end{bmatrix}^{\mathrm{T}}$. Finally, solve

$$
\underbrace{\begin{bmatrix} 3 & -2 & 3 \\ 0 & \frac{7}{3} & -3 \\ 0 & 0 & -\frac{17}{7} \end{bmatrix} \begin{bmatrix} x_1 \\ x_2 \\ x_3 \end{bmatrix}}_{\mathbf{Ux}} = \underbrace{\begin{bmatrix} 8 \\ -\frac{13}{3} \\ -\frac{51}{7} \end{bmatrix}}_{\mathbf{y}},
$$

with BS to find $\mathbf{x} \equiv [x_1 \ x_2 \ x_3]^{\mathrm{T}} \equiv [1 \ 2 \ 3]^{\mathrm{T}}$. The total cost of the two triangular systems is certainly $O(n^2)$. However, we should also include the initial $O(n^3)$ computational cost of the LU-factorization of $\mathbf{A}$ required to obtain $\mathbf{L}$ and $\mathbf{U}$. At first glance, it may seem that the whole process of factorization and the use of two triangular solvers has essentially the same cost as GEM. So the question is: why not simply use GEM?

The answer is simple. Imagine that, after solving $\mathbf{Ax} = \mathbf{b}$ with GEM, you need to solve $\mathbf{Ax} = \mathbf{c}$, that is, *only the right-hand side changes*. In practical applications, it is quite frequent to solve many linear systems with the same matrix $\mathbf{A}$ but different right-hand sides $\mathbf{b}$. In GEM we should start from scratch in order to triangularize the matrix, overall involving $O(n^3)$ operations again. Instead, once we have the LU-factorization of $\mathbf{A}$, solving $\mathbf{Ax} = \mathbf{c}$ only requires $O(n^2)$, since we only need to solve $\mathbf{Ly} = \mathbf{c}$ and then $\mathbf{Ux} = \mathbf{y}$, this time just requiring $O(n^2)$ operations.

5.3.2 Accuracy of LU

We have previously warned that LU-factorization (which is essentially GEM) should be avoided in practice as a direct solver. We will justify this warning by means of a very simple experiment. Suppose we want to solve the system

$$
\mathbf{Ax} = \mathbf{b}: \quad \begin{bmatrix} \varepsilon & 1 \\ 1 & 1 \end{bmatrix} \begin{bmatrix} x_1 \\ x_2 \end{bmatrix} \equiv \begin{bmatrix} 1 \\ 0 \end{bmatrix}, \tag{5.71}
$$

where $0 < \varepsilon \ll 1$, i.e. assume that ε is a very small positive quantity. Before we start our analysis, the reader may check that the exact solution of (5.71) is

$$
\begin{bmatrix} x_1 \\ x_2 \end{bmatrix} \equiv \frac{1}{1 - \varepsilon} \begin{bmatrix} -1 \\ 1 \end{bmatrix}. \tag{5.72}
$$

A simple calculation also shows that the exact LU-factorization of matrix $\mathbf{A}$ is

$$\begin{bmatrix} \varepsilon & 1 \\ 1 & 1 \end{bmatrix} = \underbrace{\begin{bmatrix} 1 & 0 \\ \varepsilon^{-1} & 1 \end{bmatrix}}_{\mathbf{L}} \underbrace{\begin{bmatrix} \varepsilon & 1 \\ 0 & 1 - \varepsilon^{-1} \end{bmatrix}}_{\mathbf{U}}. \tag{5.73}$$

For any ε smaller than 10^{-16}, double precision computation of the fourth entry of $\mathbf{U}$, with exact value $1 - \varepsilon^{-1}$, is rounded (and stored) as $-\varepsilon^{-1}$ (since 1 is negligible in front of ε^{-1}, to machine precision). As a result, the actual stored matrix $\mathbf{U}$ is

$$\widehat{\mathbf{U}} = \begin{bmatrix} \varepsilon & 1 \\ 0 & -\varepsilon^{-1} \end{bmatrix}. \tag{5.74}$$

Therefore, we solve

$$\begin{bmatrix} 1 & 0 \\ \varepsilon^{-1} & 1 \end{bmatrix} \underbrace{\begin{bmatrix} \varepsilon & 1 \\ 0 & -\varepsilon^{-1} \end{bmatrix} \begin{bmatrix} x_1 \\ x_2 \end{bmatrix}}_{[y_1 \; y_2]^{\mathrm{T}}} = \begin{bmatrix} 1 \\ 0 \end{bmatrix}, \tag{5.75}$$

to obtain $\begin{bmatrix} y_1 \\ y_2 \end{bmatrix} = \begin{bmatrix} 1 \\ -\varepsilon^{-1} \end{bmatrix}$, which is used as right-hand side in

$$\begin{bmatrix} \varepsilon & 1 \\ 0 & -\varepsilon^{-1} \end{bmatrix} \begin{bmatrix} x_1 \\ x_2 \end{bmatrix} = \begin{bmatrix} 1 \\ -\varepsilon^{-1} \end{bmatrix}, \tag{5.76}$$

to finally find the solution

$$\begin{bmatrix} x_1 \\ x_2 \end{bmatrix} = \begin{bmatrix} 0 \\ 1 \end{bmatrix}. \tag{5.77}$$

However, the limit of the exact solution (5.72) for $\varepsilon \to 0$ is

$$\lim_{\varepsilon \to 0} \frac{1}{1 - \varepsilon} \begin{bmatrix} -1 \\ 1 \end{bmatrix} = \begin{bmatrix} -1 \\ 1 \end{bmatrix}, \tag{5.78}$$

so that the LU solution (5.77) is completely misleading. We recommend the reader to experiment numerically with Matlab, taking for example $\varepsilon = 10^{-17}$ and repeating the analysis shown above to confirm that the obtained solution is (5.77) and not (5.78).

The origin of the problem is that the third entry, 1, of $\mathbf{A}$ in (5.71) is divided by a much smaller pivot ε to obtain the multiplier $m_{21} = \varepsilon^{-1}$. This division by a much smaller quantity leads to a huge multiplier that amplifies round-off errors. The reader may think that finding pivots of such a small order of magnitude is not the generic situation. That is usually the case. However, what really matters is not the absolute magnitude of the pivot itself, but its relative size with respect to the matrix element it is dividing. In general, the absolute size of the multiplier $m_{ik} = a_{ik}^{(k)}/a_{kk}^{(k)}$ $(i \geq k+1)$ will be large whenever $|a_{ik}^{(k)}| \gg |a_{kk}^{(k)}|$. In addition, we must emphasize that our example just involves a simple 2×2 matrix. For matrices in $\mathbb{M}_n(\mathbb{R})$, the number of computed multipliers is of order n^2. Every time a moderately large multiplier arises, the accuracy of the resulting modified rows is slightly deteriorated. These slightly inaccurate rows will serve in the incoming Gauss transformations to generate new multipliers (some of them probably large as well), thus amplifying onward the inaccuracies generated in the previous stages. This amplification is performed over and over throughout the $O(n^3)$ operations required by the LU-factorization, typically leading to catastrophic results[7].

All these complications could have been avoided from the very beginning simply by changing the order of the equations in system (5.71):

$$\begin{bmatrix} 1 & 1 \\ \varepsilon & 1 \end{bmatrix} \begin{bmatrix} x_1 \\ x_2 \end{bmatrix} = \begin{bmatrix} 0 \\ 1 \end{bmatrix}. \tag{5.79}$$

In this case, the LU-factorization of the matrix is

$$\begin{bmatrix} 1 & 1 \\ \varepsilon & 1 \end{bmatrix} = \underbrace{\begin{bmatrix} 1 & 0 \\ \varepsilon & 1 \end{bmatrix}}_{\mathbf{L}} \underbrace{\begin{bmatrix} 1 & 1 \\ 0 & 1-\varepsilon \end{bmatrix}}_{\mathbf{U}}, \tag{5.80}$$

where we must note that the resulting multiplier $m_{21} = \varepsilon$ is a small quantity. As before, we first solve

$$\begin{bmatrix} 1 & 0 \\ \varepsilon & 1 \end{bmatrix} \begin{bmatrix} y_1 \\ y_2 \end{bmatrix} = \begin{bmatrix} 0 \\ 1 \end{bmatrix}, \tag{5.81}$$

to obtain $\begin{bmatrix} y_1 \\ y_2 \end{bmatrix} = \begin{bmatrix} 0 \\ 1 \end{bmatrix}$, to be used as right-hand side in

$$\begin{bmatrix} 1 & 1 \\ 0 & 1-\varepsilon \end{bmatrix} \begin{bmatrix} x_1 \\ x_2 \end{bmatrix} = \begin{bmatrix} 0 \\ 1 \end{bmatrix}, \tag{5.82}$$

leading to the correct result

$$\begin{bmatrix} x_1 \\ x_2 \end{bmatrix} = \frac{1}{1-\varepsilon} \begin{bmatrix} -1 \\ 1 \end{bmatrix}. \tag{5.83}$$

Note that if we neglect ε in the fourth entry of the $\mathbf{U}$ matrix of (5.82) and replace $1-\varepsilon$ by 1, the rounded result is $[-1 \ 1]^{\mathrm{T}}$, which is consistent with the limit $\varepsilon \to 0$ of (5.83).

---

7 For a precise analysis of this amplification phenomenon or *growing factors* we refer the reader to the "Complementary Reading" section at the end of the chapter.

5.4 LU with Partial Pivoting

The example just seen in Section 5.3 illustrates that small pivots are the origin of potential inaccuracies when using LU to solve linear systems. As seen in that example, permuting rows allows for replacement of small pivots by moderate ones, which prevents the generation of large multipliers. The conclusion to be drawn from that elementary example is that our LU-factorization algorithm needs to incorporate the possibility of permuting rows. This new strategy of interchanging rows is usually known as *pivoting* or, more precisely, *partial pivoting*, where the partial qualifier will be clarified at the end of this section.

The pivoting technique appears for the first time in basic linear algebra courses, particularly whenever an exact zero pivotal element arises during GEM. In that situation, the row including the zero pivotal element is interchanged by *any* of the rows underneath, provided that such problematic pivot is replaced by a nonzero quantity. Numerically, however, we have just seen that even when pivots are not exactly zero, the resulting multipliers may be considerably large and lead to inaccuracies. In fact, we saw that the absolute size of the pivot is not the problem but its magnitude relative to the other matrix elements underneath. Therefore, choosing the row including the element of *maximal* magnitude will guarantee that the resulting multipliers will be smaller than unity, so that round-off errors are not amplified.

In what follows, we will see how to implement an algorithm encompassing GEM along with pivoting that is usually numerically accurate. We will also see that the combination of both techniques leads to a new type of matrix factorization. We start with a couple of definitions:

Definition: Permutation Matrix

Let $I = \{i_1, i_2, \ldots, i_n\}$ be a reordering of the first integers $\{1, 2, \ldots, n\}$ leading to the indexing $I(1) = i_1$, $I(2) = i_2$, $\ldots, I(n) = i_n$, and let $\{e_1, e_2, \ldots, e_n\}$ be the canonical basis of $\mathbb{R}^n$. The *permutation matrix* associated with the indexing I is

$$\mathbf{P}_I = \left[\, \mathbf{e}_{I(1)} \,\middle|\, \mathbf{e}_{I(2)} \,\middle|\, \cdots \,\middle|\, \mathbf{e}_{I(n)} \,\right] = \left[\, \mathbf{e}_{i_1} \,\middle|\, \mathbf{e}_{i_2} \,\middle|\, \cdots \,\middle|\, \mathbf{e}_{i_n} \,\right]. \qquad (5.84)$$

Since $\mathbf{P}_I$ is a reordering of the canonical basis, we have

$$\det \mathbf{P}_I \neq 0 \quad \text{and} \quad \mathbf{P}_I^{\mathrm{T}} \mathbf{P}_I = \mathbf{P}_I \mathbf{P}_I^{\mathrm{T}} = \mathbf{I}_n.$$

For example, consider the reordering $I = \{3, 1, 2\}$ that leads to the indexing $I(1) = 3$, $I(2) = 1$, and $I(3) = 2$. In this case, the permutation matrix is

$$\mathbf{P}_I = \mathbf{P}_{\{3,1,2\}} = \left[\, \mathbf{e}_3 \,\middle|\, \mathbf{e}_1 \,\middle|\, \mathbf{e}_2 \,\right] = \begin{bmatrix} 0 & 1 & 0 \\ 0 & 0 & 1 \\ 1 & 0 & 0 \end{bmatrix}.$$

Definition: Exchange Matrix

Let $I = \{i_1, i_2, \ldots, i_n\}$ be a reordering of the first integers $\{1, 2, \ldots, n\}$ such that

$$I(i) = \begin{cases} k_2 & (i = k_1), \\ k_1 & (i = k_2), \\ i & (\text{otherwise}) \end{cases} \quad (5.85)$$

where $1 \leq k_1 \leq k_2 \leq n$. That is, the indexing has the form

$$I = \{1, 2, \ldots, k_1 - 1, k_2, k_1 + 1, \ldots, k_2 - 1, k_1, k_2 + 1, \ldots, n\}.$$

The permutation matrix associated with the indexing (5.85) is called an *exchange matrix* and has the form

$$(5.86)$$

The two subscripts in $\mathbf{E}_{k_1 k_2}$ indicate which rows or, equivalently, columns of the identity matrix $\mathbf{I}_n$ have been exchanged (vertical boxes above). An arbitrary exchange matrix $\mathbf{E}$ is, by construction, *symmetric*, i.e. $\mathbf{E}^{\mathrm{T}} = \mathbf{E}$.

Before illustrating the functionality of exchange matrices, it is convenient to review the concept of matrix-vector product in a slightly different way.

Take for example a matrix $\mathbf{A} = [a_{ij}] \in M_n(\mathbb{R})$ and a vector $\mathbf{b} = [b_i] \in \mathbb{R}^n$. The matrix–vector product $\mathbf{A}\,\mathbf{b}$ is

$$
\begin{bmatrix} a_{11} & a_{12} & \cdots & a_{1n} \\ a_{21} & a_{22} & \cdots & a_{2n} \\ \vdots & \vdots & \ddots & \vdots \\ a_{n1} & a_{n2} & \cdots & a_{nn} \end{bmatrix} \begin{bmatrix} b_1 \\ b_2 \\ \vdots \\ b_n \end{bmatrix} = b_1 \begin{bmatrix} a_{11} \\ a_{21} \\ \vdots \\ a_{n1} \end{bmatrix} + b_2 \begin{bmatrix} a_{12} \\ a_{22} \\ \vdots \\ a_{n2} \end{bmatrix} + \cdots + b_n \begin{bmatrix} a_{1n} \\ a_{2n} \\ \vdots \\ a_{nn} \end{bmatrix}, \quad (5.87)
$$

where we have included vertical lines between the columns of $\mathbf{A}$ to emphasize that the result of the product is simply the linear combination of these columns, and where the coefficients of the combination are precisely the components of the vector $\mathbf{b}$. For example, take the exchange matrix $\mathbf{E}_{32} \in M_3(\mathbb{R})$ and apply it on an arbitrary vector $\mathbf{b} \in \mathbb{R}^3$

$$
\mathbf{E}_{32}\mathbf{b} = \begin{bmatrix} 1 & 0 & 0 \\ 0 & 0 & 1 \\ 0 & 1 & 0 \end{bmatrix} \begin{bmatrix} b_1 \\ b_2 \\ b_3 \end{bmatrix} = b_1 \begin{bmatrix} 1 \\ 0 \\ 0 \end{bmatrix} + b_2 \begin{bmatrix} 0 \\ 0 \\ 1 \end{bmatrix} + b_3 \begin{bmatrix} 0 \\ 1 \\ 0 \end{bmatrix} = \begin{bmatrix} b_1 \\ b_3 \\ b_2 \end{bmatrix}.
$$

The result is that $\mathbf{E}_{32}$ exchanges the third and second rows of $\mathbf{b}$. From the previous result, we conclude the following:

Property: Action of an Exchange Matrix

Let $\mathbf{E}_{k_1 k_2} \in M_n(\mathbb{R})$ be an exchange matrix, $\mathbf{A} \in M_n(\mathbb{R})$ an arbitrary matrix. Then,

1. The product $\mathbf{E}_{k_1 k_2}\mathbf{A}$ exchanges rows k_1 and k_2 of $\mathbf{A}$;
2. The product $\mathbf{A}\mathbf{E}_{k_1 k_2}$ exchanges columns k_1 and k_2 of $\mathbf{A}$;
3. $\mathbf{E}_{k_1 k_2}^{-1} = \mathbf{E}_{k_1 k_2}^{T} = \mathbf{E}_{k_1 k_2}$ since $\mathbf{E}_{k_1 k_2}\mathbf{E}_{k_1 k_2} = \mathbf{I}_n$.

The first case is a direct consequence of our previous example. The second case is left as a simple exercise for the reader. The third is the result of applying twice $\mathbf{E}_{k_1 k_2}$ to a matrix, leaving that matrix invariant.

With these new elements, we are ready to implement pivoting in the solution of linear systems. At stage k of the GEM algorithm, we search for the element $a_{ik}^{(k)}$, $(i = k+1, \ldots, n)$ with maximum absolute value within the rectangular box of the $\mathbf{A}^{(k)}$ matrix below:

$$
\begin{bmatrix} a_{11}^{(1)} & a_{12}^{(1)} & \cdots & a_{1k}^{(1)} & \cdots & a_{1n}^{(1)} \\ 0 & a_{22}^{(2)} & \cdots & a_{2k}^{(2)} & \cdots & a_{2n}^{(2)} \\ \vdots & 0 & \ddots & \vdots & & \vdots \\ \vdots & \vdots & 0 & \boxed{a_{kk}^{(k)}} & \cdots & a_{kn}^{(k)} \\ \vdots & \vdots & \vdots & \vdots & \ddots & \vdots \\ 0 & \cdots & 0 & a_{nk}^{(k)} & \cdots & a_{nn}^{(k)} \end{bmatrix}. \quad (5.88)
$$

Let $\sigma(k)$ be the index such that[8]

$$|a_{\sigma(k)k}^{(k)}| \geq \max_{\ell \geq k} |a_{\ell k}^{(k)}|. \tag{5.89}$$

In other words, $\sigma(k)$ is the row index where the maximal absolute element is located. Now consider the exchange matrix $\mathbf{E}_{k\sigma(k)}$, which, to avoid cumbersome notation, we will denote simply by $\mathbf{E}_k$. The product $\mathbf{E}_k \mathbf{A}^{(k)}$ exchanges rows k and $\sigma(k)$ of $\mathbf{A}^{(k)}$ as shown in the schematics below:

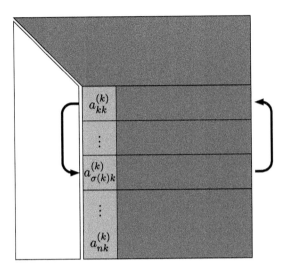

where the white trapezoidal region only contains null elements resulting from the previous Gauss transformations. As a result, swapping rows k and $\sigma(k)$ preserves that trapezoidal shape, and only affects the two horizontal boxes corresponding to the exchanged rows. After performing the exchange of rows, the resulting elements of $\mathbf{E}_k \mathbf{A}^{(k)}$ are

$$[\mathbf{E}_k \mathbf{A}^{(k)}]_{ij} = \begin{cases} a_{\sigma(k)j} & (i = k) \\ a_{kj} & (i = \sigma(k)) \\ a_{ij}^{(k)} & \text{(otherwise)} \end{cases}. \tag{5.90}$$

Once the pivoting has been performed, we proceed with the corresponding Gauss transformation leading to the $(k+1)$th stage of GEM, so that overall,

$$\mathbf{A}^{(k+1)} = \mathbf{M}_k \mathbf{E}_k \mathbf{A}^{(k)}, \tag{5.91}$$

where $\mathbf{M}_k$ is built from the multipliers $m_{ik} = a_{ik}^{(k)}/a_{kk}^{(k)}$ $(i \geq k+1)$ calculated from the transformed (and accordingly renamed) elements of $\mathbf{A}^{(k)}$. As a result,

---

8 Since $\mathbf{A}$ is Non-singular, $|a_{\sigma(k)k}^{(k)}| > 0$.

the composition of successive row exchanges and Gauss transformations reads

$$\mathbf{A}^{(2)} \equiv \mathbf{M}_1 \mathbf{E}_1 \mathbf{A}^{(1)} \;\rightarrow\; \mathbf{A}^{(3)} \equiv \mathbf{M}_2 \mathbf{E}_2 \mathbf{A}^{(2)} \equiv \mathbf{M}_2 \mathbf{E}_2 \mathbf{M}_1 \mathbf{E}_1 \mathbf{A}^{(1)} \;\rightarrow\; \cdots$$

$$(5.92)$$

The resulting upper triangular matrix is therefore

$$\mathbf{U} \equiv \mathbf{A}^{(n)} \equiv \mathbf{M}_{n-1} \mathbf{E}_{n-1} \mathbf{M}_{n-2} \mathbf{E}_{n-2} \cdots \mathbf{M}_2 \mathbf{E}_2 \mathbf{M}_1 \mathbf{E}_1 \mathbf{A}^{(1)}. \quad (5.93)$$

For convenience, we define

$$\mathbf{M} \doteq \mathbf{M}_{n-1} \mathbf{E}_{n-1} \mathbf{M}_{n-2} \mathbf{E}_{n-2} \cdots \mathbf{M}_2 \mathbf{E}_2 \mathbf{M}_1 \mathbf{E}_1, \quad (5.94)$$

and

$$\mathbf{P} \doteq \mathbf{E}_{n-1} \mathbf{E}_{n-2} \cdots \mathbf{E}_2 \mathbf{E}_1, \quad (5.95)$$

so that we can write (5.93) as

$$\mathbf{U} \equiv \mathbf{M}\mathbf{A}^{(1)} \equiv \mathbf{M}\mathbf{A}. \quad (5.96)$$

The matrix $\mathbf{P}$ in (5.95) is the composition of the $n-1$ row exchanges that have been performed throughout the triangularization and, as a result, is a *permutation* matrix. Introducing the neutral factor $\mathbf{P}^{-1}\mathbf{P}$ in (5.96) we obtain

$$\mathbf{M}\mathbf{P}^{-1}\mathbf{P}\mathbf{A} \equiv \mathbf{U}, \quad (5.97)$$

which, after left multiplying first by $\mathbf{M}^{-1}$ and after by $\mathbf{P}$ becomes

$$\mathbf{P}\mathbf{A} \equiv \mathbf{P}\mathbf{M}^{-1}\mathbf{U}. \quad (5.98)$$

In what follows we will prove that (5.98) can be written as

$$\mathbf{P}\mathbf{A} \equiv \mathbf{L}\mathbf{U}, \quad (5.99)$$

where $\mathbf{L} \equiv \mathbf{P}\mathbf{M}^{-1}$ is a lower triangular matrix. Since the proof is a bit involved, we start by two simple properties:

> Let $\mathbf{E}_k \in \mathbb{M}_n(\mathbb{R})$ be an exchange matrix, $\mathbf{m}_k \in \mathbb{R}^n$ the multiplier vector defined in (5.45), and $\mathbf{e}_k^{\mathsf{T}}$ the kth canonical basis vector of $\mathbb{R}^n$. Then,
>
> $$\mathbf{e}_k^{\mathsf{T}}\mathbf{E}_\ell = \mathbf{e}_k^{\mathsf{T}} \quad (\forall \ell > k). \quad (5.100)$$
>
> Similarly, the matrix resulting from the product
>
> $$\mathbf{E}_k \mathbf{m}_\ell \mathbf{e}_\ell^{\mathsf{T}} \quad (5.101)$$
>
> is lower triangular if $k \geq \ell + 1$.

Recall that, for brevity, we wrote $\mathbf{E}_\ell = \mathbf{E}_{\ell\sigma(\ell)}$, i.e. the product $\mathbf{e}_k^{\mathsf{T}}\mathbf{E}_\ell$ exchanges columns ℓ and $\sigma(\ell) \geq \ell$ of the canonical row vector $\mathbf{e}_k^{\mathsf{T}}$. If $\ell > k$,

$\mathbf{E}_\ell$ only exchanges null components of $\mathbf{e}_k^T$, leaving it invariant and leading to (5.100). Similarly, (5.101) comes from the fact that the dyadic product $\mathbf{m}_\ell \mathbf{e}_\ell^T$ is a lower triangular matrix whose nonzero elements are located along the ℓth column and only from the $(\ell+1)$th row downwards. Therefore, $\mathbf{E}_k$ with $k \geq \ell + 1$ only exchanges elements from column ℓ and row $\ell + 1$ downwards (i.e. below the diagonal) leaving that dyadic product still lower triangular.

From the definitions of $\mathbf{M}$ and $\mathbf{P}$ in (5.94) and (5.95), we have that

$$\mathbf{PM}^{-1} \equiv \mathbf{E}_{n-1}\,\mathbf{E}_{n-2}\cdots\mathbf{E}_2\,\mathbf{E}_1(\mathbf{M}_{n-1}\,\mathbf{E}_{n-1}\,\mathbf{M}_{n-2}\,\mathbf{E}_{n-2}\cdots\mathbf{M}_2\,\mathbf{E}_2\,\mathbf{M}_1\,\mathbf{E}_1)^{-1}$$
$$= \mathbf{E}_{n-1}\,\mathbf{E}_{n-2}\cdots\mathbf{E}_2\,\mathbf{E}_1\mathbf{E}_1^{-1}\mathbf{M}_1^{-1}\mathbf{E}_2^{-1}\mathbf{M}_2^{-1}\cdots\mathbf{E}_{n-2}^{-1}\mathbf{M}_{n-2}^{-1}\,\mathbf{E}_{n-1}^{-1}\mathbf{M}_{n-1}^{-1}, \tag{5.102}$$

which, since $\mathbf{E}_k^{-1} \equiv \mathbf{E}_k$, reads

$$\mathbf{PM}^{-1} \equiv \mathbf{E}_{n-1}\mathbf{E}_{n-2}\cdots\mathbf{E}_3\,\mathbf{E}_2\,\mathbf{M}_1^{-1}\mathbf{E}_2\mathbf{M}_2^{-1}\cdots\mathbf{E}_{n-2}\mathbf{M}_{n-2}^{-1}\,\mathbf{E}_{n-1}\mathbf{M}_{n-1}^{-1}. \tag{5.103}$$

In order to analyze the structure of the right-hand side of the last expression, let us define $\mathbf{L}_1 \doteq \mathbf{M}_1^{-1}$ and consider the recurrences

$$\widetilde{\mathbf{L}}_i \equiv \mathbf{E}_{i+1}\,\mathbf{L}_i\,\mathbf{E}_{i+1}, \tag{5.104}$$

and

$$\mathbf{L}_{i+1} \equiv \widetilde{\mathbf{L}}_i\,\mathbf{M}_{i+1}^{-1}, \tag{5.105}$$

for $i = 1, 2, \ldots, n-2$, so that we can easily group terms within the products of (5.103):

$$\mathbf{E}_{n-1}\,\mathbf{E}_{n-2}\cdots\mathbf{E}_3\underbrace{\overbrace{\mathbf{E}_2\,\underbrace{\mathbf{L}_1}_{\widetilde{\mathbf{L}}_1}\,\mathbf{E}_2\mathbf{M}_2^{-1}}^{\mathbf{L}_2}}_{\widetilde{\mathbf{L}}_2}\,\mathbf{E}_3\cdots\mathbf{E}_{n-2}\mathbf{M}_{n-2}^{-1}\,\mathbf{E}_{n-1}\mathbf{M}_{n-1}^{-1}. \tag{5.106}$$

Let us start with $\widetilde{\mathbf{L}}_1 \equiv \mathbf{E}_2\mathbf{L}_1\mathbf{E}_2$. Since $\mathbf{L}_1 \equiv \mathbf{M}_1^{-1}$, we can write $\widetilde{\mathbf{L}}_1$ in terms of the dyadic products (5.52):

$$\widetilde{\mathbf{L}}_1 = \mathbf{E}_2(\mathbf{I}_n + \mathbf{m}_1\mathbf{e}_1^T)\mathbf{E}_2 = \mathbf{I}_n + \mathbf{E}_2\mathbf{m}_1\mathbf{e}_1^T\mathbf{E}_2 = \mathbf{I}_n + \mathbf{E}_2\mathbf{m}_1\mathbf{e}_1^T, \tag{5.107}$$

since, according to (5.100), $\mathbf{e}_1^T\mathbf{E}_2 \equiv \mathbf{e}_1^T$.

Following recurrence (5.105), we have

$$\mathbf{L}_2 = \widetilde{\mathbf{L}}_1\mathbf{M}_2^{-1} \equiv (\mathbf{I}_n + \mathbf{E}_2\mathbf{m}_1\mathbf{e}_1^T)(\mathbf{I}_n + \mathbf{m}_2\mathbf{e}_2^T) \equiv \mathbf{I}_n + \mathbf{m}_2\mathbf{e}_2^T + \mathbf{E}_2\mathbf{m}_1\mathbf{e}_1^T, \tag{5.108}$$

since $\mathbf{E}_2\mathbf{m}_1\mathbf{e}_1^T\mathbf{m}_2\mathbf{e}_2^T \equiv 0$, according to (5.55).

In order to obtain $\widetilde{\mathbf{L}}_2$, we use recurrence (5.104) and property (5.100):

$$\widetilde{\mathbf{L}}_2 = \mathbf{E}_3 \mathbf{L}_2 \mathbf{E}_3 = \mathbf{I}_n + \mathbf{E}_3 \mathbf{m}_2 \mathbf{e}_2^{\mathrm{T}} + \mathbf{E}_3 \mathbf{E}_2 \mathbf{m}_1 \mathbf{e}_1^{\mathrm{T}}. \tag{5.109}$$

Matrices $\mathbf{L}_3$, $\mathbf{L}_4$, etc. can similarly be obtained by recursively using (5.104) and (5.105), along with properties (5.55), (5.100), and (5.101). After a bit of algebra, the reader may check that

$$\mathbf{L}_3 = \mathbf{I}_n + \mathbf{m}_3 \mathbf{e}_3^{\mathrm{T}} + \mathbf{E}_3 \mathbf{m}_2 \mathbf{e}_2^{\mathrm{T}} + \mathbf{E}_3 \mathbf{E}_2 \mathbf{m}_1 \mathbf{e}_1^{\mathrm{T}}, \tag{5.110}$$

and that

$$\mathbf{L}_4 = \mathbf{I}_n + \mathbf{m}_4 \mathbf{e}_4^{\mathrm{T}} + \mathbf{E}_4 \mathbf{m}_3 \mathbf{e}_3^{\mathrm{T}} + \mathbf{E}_4 \mathbf{E}_3 \mathbf{m}_2 \mathbf{e}_2^{\mathrm{T}} + \mathbf{E}_4 \mathbf{E}_3 \mathbf{E}_2 \mathbf{m}_1 \mathbf{e}_1^{\mathrm{T}}. \tag{5.111}$$

The reader has probably noticed that the exchange operators $\mathbf{E}_k$ appearing in expressions (5.108), (5.110), and (5.111), always act on the left of dyadic products $\mathbf{m}_\ell \mathbf{e}_\ell^{\mathrm{T}}$ with $k > \ell$, i.e. preserving their lower triangular structure according to (5.101). Therefore, we conclude that $\mathbf{L}_2$, $\mathbf{L}_3$, $\mathbf{L}_4$, and, in general, the last term of the iteration

$$\mathbf{L}_{n-1} = \mathbf{I}_n + \mathbf{m}_{n-1} \mathbf{e}_{n-1}^{\mathrm{T}} + \sum_{k=1}^{n-2} \mathbf{E}_{n-1} \mathbf{E}_{n-2} \cdots \mathbf{E}_{k+2} \mathbf{E}_{k+1} \mathbf{m}_k \mathbf{e}_k^{\mathrm{T}}, \tag{5.112}$$

are lower triangular matrices. The last matrix $\mathbf{L}_{n-1}$, henceforth referred to simply as $\mathbf{L}$, is the result of the right-hand side product $\mathbf{P}\mathbf{M}^{-1}$ of (5.103). Therefore, from (5.98) we conclude that

$$\mathbf{P}\mathbf{A} = \mathbf{L}\mathbf{U}. \tag{5.113}$$

In other words, GEM applied to $\mathbf{A}$ with partial pivoting is equivalent to LU-factorization of $\mathbf{P}\mathbf{A}$ without pivoting. The only problem with this interpretation is that we do not know the permutation matrix $\mathbf{P}$ in advance.

Since $\mathbf{P}$ is a permutation matrix, Eq. (5.113) is also expressed as

$$\mathbf{A} = \mathbf{P}^{\mathrm{T}} \mathbf{L}\mathbf{U}. \tag{5.114}$$

In numerical mathematics, Eq. (5.113) or (5.114) are referred to as the LU-factorization of matrix $\mathbf{A}$ with *partial pivoting*.[9] This is because, at stage k of the factorization, our pivoting strategy has searched maximal pivots just along the kth column of the matrix in (5.88). We have not considered the possibility of finding better pivots within the remaining columns $k + 1, \ldots, n$,

---

9 Or simply LU-factorization.

that is, within the boxed region of the matrix

$$
\begin{bmatrix}
a_{11}^{(1)} & a_{12}^{(1)} & \cdots & a_{1k}^{(1)} & \cdots & a_{1n}^{(1)} \\
0 & a_{22}^{(2)} & \cdots & a_{2k}^{(2)} & \cdots & a_{2n}^{(2)} \\
\vdots & 0 & \ddots & \vdots & & \vdots \\
\vdots & \vdots & 0 & \boxed{\begin{matrix} a_{kk}^{(k)} & \cdots & a_{kn}^{(k)} \\ \vdots & \ddots & \vdots \\ a_{nk}^{(k)} & \cdots & a_{nn}^{(k)} \end{matrix}}
\end{bmatrix}
, \tag{5.115}
$$

and, if the search is successful, proceeding with the corresponding column exchange accordingly. This second strategy, known as *complete pivoting*, is always numerically stable, although the complete search is time consuming. In practice, GEM with partial pivoting works fine,[10] and is the standard procedure for solving linear systems.

In order to solve a general system

$$
\mathbf{Ax} = \mathbf{b}, \tag{5.116}
$$

we first use (5.114) so that it reads

$$
\mathbf{P}^{\mathsf{T}}\mathbf{LUx} = \mathbf{b}, \tag{5.117}
$$

or

$$
\mathbf{LUx} = \mathbf{Pb}. \tag{5.118}
$$

From here, we proceed as in Section 5.3, by introducing $\mathbf{y} = \mathbf{Ux}$ so we first solve

$$
\mathbf{Ly} = \mathbf{Pb} \tag{5.119}
$$

using FS, and we use the solution $\mathbf{y}$ as the right-hand side of the system

$$
\mathbf{Ux} = \mathbf{y}, \tag{5.120}
$$

to be solved using BS.

In practice, the permutation matrix $\mathbf{P}$ is never computed. Instead, a vector storing the index reordering throughout the pivoting will suffice. Since the matrix $\mathbf{A}$ is no longer needed after the factorization, it is common practice to economize memory storage by saving matrices $\mathbf{L}$ and $\mathbf{U}$ precisely in the lower and upper parts of the original matrix $\mathbf{A}$, respectively (taking advantage of the

---

10 Although some rare unstable cases have been found (see references in the Complementary Reading section, at the end of the chapter).

fact that the diagonal entries in **L** are all equal to 1, so they do not need to be stored). However, we will not proceed in that way. Code 13 provides a simple implementation of the LU-factorization of an arbitrary matrix **A** using partial pivoting. The code also provides a vector P containing the row reordering.

Code 14 solves the linear system $\mathbf{Ax} = \mathbf{b}$ for an arbitrary right-hand side **b** using the LU-factorization of **A** provided by Code 13. This code makes use of FS and BS, invoking Codes 11 and 12.

```
% Code 13: PA = LU factorization (partial pivoting)
% Input: A (non-singular square matrix)
% Output: L (unit lower triangular matrix)
%          U (upper triangular matrix)
%          P (reordering vector)
function [P,L,U] = pplu(A)
[m,n] = size(A); if m ~= n; error('not square matrix'); end
U = A; L = eye(n); P = [1:n]';

for k = 1:n-1
  ["",imax] = max(abs(U(k:end,k)));
  imax = imax + k - 1; i1 = [k,imax]; i2 = [imax,k];
  U(i1,:) = U(i2,:); P(k) = imax;
  L(i1,1:k-1) = L(i2,1:k-1);
   for j = [k+1:n]
     L(j,k) = U(j,k)/U(k,k);
     U(j,k:n) = U(j,k:n) - L(j,k)*U(k,k:n);
   end
end
```

```
% Code 14: PA = LU (Solver for Ax = b)
% Input:  L (unit lower triangular matrix)
%          U (upper triangular matrix)
%          P (reordering vector)
%          b (right-hand side)
% Output: solution x
% Note: requires fs.m and bs.m (Codes 11 & 12)
function x = plusolve(L,U,P,b)
n = length(b);
for k = 1:n-1; b([k P(k)]) = b([P(k) k]); end
y = fs(L,b); x = bs(U,y);
```

Practical 5.1: Equivalent Resistance

The figure below shows a series of electrical circuits consisting of identical resistors ($R = 3 \; \Omega$) connected to a battery $V = 1$ V. The circuit at the bottom is a generalization of the first three circuits.

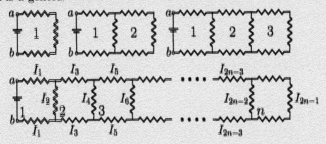

Use Kirchoff's laws[11] and confirm that the $2n - 1$ intensities satisfy:

$$
\begin{aligned}
2R\,I_1 \;+\; & R\,I_2 & & = V \\
I_1 \;-\; & I_2 \;-\; & I_3 & = 0 \\
-R\,I_2 \;+\; & 2R\,I_3 \;+\; & R\,I_4 & = 0 \\
I_3 \;-\; & I_4 \;-\; & I_5 & = 0 \\
& \vdots & & \vdots \\
-R\,I_{2n-4} \;+\; & 2R\,I_{2n-3} \;+\; & R\,I_{2n-2} & = 0 \\
I_{2n-3} \;-\; & I_{2n-2} \;-\; & I_{2n-1} & = 0 \\
& -R\,I_{2n-2} \;+\; & 3R\,I_{2n-1} & = 0,
\end{aligned}
$$

or, in matricial form,

$$
\begin{bmatrix}
2R & R & 0 & 0 & 0 & 0 & \cdots \\
1 & -1 & -1 & 0 & 0 & 0 & \cdots \\
0 & -R & 2R & R & 0 & 0 & \cdots \\
0 & 0 & 1 & -1 & -1 & 0 & \cdots \\
\vdots & \vdots & 0 & \ddots & \ddots & \ddots & \\
\vdots & \vdots & \vdots & \vdots & -R & 2R & R & 0 \\
\vdots & \vdots & \vdots & \vdots & & 1 & -1 & -1 \\
\vdots & \vdots & \vdots & \vdots & & & 0 & -R & 3R
\end{bmatrix}
\begin{bmatrix}
I_1 \\ I_2 \\ I_3 \\ I_4 \\ I_5 \\ \vdots \\ I_{2n-3} \\ I_{2n-2} \\ I_{2n-1}
\end{bmatrix}
=
\begin{bmatrix}
V \\ 0 \\ 0 \\ 0 \\ 0 \\ \vdots \\ 0 \\ 0 \\ 0
\end{bmatrix}.
$$

(a) Build the matrix above and solve the system for $n = 5, 10, \ldots, 30$. For $n = 10$, provide the values I_1, I_2, and I_{15} and plot (k, I_k) for $k = 1, 2, \ldots, 2n - 1$. How do the intensities decay?

(b) Show that the equivalence resistance $R_e(n)$ between a and b satisfies $\lim\limits_{n \to \infty} R_e(n) = R(1 + \sqrt{3})$. Do you approach this limit in (a)?

---

11 G. R. Kirchhoff (1824–1887), German physicist, known for his contributions to the theory of electrical circuits, spectroscopy, and black-body radiation.

5.5   The Least Squares Problem

In Part I, we studied different interpolatory methods to find the polynomial that passes through a set of given points $\{(x_0, y_0),\ (x_1, y_1),\ ,\ldots, (x_n, y_n)\}$. Under certain circumstances, this polynomial exists and it is unique. In particular, the method of undetermined coefficients that leads to the system (2.5) of $n+1$ linear equations has a unique solution if (i) all the abscissas x_j are different and (ii) the polynomial has degree n.

In many experimental branches of science and technology it is frequently needed to *fit* a set of empirical data (previously collected in the laboratory, for example) to a simple mathematical rule. The most common fitting is *linear regression*, that is, for a given set of experimental data $\{(x_0, y_0),\ (x_1, y_1),\ldots,$ $(x_n, y_n)\}$, we want to find the straight line $y = \alpha x + \beta$ that *best fits* the points. For example, consider a complex network of resistors such as the one seen in Practical 5.1. Let us imagine that we ignore its internal structure but still we need to determine its equivalent resistance by experimental techniques. This could be done by, for example, connecting the terminals a and b of any of the circuits of Practical 5.1 to a battery of known voltage V and then measuring the intensity I with a multimeter. According to Ohm's law, the equivalent resistance of the circuit is simply $R = V/I$. However, just one measure is usually not enough. To have a better estimation of R, we need to make different measurements due to inaccuracies in the experiment, such as variations in the nominal voltage of the battery used or the lack of precision in the multimeter. Therefore, we repeat the experiment with batteries providing different known voltages $\{V_0,\ V_1,\ \ldots, V_n\}$ and measuring the corresponding intensities $\{I_0,\ I_1,\ \ldots, I_n\}$ provided by the readings of the multimeter. If we plot the empirical data in the (V, I) plane, we will typically observe that the points are scattered as shown in Figure 5.1, but aligned with an imaginary straight

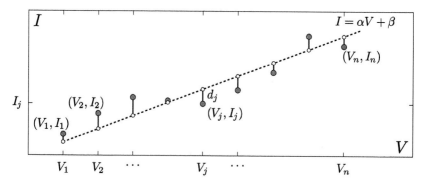

Figure 5.1   Linear fitting. The gray disks are the experimental observations.

line $I = \alpha V + \beta$ (dashed). Linear regression consists precisely in identifying the coefficients α and β, which makes $\alpha V + \beta$ fit the points in an *optimal* way. To be more precise, linear regression provides the values of α and β such that the sum of the squares of the differences $d_j = I_j - \alpha V_j - \beta$ $(j = 0, 1, \ldots, n)$ between the expected ideal results (white bullets) and the measured quantities (gray disks)

$$\sum_{j=0}^{n} (I_j - \alpha V_j - \beta)^2 \tag{5.121}$$

is a *minimum*. For this reason, linear regression is commonly known as a *least squares approximation*. It that sense, it is said that the straight line $\alpha V + \beta$ is the best fit for the given data. Once α is determined, the sought equivalent resistance is $R = 1/\alpha$.

Any textbook on statistical data analysis provides explicit expressions of α and β for the linear fitting $y = \alpha x + \beta$ of a given arbitrary set of collected data $\{(x_0, y_0), (x_1, y_1), , \ldots, (x_n, y_n)\}$. The standard procedure is to impose the function

$$S(\alpha, \beta) = \sum_{j=0}^{n} (y_j - \alpha x_j - \beta)^2 \tag{5.122}$$

to have a *local minimum* at some point $(\alpha, \beta) \in \mathbb{R}^2$, by imposing $\partial_\alpha S = \partial_\beta S = 0$. From the two previous conditions we can find explicit expressions for α and β. We do not include here these expressions because later on in this section we will arrive at the same result by means of a more general framework.

The minimization procedure used by linear regression to fit the data is very different from the interpolation method mentioned at the beginning of this introduction. On the one hand, it is impossible to require a first order polynomial to pass through more than two points simultaneously. In other words, by imposing a first degree polynomial $y = a_0 + a_1 x$ to pass through an arbitrary set of $n + 1$ points $\{(x_0, y_0), (x_1, y_1), , \ldots, (x_n, y_n)\}$ we obtain the linear system of equations

$$
\begin{aligned}
a_0 + a_1 x_0 &= y_0, \\
a_0 + a_1 x_1 &= y_1, \\
&\vdots \\
a_0 + a_1 x_n &= y_n
\end{aligned}
\quad \text{or} \quad
\begin{bmatrix} 1 & x_0 \\ 1 & x_1 \\ \vdots & \vdots \\ 1 & x_n \end{bmatrix}
\begin{bmatrix} a_0 \\ a_1 \end{bmatrix}
=
\begin{bmatrix} y_0 \\ y_1 \\ \vdots \\ y_n \end{bmatrix},
\tag{5.123}
$$

which is *overdetermined* for $n > 1$. On the other hand, we may have two or more different experimental measurements y_j (in our previous example, different readings from the multimeter) for the same x_j (using for example two or more identical batteries of, presumably, the same nominal voltage). This would violate assumption (1) at the beginning of this introduction.

Linear regression is just a particular case of least squares problem, where the curve to be adjusted to the experimental points is a straight line, i.e. a polynomial of first degree. However, the theory behind least squares problems is more general and can be applied to analyze data arising from physical phenomena of very different nature, such as the linear Ohm's law seen in our example, or periodic behaviors arising in mechanical oscillations and waves, for example. The least squares approach allows one to fit sets of arbitrary data (statistical or deterministic) to linear combinations of mathematical functions of different type, ranging from monomials $\{1, x, x^2, \ldots\}$ that lead to polynomial fittings of the form

$$p(x) \equiv a_0 + a_1 x + a_2 x^2 + \cdots ,$$

to periodic functions $\{1, \sin x, \sin 2x, \ldots, \cos x, \cos 2x, \ldots\}$ used in trigonometric expansions

$$F(x) \equiv a_0 + a_1 \sin x + a_2 \sin 2x + \cdots + b_1 \cos x + b_2 \cos 2x + \cdots ,$$

for example.

In this section we will study a new type of linear transformations that lead to suitable matrix factorization, and provide the least squares solution of overdetermined systems such as (5.123).

5.5.1 QR Factorization

We begin with a fundamental definition:

Orthogonal Matrix

A square matrix $\mathbf{Q} \in M_n(\mathbb{R})$ is said to be *orthogonal* if

$$\mathbf{Q}^T \mathbf{Q} = \mathbf{Q} \mathbf{Q}^T = \mathbf{I}_n. \tag{5.124}$$

As a result, orthogonal matrices are invertible, with $\mathbf{Q}^{-1} \equiv \mathbf{Q}^T$. Another consequence is that the product of k orthogonal matrices $\mathbf{Q}_1 \mathbf{Q}_2 \cdots \mathbf{Q}_k$ is also an orthogonal matrix, since

$$(\mathbf{Q}_1 \mathbf{Q}_2 \cdots \mathbf{Q}_k)^T \mathbf{Q}_1 \mathbf{Q}_2 \cdots \mathbf{Q}_k \equiv \mathbf{Q}_k^T \cdots \mathbf{Q}_2^T \underbrace{\overbrace{\mathbf{Q}_1^T \mathbf{Q}_1}^{\mathbf{I}_n} \mathbf{Q}_2 \cdots \mathbf{Q}_k}_{\mathbf{I}_n} \equiv \mathbf{I}_n.$$

Geometrically, orthogonal matrices represent transformations in $\mathbb{R}^n$ that *preserve norms and angles*. Consider the standard *scalar product* and *norm* in $\mathbb{R}^n$:

$$\langle \mathbf{x}, \mathbf{y} \rangle \doteq \mathbf{x}^T \mathbf{y} = \mathbf{y}^T \mathbf{x} = \sum_{j=1}^{n} x_j y_j, \tag{5.125}$$

and

$$\|\mathbf{x}\|_2 \equiv \sqrt{\langle \mathbf{x}, \mathbf{x} \rangle}, \tag{5.126}$$

for any two vectors $\mathbf{x}, \mathbf{y} \in \mathbb{R}^n$. Then, for any orthogonal matrix $\mathbf{Q} \in M_n(\mathbb{R})$ we have that

$$\langle \mathbf{Qx}, \mathbf{Qy} \rangle \equiv (\mathbf{Qx})^T \mathbf{Qy} \equiv \mathbf{x}^T \mathbf{Q}^T \mathbf{Qy} \equiv \mathbf{x}^T \mathbf{y} \equiv \langle \mathbf{x}, \mathbf{y} \rangle, \tag{5.127}$$

and

$$\|\mathbf{Qx}\|_2 \equiv \sqrt{(\mathbf{Qx})^T \mathbf{Qx}} \equiv \sqrt{\mathbf{x}^T \mathbf{Q}^T \mathbf{Qx}} \equiv \sqrt{\mathbf{x}^T \mathbf{x}} \equiv \|\mathbf{x}\|_2. \tag{5.128}$$

In what follows, consider two nonzero orthogonal vectors $\mathbf{u}$ and $\mathbf{v}$, with $\|\mathbf{u}\| \equiv 1$. Let L be the straight line passing through the origin and parallel to $\mathbf{v}$ as shown in Figure 5.2 and consider the linear operator

$$\mathbf{Q} = \mathbf{I} - 2\mathbf{u}\mathbf{u}^T, \tag{5.129}$$

built from the identity and the dyadic product of $\mathbf{u}$ with itself. This operator has two very interesting properties. On the one hand,

$$\mathbf{Qu} \equiv (\mathbf{I} - 2\mathbf{u}\mathbf{u}^T)\mathbf{u} \equiv \mathbf{u} - 2\mathbf{u}\mathbf{u}^T \mathbf{u} \equiv -\mathbf{u}, \tag{5.130}$$

since $\mathbf{u}^T \mathbf{u} \equiv \|\mathbf{u}\|^2 \equiv 1$. On the other hand,

$$\mathbf{Qv} \equiv (\mathbf{I} - 2\mathbf{u}\mathbf{u}^T)\mathbf{v} \equiv \mathbf{v} - 2\mathbf{u}^T \mathbf{v} \equiv \mathbf{v}, \tag{5.131}$$

since $\mathbf{u}^T \mathbf{v} \equiv \langle \mathbf{u}, \mathbf{v} \rangle \equiv 0$. Take any vector $\mathbf{x}$ lying within the plane $\mathbf{u} - \mathbf{v}$, so that it can be expressed as the linear combination $\mathbf{x} \equiv \alpha \mathbf{u} + \beta \mathbf{v}$. Let us analyze the effect of the operator $\mathbf{Q}$ on $\mathbf{x}$. By virtue of (5.130) and (5.131) we have

$$\mathbf{Qx} \equiv \mathbf{Q}(\alpha \mathbf{u} + \beta \mathbf{v}) \equiv -\alpha \mathbf{u} + \beta \mathbf{v}. \tag{5.132}$$

This effect is better illustrated in Figure 5.2, where we can clearly see that $\mathbf{x}$ has been *reflected* across the line L orthogonal to $\mathbf{u}$ through the action of $\mathbf{Q}$. That is, the linear operator $\mathbf{Q}$ defined in (5.129) *reflects* vectors across the line L orthogonal to $\mathbf{u}$. Furthermore, since

$$\mathbf{Q}^T = (\mathbf{I} - 2\mathbf{u}\mathbf{u}^T)^T = \mathbf{I} - 2\mathbf{u}\mathbf{u}^T \equiv \mathbf{Q}, \tag{5.133}$$

and

$$\mathbf{Q}^T \mathbf{Q} = \mathbf{QQ} = (\mathbf{I} - 2\mathbf{u}\mathbf{u}^T)(\mathbf{I} - 2\mathbf{u}\mathbf{u}^T) = \mathbf{I} - 4\mathbf{u}\mathbf{u}^T + 4\mathbf{u}\mathbf{u}^T\mathbf{u}\mathbf{u}^T = \mathbf{I}, \tag{5.134}$$

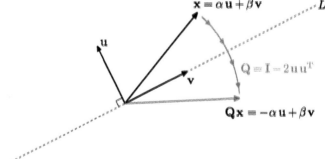

Figure 5.2 The resulting vector $\mathbf{Qx}$ is the reflection of $\mathbf{x}$ across the line L.

$\mathbf{Q}$ is orthogonal with inverse $\mathbf{Q}^{-1} = \mathbf{Q}$. The operator $\mathbf{Q}$ defined in (5.129) is called *reflector* or *Householder transformation*,[12] whereas $\mathbf{u}$ is usually known as the *Householder vector*. In general, if $\mathbf{u}$ is not of unit length, reflectors are defined as follows:

Householder Transformation (Reflector)

For a given nonzero vector $\mathbf{u}$, its corresponding reflector or Householder transformation is[13]

$$\mathbf{Q} = \mathbf{I} - \frac{2}{\|\mathbf{u}\|^2}\mathbf{u}\mathbf{u}^{\mathrm{T}}. \tag{5.135}$$

It is left as a simple exercise for the reader to check that the reflector defined in (5.135) satisfies $\mathbf{Q}\mathbf{u} = -\mathbf{u}$ and $\mathbf{Q}\mathbf{v} = \mathbf{v}$, for any $\mathbf{v} \perp \mathbf{u}$.

Property

Let $\mathbf{x}$ and $\mathbf{y}$ be two different nonzero vectors in $\mathbb{R}^n$ with same norm. Then the reflector

$$\mathbf{Q} = \mathbf{I}_n - \gamma\mathbf{u}\,\mathbf{u}^{\mathrm{T}}, \tag{5.136}$$

with $\mathbf{u} = \mathbf{x} - \mathbf{y}$ and $\gamma = 2/\|\mathbf{u}\|^2$ satisfies

$$\mathbf{Q}\mathbf{x} = \mathbf{y}, \quad \text{and} \quad \mathbf{Q}\mathbf{y} = \mathbf{x}. \tag{5.137}$$

To prove the property above, first notice that

$$\mathbf{Q}(\mathbf{x} - \mathbf{y}) = \mathbf{Q}\mathbf{u} = -\mathbf{u} = -\mathbf{x} + \mathbf{y}, \tag{5.138}$$

and that

$$\begin{aligned}
\mathbf{Q}(\mathbf{x} + \mathbf{y}) &= \mathbf{x} + \mathbf{y} - \gamma\mathbf{u}\langle\mathbf{u}, \mathbf{x}\rangle - \gamma\mathbf{u}\langle\mathbf{u}, \mathbf{y}\rangle \\
&= \mathbf{x} + \mathbf{y} - \gamma\mathbf{u}\langle\mathbf{x} - \mathbf{y}, \mathbf{x}\rangle - \gamma\mathbf{u}\langle\mathbf{x} - \mathbf{y}, \mathbf{y}\rangle \\
&= \mathbf{x} + \mathbf{y} - \gamma\mathbf{u}\|\mathbf{x}\|^2 + \gamma\mathbf{u}\langle\mathbf{y}, \mathbf{x}\rangle - \gamma\mathbf{u}\langle\mathbf{x}, \mathbf{y}\rangle + \gamma\mathbf{u}\|\mathbf{y}\|^2 \\
&= \mathbf{x} + \mathbf{y},
\end{aligned} \tag{5.139}$$

since $\langle\mathbf{y}, \mathbf{x}\rangle = \langle\mathbf{x}, \mathbf{y}\rangle$ and $\|\mathbf{y}\|^2 = \|\mathbf{x}\|^2$. Addition and subtraction of Eqs. (5.138) and (5.139) leads to the first and second equation of (5.137), respectively. Figure 5.3a outlines graphically what we have just proved algebraically.

---

12 After A. S. Householder, American mathematician (1904–1993), who first employed it in matrix computations.
13 Unless stated otherwise, $\|\mathbf{u}\|$ stands for the Euclidean 2-norm of $\mathbf{u}$.

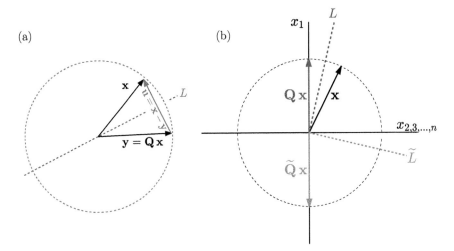

Figure 5.3 (a) Properties (5.137). (b) There are two possible reflectors (across lines L or $\widetilde{L}$) capable of annihilating the components $x_{2:n}$ of **x**.

The motivation for the use of reflectors comes from the fact that any vector $\mathbf{x} = [x_1\ x_2\ x_3\ \cdots\ x_n]^\mathrm{T}$, with arbitrary nonzero components, can be reflected across a suitable hyperplane L (dashed gray line in Figure 5.3b) leading to a new vector $\mathbf{y} = \mathbf{Q}\,\mathbf{x} = [y_1\ 0\ 0\ \cdots\ 0]^\mathrm{T}$ (gray vector pointing upwards in Figure 5.3b) with the same norm and all its components canceled but the first. Equations (5.136) and (5.137) provide the right reflector for this purpose. Our aim is to find a reflector **Q** such that

$$\mathbf{Q}\,\mathbf{x} = \mathbf{Q}\begin{bmatrix} x_1 \\ x_2 \\ x_3 \\ \vdots \\ x_n \end{bmatrix} = \begin{bmatrix} y_1 \\ 0 \\ 0 \\ \vdots \\ 0 \end{bmatrix} = \mathbf{y},$$

with $\|\mathbf{y}\| = \|\mathbf{x}\|$. This can be done if we define[14] $\tau = \pm\|\mathbf{x}\|$ and take

$$\mathbf{y} = \begin{bmatrix} -\tau \\ 0 \\ 0 \\ \vdots \\ 0 \end{bmatrix},$$

---

14 The two possible signs of τ will be justified in a moment.

so that the Householder vector $\mathbf{u}$ of the reflector (5.136) is

$$\mathbf{u} \equiv \mathbf{x} - \mathbf{y} \equiv \begin{bmatrix} x_1 \\ x_2 \\ x_3 \\ \vdots \\ x_n \end{bmatrix} - \begin{bmatrix} -\tau \\ 0 \\ 0 \\ \vdots \\ 0 \end{bmatrix} = \begin{bmatrix} x_1 + \tau \\ x_2 \\ x_3 \\ \vdots \\ x_n \end{bmatrix},$$

and, by virtue of Eq. (5.137), $\mathbf{Q}\,\mathbf{x} \equiv \mathbf{y}$. The vector $\mathbf{u}$ can be normalized (without affecting the reflector[15]) with respect to its first component so that we take

$$\mathbf{u} \equiv \begin{bmatrix} 1 \\ x_2/(x_1 + \tau) \\ x_3/(x_1 + \tau) \\ \vdots \\ x_n/(x_1 + \tau) \end{bmatrix},$$

and therefore its norm now reads

$$\begin{aligned}
\|\mathbf{u}\|^2 &\equiv 1 + \frac{x_2^2}{(x_1 + \tau)^2} + \frac{x_3^2}{(x_1 + \tau)^2} + \cdots + \frac{x_n^2}{(x_1 + \tau)^2} \\
&= \frac{1}{(x_1 + \tau)^2}\left\{(x_1 + \tau)^2 + x_2^2 + \cdots + x_n^2\right\} \\
&= \frac{1}{(x_1 + \tau)^2}\left\{\tau^2 + 2x_1\tau + \overbrace{x_1^2 + x_2^2 + \cdots + x_n^2}^{\|\mathbf{x}\|^2 = \tau^2}\right\} \\
&= \frac{1}{(x_1 + \tau)^2}\left\{2\tau^2 + 2x_1\tau\right\} \equiv \frac{2\tau}{x_1 + \tau}.
\end{aligned} \tag{5.140}$$

As a result, the γ factor in (5.136) now reads

$$\gamma = \frac{2}{\|\mathbf{u}\|^2} = 1 + \frac{x_1}{\tau}. \tag{5.141}$$

The two possible signs of τ come from the fact that there are two possible reflectors capable of annihilating the $n - 1$ components $x_{2:n}$ of $\mathbf{x}$. This is better understood by looking at Figure 5.3b. We can see that reflection of $\mathbf{x}$ across $\widetilde{L}$ (light gray dashed line) leads to the vector $\widetilde{\mathbf{y}} = \widetilde{\mathbf{Q}}\,\mathbf{x}$ (light gray vector pointing downwards), antiparallel to $\mathbf{Q}\,\mathbf{x}$, but also with $\widetilde{y}_{2:n} = 0$. This arbitrariness is solved by choosing the sign that avoids the normalizing factor $x_1 + \tau$ from being zero or numerically very small in magnitude.[16] This is accomplished by defining

$$\tau \doteq \mathrm{sgn}(x_1)\|\mathbf{x}\|. \tag{5.142}$$

With this choice, $y_1 = -\tau = -\mathrm{sgn}(x_1)\|\mathbf{x}\|$, so that we reflect $\mathbf{x}$ across the line $\widetilde{L}$ depicted in Figure 5.3b. This strategy maximizes the norm of the Householder

---

15 $\forall \beta \in \mathbb{R}$, $\beta\mathbf{u}$ generates the same reflector: $\mathbf{Q} = \mathbf{I} - (2/\beta^2|\mathbf{u}|^2)\beta^2\mathbf{u}\mathbf{u}^\mathrm{T} = \mathbf{I} - (2/|\mathbf{u}|^2)\mathbf{u}\mathbf{u}^\mathrm{T}$.
16 In that sense, we choose the *well-conditioned* reflector.

vector, $\|\mathbf{u}\| \equiv \|\mathbf{x} - \tilde{\mathbf{Q}}\,\mathbf{x}\|$, is maximized, so that the denominator in $\gamma = 2/\|\mathbf{u}\|^2$ is made as large as possible in order to avoid numerical instabilities.

Summary (Vector Reduction Using Reflectors)

Let $\mathbf{x} = [x_1\ x_2\ \cdots\ x_n]^{\mathrm{T}}$ be an arbitrary nonzero vector and the reflector $\mathbf{Q} = \mathbf{I}_n - \gamma\mathbf{u}\mathbf{u}^{\mathrm{T}}$, with

$$\mathbf{u} = \left[1\ \frac{x_2}{\tau + x_1}\ \cdots\ \frac{x_n}{\tau + x_1}\right]^{\mathrm{T}}, \tag{5.143}$$

$$\tau \doteq \mathrm{sgn}(x_1)\|\mathbf{x}\|, \tag{5.144}$$

and

$$\gamma \equiv 1 + \frac{x_1}{\tau}. \tag{5.145}$$

Then,

$$\mathbf{Q}\mathbf{x} = \mathbf{Q}\begin{bmatrix} x_1 \\ x_2 \\ \vdots \\ x_n \end{bmatrix} = \begin{bmatrix} -\tau \\ 0 \\ \vdots \\ 0 \end{bmatrix}, \tag{5.146}$$

The result above opens the door for a new methodology to triangularize matrices, which leads to a new type of factorization:

Theorem (QR-Factorization)

Every matrix $\mathbf{A} \in \mathbb{M}_n(\mathbb{R})$ can be expressed as a product $\mathbf{A} = \mathbf{Q}\,\mathbf{R}$, where $\mathbf{Q}$ and $\mathbf{R}$ are orthogonal and upper triangular matrices, respectively.

The proof is by induction on n. For $n = 1$, we take $\mathbf{A} \equiv [a_{11}] \equiv \mathbf{Q}\,\mathbf{R}$ with $\mathbf{Q} \equiv [1]$ and $\mathbf{R} \equiv [a_{11}]$. Assuming that the theorem holds for arbitrary matrices in $\mathbb{M}_{n-1}(\mathbb{R})$, we must show that it also holds for arbitrary matrices in $\mathbb{M}_n(\mathbb{R})$ with $n \geq 2$.

We begin by exploiting property (5.146), where $\mathbf{x}$ is the first column of the original matrix $\mathbf{A}$. Let $\mathbf{Q}_1$ be the reflector that annihilates the elements $a_{2:n,1}$, that is,

$$\mathbf{Q}_1\begin{bmatrix} a_{11} \\ a_{21} \\ \vdots \\ a_{n1} \end{bmatrix} = \begin{bmatrix} -\tau_1 \\ 0 \\ \vdots \\ 0 \end{bmatrix}. \tag{5.147}$$

Since $\mathbf{Q}_1^T = \mathbf{Q}_1$, we may write

$$\mathbf{Q}_1^T \mathbf{A} = \mathbf{Q}_1 \mathbf{A} = \begin{bmatrix} -\tau_1 & \hat{a}_{12} & \cdots & \hat{a}_{1n} \\ 0 & & & \\ \vdots & & \hat{\mathbf{A}}_2 & \\ 0 & & & \end{bmatrix},$$

where the last $n - 1$ elements of the first column have been reduced to zero, at the price of transforming the remaining columns of $\mathbf{A}$, whose elements are now denoted by $\hat{a}_{ij}$. The block matrix denoted by $\hat{\mathbf{A}}_2 \in \mathbb{M}_{n-1}(\mathbb{R})$ admits, by the induction hypothesis, the factorization $\hat{\mathbf{A}}_2 = \hat{\mathbf{Q}}_2 \hat{\mathbf{R}}_2$, with $\hat{\mathbf{Q}}_2$ and $\hat{\mathbf{R}}_2$ orthogonal and upper triangular matrices in $\mathbb{M}_{n-1}(\mathbb{R})$, respectively. To end the proof, we finally introduce the matrix

$$\mathbf{Q}_2 \doteq \begin{bmatrix} 1 & 0 & \cdots & 0 \\ 0 & & & \\ \vdots & & \hat{\mathbf{Q}}_2 & \\ 0 & & & \end{bmatrix}, \tag{5.148}$$

containing the orthogonal matrix $\hat{\mathbf{Q}}_2$ of the factorization $\hat{\mathbf{A}}_2 = \hat{\mathbf{Q}}_2 \hat{\mathbf{R}}_2$. Matrix $\mathbf{Q}_2$ is, by construction, orthogonal:

$$\mathbf{Q}_2^T \mathbf{Q}_2 = \begin{bmatrix} 1 & 0 & \cdots & 0 \\ 0 & & & \\ \vdots & & \hat{\mathbf{Q}}_2^T \hat{\mathbf{Q}}_2 & \\ 0 & & & \end{bmatrix} = \begin{bmatrix} 1 & 0 & \cdots & 0 \\ 0 & & & \\ \vdots & & \mathbf{I}_{n-1} & \\ 0 & & & \end{bmatrix} = \mathbf{I}_n.$$

Therefore, we may write

$$\mathbf{Q}_2^T \mathbf{Q}_1^T \mathbf{A} = \begin{bmatrix} 1 & 0 & \cdots & 0 \\ 0 & & & \\ \vdots & & \hat{\mathbf{Q}}_2^T & \\ 0 & & & \end{bmatrix} \begin{bmatrix} -\tau_1 & \hat{a}_{12} & \cdots & \hat{a}_{1n} \\ 0 & & & \\ \vdots & & \hat{\mathbf{A}}_2 & \\ 0 & & & \end{bmatrix}$$

$$= \begin{bmatrix} -\tau_1 & \hat{a}_{12} & \cdots & \hat{a}_{1n} \\ 0 & & & \\ \vdots & & \hat{\mathbf{R}}_2 & \\ 0 & & & \end{bmatrix}, \tag{5.149}$$

where $\hat{\mathbf{R}}_2$ is upper triangular, so that the resulting matrix in (5.149) is also upper triangular. Consequently, the last expression reads

$$\mathbf{Q}_2^T \mathbf{Q}_1^T \mathbf{A} = \mathbf{R},$$

with $\mathbf{R}$ being an upper triangular matrix. Left multiplication on both sides of the last expression, first by $\mathbf{Q}_2$ and then by $\mathbf{Q}_1$, leads to

$$\mathbf{Q}_1 \underbrace{\mathbf{Q}_2 \mathbf{Q}_2^T}_{\mathbf{I}_n} \overbrace{\mathbf{Q}_1^T}^{\mathbf{I}_n} \mathbf{A} = \mathbf{Q}_1 \, \mathbf{Q}_2 \mathbf{R},$$

or simply

$$\mathbf{A} = \mathbf{Q} \, \mathbf{R},$$

where we have implicitly defined $\mathbf{Q} \doteq \mathbf{Q}_1 \, \mathbf{Q}_2$, which is an orthogonal matrix resulting from the product of two orthogonal matrices.

The previous constructive proof envisages how to proceed to triangularize any matrix using reflectors. Recall that the first transformation $\mathbf{Q}_1$ in (5.147) is given by the reflector

$$\mathbf{Q}_1 = \mathbf{I}_n - \gamma_1 \mathbf{u}_1 \mathbf{u}_1^T, \tag{5.150}$$

where the corresponding Householder vector $\mathbf{u}_1$ and its associated factor γ_1 are obtained from expressions (5.143)–(5.145), with $\mathbf{x}$ being the first column of $\mathbf{A}$:

$$\mathbf{x} = \begin{bmatrix} x_1 \\ x_2 \\ \vdots \\ x_n \end{bmatrix} = \begin{bmatrix} a_{11} \\ a_{21} \\ \vdots \\ a_{n1} \end{bmatrix}.$$

This transformation led to

$$\mathbf{Q}_1^T \mathbf{A} = \mathbf{Q}_1 \mathbf{A} = \begin{bmatrix} -\tau_1 & \widehat{a}_{12} & \cdots & \widehat{a}_{1n} \\ \hline 0 & & & \\ \vdots & & \widehat{\mathbf{A}}_2 & \\ 0 & & & \end{bmatrix} = \begin{bmatrix} -\tau_1 & \widehat{a}_{12} & \cdots & \widehat{a}_{1n} \\ \hline 0 & \widehat{a}_{22} & \cdots & \widehat{a}_{2n} \\ \vdots & \vdots & & \vdots \\ 0 & \widehat{a}_{n2} & \cdots & \widehat{a}_{nn} \end{bmatrix} \tag{5.151}$$

In the second transformation, $\mathbf{Q}_2$ is given by (5.148); the reflector $\widehat{\mathbf{Q}}_2 \in \mathrm{M}_{n-1}(\mathbb{R})$ is

$$\widehat{\mathbf{Q}}_2 = \mathbf{I}_{n-1} - \gamma_2 \mathbf{u}_2 \mathbf{u}_2^T, \tag{5.152}$$

but in this case its associated Householder vector $\mathbf{u}_2 \in \mathbb{R}^{n-1}$ is built from the first column of the matrix $\widehat{\mathbf{A}}_2$, boxed in (5.151):

$$\mathbf{x} = \begin{bmatrix} x_1 \\ x_2 \\ \vdots \\ x_{n-1} \end{bmatrix} = \begin{bmatrix} \widehat{a}_{22} \\ \widehat{a}_{32} \\ \vdots \\ \widehat{a}_{n2} \end{bmatrix}.$$

This second transformation led to

$$
\mathbf{Q}_2^{\mathrm{T}}\,\mathbf{Q}_1^{\mathrm{T}}\,\mathbf{A} \equiv
\begin{bmatrix}
1 & 0 & \cdots & 0 \\
\hline
0 & & & \\
\vdots & & \widehat{\mathbf{Q}}_2^{\mathrm{T}} & \\
0 & & &
\end{bmatrix}
\begin{bmatrix}
-\tau_1 & \widehat{a}_{12} & \cdots & \widehat{a}_{1n} \\
\hline
0 & \widehat{a}_{22} & \cdots & \widehat{a}_{2n} \\
\vdots & \vdots & & \vdots \\
0 & \widehat{a}_{n2} & \cdots & \widehat{a}_{nn}
\end{bmatrix}
$$

$$
\equiv
\begin{bmatrix}
-\tau_1 & \widehat{a}_{12} & \widehat{a}_{13} & \cdots & \widehat{a}_{1n} \\
\hline
0 & -\tau_2 & \widetilde{a}_{22} & \cdots & \widetilde{a}_{2n} \\
0 & 0 & & & \\
\vdots & \vdots & & \widetilde{\widetilde{\mathbf{A}}}_3 & \\
0 & 0 & & &
\end{bmatrix} ,
\tag{5.153}
$$

Note that the elements of the second row and below have suffered *two* transformations. Accordingly, they have been denoted as $\widetilde{a}_{ij}$ (instead of $\widehat{a}_{ij}$) and by the $(n-2) \times (n-2)$ matrix $\widetilde{\widetilde{\mathbf{A}}}_3$. The third orthogonal transformation $\mathbf{Q}_3$ required to reduce the elements of the first column of $\widetilde{\widetilde{\mathbf{A}}}_3$ is therefore of the form

$$
\mathbf{Q}_3 \equiv
\begin{bmatrix}
1 & 0 & 0 & \cdots & 0 \\
\hline
0 & 1 & 0 & \cdots & 0 \\
\hline
0 & 0 & & & \\
\vdots & \vdots & & \widetilde{\widetilde{\mathbf{Q}}}_3 & \\
0 & 0 & & &
\end{bmatrix} ,
\tag{5.154}
$$

where the reflector $\widetilde{\widetilde{\mathbf{Q}}}_3$ is

$$
\widetilde{\widetilde{\mathbf{Q}}}_3 \equiv \mathbf{I}_{n-2} - \gamma_3 \mathbf{u}_3 \mathbf{u}_3^{\mathrm{T}},
\tag{5.155}
$$

and where $\mathbf{u}_3 \in \mathbb{R}^{n-2}$ is built from the vector $\mathbf{x} \equiv [\widetilde{a}_{33}\ \widetilde{a}_{43}\ \cdots\ \widetilde{a}_{n3}]^{\mathrm{T}}$, i.e. the first column of matrix $\widetilde{\widetilde{\mathbf{A}}}_3$. In general, the jth transformation is given by the orthogonal matrix

$$
\mathbf{Q}_j \equiv
\begin{bmatrix}
\mathbf{I}_{j-1} & 0 \\
\hline
0 & \mathbf{I}_{n-j+1} - \gamma_j \mathbf{u}_j \mathbf{u}_j^{\mathrm{T}}
\end{bmatrix} ,
\tag{5.156}
$$

for $j = 1, 2, \ldots, n-1$, where the reflectors are built from the vectors $\mathbf{u}_j \in \mathbb{R}^{n-j+1}$ and factors γ_j obtained from the first column of the submatrix $\mathbf{A}_j$ whose first column needs to be reduced. Overall, after $n-1$ Householder transformations of the form (5.156), the upper triangular matrix $\mathbf{R}$ is obtained:

$$
\mathbf{Q}_{n-1}^{\mathrm{T}}\,\mathbf{Q}_{n-2}^{\mathrm{T}}\,\cdots\,\mathbf{Q}_2^{\mathrm{T}}\,\mathbf{Q}_1^{\mathrm{T}}\,\mathbf{A} \equiv \mathbf{R}.
\tag{5.157}
$$

After left multiplication of both sides of (5.157) by $\mathbf{Q}_1\,\mathbf{Q}_2\,\cdots\,\mathbf{Q}_{n-2}\,\mathbf{Q}_{n-1}$

$$
\mathbf{Q}_1\,\mathbf{Q}_2\,\cdots\,\mathbf{Q}_{n-2}\,\underbrace{\mathbf{Q}_{n-1}\,\overbrace{\mathbf{Q}_{n-1}^{\mathrm{T}}}^{\mathbf{I}_n}\,\mathbf{Q}_{n-2}^{\mathrm{T}}}_{\mathbf{I}_n}\,\cdots\,\mathbf{Q}_2^{\mathrm{T}}\,\mathbf{Q}_1^{\mathrm{T}}\,\mathbf{A} \equiv \mathbf{Q}_1\,\mathbf{Q}_2\,\cdots\,\mathbf{Q}_{n-2}\,\mathbf{Q}_{n-1}\,\mathbf{R},
$$

Eq. (5.157) reads $\mathbf{A} \equiv \mathbf{Q}_1 \, \mathbf{Q}_2 \, \cdots \, \mathbf{Q}_{n-2} \, \mathbf{Q}_{n-1} \, \mathbf{R}$, or simply

$$\mathbf{A} = \mathbf{Q} \, \mathbf{R}, \tag{5.158}$$

where $\mathbf{Q} \doteq \mathbf{Q}_1 \, \mathbf{Q}_2 \, \cdots \, \mathbf{Q}_{n-2} \, \mathbf{Q}_{n-1}$ is an orthogonal matrix, since it is the product of orthogonal matrices. Expression (5.158) is better known as the QR-factorization of the matrix $\mathbf{A}$, and it can be shown that its computational cost is $4n^3/3$, that is, twice the computational cost of LU. However, QR-factorization has a large variety of useful applications, ranging from the computation of eigenvalues, to the orthonormalization of vectors or the least squares approximation of overdetermined systems, the last two being addressed later on in this chapter. These facts vindicate the use of QR-factorization in many areas of numerical mathematics, and this is the reason why many numerical analysts consider QR as the most important matrix factorization of numerical linear algebra.

Factorization (5.158) can be used to solve the linear system

$$\mathbf{A}\mathbf{x} = \mathbf{b},$$

which can be written as

$$\mathbf{Q} \, \mathbf{R} \, \mathbf{x} = \mathbf{b},$$

or

$$\mathbf{R} \, \mathbf{x} = \mathbf{Q}^{\mathrm{T}} \, \mathbf{b}. \tag{5.159}$$

By introducing $\mathbf{e} \doteq \mathbf{Q}^{\mathrm{T}} \, \mathbf{b}$, system (5.158) now reads

$$\mathbf{R} \, \mathbf{x} = \mathbf{e},$$

and it can be solved using BS in $O(n^2)$ operations. As with LU, changing the right-hand side $\mathbf{b}$ by $\mathbf{d}$ only requires one to compute $\mathbf{Q}^{\mathrm{T}}\mathbf{d}$ and to proceed again with BS.

In practice, the matrix $\mathbf{Q}$ of (5.158) or the sequential transformation matrices $\mathbf{Q}_j$ of (5.157) are seldom needed explicitly, so they are *never* built. The reason is that we only need to compute the *action* of the matrix $\mathbf{I}_{n-j+1} - \gamma_j \mathbf{u}_j \mathbf{u}_j^{\mathrm{T}}$ appearing in (5.156) on the corresponding $(n-j+1) \times (n-j+1)$ matrix $\widehat{\mathbf{A}}_j$, that is,

$$(\mathbf{I}_{n-j+1} - \gamma_j \mathbf{u}_j \mathbf{u}_j^{\mathrm{T}})\widehat{\mathbf{A}}_j \equiv \widehat{\mathbf{A}}_j - \mathbf{u}_j \mathbf{v}_j^{\mathrm{T}} \widehat{\mathbf{A}}_j, \tag{5.160}$$

where $\mathbf{v}_j \doteq \gamma_j \, \mathbf{u}_j$. The computational cost of the product $\mathbf{u}_j \mathbf{v}_j^{\mathrm{T}} \widehat{\mathbf{A}}_j$ depends on how it is performed. An optimal way consists of pre-computing the auxiliary *row* vector $\mathbf{v}_{\mathrm{aux.}}^{\mathrm{T}} = \mathbf{v}_j^{\mathrm{T}} \widehat{\mathbf{A}}_j$ and afterwards evaluating the dyadic product $\mathbf{u}_j \mathbf{v}_{\mathrm{aux.}}^{\mathrm{T}}$, overall requiring $3n^2$ operations (see Exercise 13). Code 15 performs the

QR-factorization of an arbitrary square matrix[17] using Householder reflectors. However, Code 15 *does not* provide the matrix $\mathbf{Q}$ but the matrix

$$
\begin{bmatrix}
r_{11} & r_{12} & r_{13} & \cdots & r_{1n} \\
\mathbf{u}_{1,2} & r_{22} & r_{23} & \cdots & r_{2n} \\
\mathbf{u}_{1,3} & \mathbf{u}_{2,2} & r_{33} & & \vdots \\
\mathbf{u}_{1,4} & \mathbf{u}_{2,3} & \mathbf{u}_{3,2} & \ddots & \\
\vdots & \vdots & \vdots & & \\
\mathbf{u}_{1,n} & \mathbf{u}_{2,n-1} & \mathbf{u}_{3,n-2} & \cdots & r_{nn}
\end{bmatrix}
\tag{5.161}
$$

whose diagonal and upper diagonal elements constitute the upper triangular matrix $\mathbf{R}$. The lower triangular jth column in (5.161) is the Householder vector (5.143)

$$
[\mathbf{u}_{j,2}\ \mathbf{u}_{j,3}\ \cdots\ \mathbf{u}_{j,n-j+1}]^{\mathrm{T}} = \begin{bmatrix} \dfrac{x_2}{\tau + x_1} & \dfrac{x_3}{\tau + x_1} & \cdots & \dfrac{x_{n-j+1}}{\tau + x_1} \end{bmatrix}^{\mathrm{T}},
$$

required to generate the action of the reflector $\mathbf{Q}_j$ in (5.160). Note that the first component of the vector (which is, by construction, 1) is not stored.

```
% Code 15: QR - Householder factorization
% Input:  A (n x m matrix with n >= m and rank(A) = m)
% Output: UR matrix containing (see text):
%            1) Householder vectors u_j (low triang. part)
%            2) R (upper triangular matrix)
function UR = myqr(A)
[n,m] = size(A);
if n > m; M = m; else M = m - 1; end
for jj = 1:M
    B = A(jj:n,jj:m);
    x = B(:,1); tau = sign(x(1))*norm(x); gam = 1+x(1)/tau;
    u = [1 ; x(2:end)/(tau+x(1))]; vauxT = gam*u'*B;
    A(jj:n,jj:m) = B - u*vauxT; A(jj+1:n,jj) = u(2:end);
end
UR = A;
```

To solve a non-singular square system $\mathbf{A}\,\mathbf{x} = \mathbf{b}$, the right-hand side $\mathbf{Q}^{\mathrm{T}}\mathbf{b}$ in (5.159) is obtained by performing on the vector $\mathbf{b}$ exactly the same set of transformations (5.160) applied on the matrices $\widehat{\mathbf{A}}_j$. Code 16 does this task

---

17 It also works for full-rank $n \times m$ matrices with $n > m$.

by using the Householder vectors stored in the matrix UR previously obtained after invoking the function myqr.m of Code 15. Finally, Code 16 extracts the upper triangular matrix $\mathbf{R}$ from UR in order to solve $\mathbf{R}\,\mathbf{x} = \mathbf{Q}^T\,\mathbf{b}$ using BS. Both Codes 15 and 16 can be used jointly to solve square systems or, as we will see in Section 5.5.2, least squares problems. The reader has probably noticed that the transformation counter jj in both codes runs from 1 to M, where M is $m-1$ in the square case $n = m$. That is, we perform $m-1 = n-1$ Householder transformations to solve a square non-singular system. Section 5.5.2 deals with the least squares problem case where $n > m$, i.e. overdetermined systems. In that case, we will see that m transformations will be required.

```
% Code 16: QR - Solving Ax = b (requires bs.m)
% Input: 1) UR matrix provided by Code 15, i.e. UR = myqr(A)
%        2) b right-hand side
% Output: x solution minimizing || Ax -b ||
function x = qrsolve(UR,b);
[n,m] = size(UR);
if n > m; M = m; else M = m - 1; end
for jj = 1:M
    u = [1; UR(jj+1:n,jj)]; gam = 2/norm(u)^2;
    b(jj:n) = b(jj:n) - gam*u*(u'*b(jj:n));
end
x = bs(triu(UR(1:m,1:m)),b(1:m));
```

5.5.2   Linear Data Fitting

Linear regression was already introduced at the beginning of this section as a way to fit a straight line $y = \alpha + \beta x$ to a given set of scattered data $\{(x_1,y_1),\ \ldots,\ (x_n,y_n)\}$. Algebraically, we concluded that the system of equations

$$
\begin{bmatrix} 1 & x_1 \\ 1 & x_2 \\ \vdots & \vdots \\ 1 & x_n \end{bmatrix}
\begin{bmatrix} \alpha \\ \beta \end{bmatrix} =
\begin{bmatrix} y_1 \\ y_2 \\ \vdots \\ y_n \end{bmatrix},
\tag{5.162}
$$

was in general overdetermined and, therefore, our maximum aspiration was to look for the optimal values of the coefficients α and β such that the quantity

$$
\sum_{j=1}^{n} [y_j - (\alpha + \beta\, x_j)]^2
\tag{5.163}
$$

was a minimum. System (5.162) is a particular case of the general overdetermined system

$$
\begin{bmatrix}
a_{11} & a_{12} & \cdots & a_{1m} \\
a_{21} & a_{22} & \cdots & a_{2m} \\
\vdots & \vdots & & \vdots \\
a_{n1} & a_{n2} & \cdots & a_{nm}
\end{bmatrix}
\begin{bmatrix}
x_1 \\
\vdots \\
x_m
\end{bmatrix}
=
\begin{bmatrix}
b_1 \\
b_2 \\
\vdots \\
b_n
\end{bmatrix},
\tag{5.164}
$$

where $n > m = 2$. System (5.164) can be briefly expressed as

$$
\mathbf{A}\,\mathbf{x} = \mathbf{b}, \quad \mathbf{A} \in \mathrm{M}_{nm}(\mathbb{R}), \quad \mathbf{x} \in \mathbb{R}^m, \quad \mathbf{b} \in \mathbb{R}^n \quad (n \geqslant m), \tag{5.165}
$$

where $\mathrm{M}_{nm}(\mathbb{R})$ is the vector space of $n \times m$ real matrices. Unless stated otherwise, we will always assume that $\mathrm{rank}(\mathbf{A}) = m$. In this more general case, the main goal is to find the solution vector $\mathbf{x} = [x_0 \ \cdots \ x_m]^{\mathrm{T}} \in \mathbb{R}^m$ that minimizes the norm of the so-called *residual vector* of (5.165) defined as

$$
\mathbf{r} \doteq \mathbf{b} - \mathbf{A}\,\mathbf{x}, \tag{5.166}
$$

that is, to minimize the scalar function

$$
r(x_1, x_2, \ldots, x_m) \doteq \|\mathbf{r}\| \equiv \|\mathbf{A}\,\mathbf{x} - \mathbf{b}\| = \left\{ \sum_{i=1}^{n} [(\mathbf{A}\,\mathbf{x})_i - b_i]^2 \right\}^{1/2}. \tag{5.167}
$$

The reader may wonder whether this minimization problem always has a solution and, if there is one, whether that solution is unique. To answer these two questions we first need (as we will justify later) to see if there is a QR-factorization for non-square matrices such as the matrix $\mathbf{A}$ appearing in (5.165). That is, we wonder if an arbitrary matrix $\mathbf{A} \in \mathrm{M}_{nm}(\mathbb{R})$ can be factorized as a product $\mathbf{A} = \mathbf{QR}$, with $\mathbf{Q}$ and $\mathbf{R}$ being orthogonal and upper triangular matrices, respectively. The QR-theorem guarantees that any square matrix admits such factorization. In our rectangular case, the problem is solved by expanding the $n \times m$ matrix $\mathbf{A}$ with an *arbitrary* $n \times (n - m)$ auxiliary matrix $\mathbf{A}_{\mathrm{aux.}}$, resulting in the $n \times n$ matrix $\overline{\mathbf{A}}$, as shown in the diagram of Figure 5.4.

The contents of $\mathbf{A}_{\mathrm{aux.}}$ are irrelevant for our analysis. The key point is that the resulting enlarged matrix $\overline{\mathbf{A}}$ is square and therefore admits a QR-factorization, so that $\overline{\mathbf{A}} = \mathbf{Q}\overline{\mathbf{R}}$, where $\mathbf{Q}, \overline{\mathbf{R}} \in \mathrm{M}_n(\mathbb{R})$ are orthogonal, and upper triangular matrices, respectively. From the diagram we observe that the first m columns of

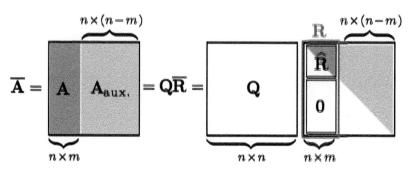

Figure 5.4 QR-factorization of a non-square matrix $\mathbf{A} \in \mathrm{M}_{nm}(\mathbb{R})$.

matrix $\overline{\mathbf{R}}$ (dark gray $n \times m$ rectangular frame on the right-hand side) constitute the matrix $\mathbf{R} \in M_{nm}(\mathbb{R})$ resulting from the triangularization of the matrix $\mathbf{A}$ (dark gray rectangle on the left-hand side).[18] The matrix $\mathbf{R}$ contains an $m \times m$ upper triangular matrix $\widehat{\mathbf{R}}$ on top, along with a zero $(n - m) \times m$ matrix underneath. Summing up, any arbitrary matrix $\mathbf{A} \in M_{nm}(\mathbb{R})$ admits a factorization of the form

$$\mathbf{A} = \mathbf{Q}\mathbf{R} = \mathbf{Q}\begin{bmatrix} \widehat{\mathbf{R}} \\ 0 \end{bmatrix}, \quad \mathbf{Q} \in M_n(\mathbb{R}), \ \mathbf{R} \in M_{nm}(\mathbb{R}), \ \widehat{\mathbf{R}} \in M_m(\mathbb{R}), \quad (5.168)$$

where $\mathbf{Q}$ is an orthogonal transformation and $\widehat{\mathbf{R}}$ is upper triangular. Equation (5.168) can also be expressed as

$$\mathbf{R} = \mathbf{Q}^T \mathbf{A}, \tag{5.169}$$

where $\mathbf{Q}^T$ is the result of applying m (not $m = 1$, as it would be in the square case $m \equiv n$) Householder transformations, that is,

$$\mathbf{Q}^T \mathbf{A} = \mathbf{Q}_m^T \mathbf{Q}_{m-1}^T \cdots \mathbf{Q}_2^T \mathbf{Q}_1^T \mathbf{A},$$

so that $\mathbf{Q} = \mathbf{Q}_1 \mathbf{Q}_2 \cdots \mathbf{Q}_{m-1} \mathbf{Q}_m$. Left multiplying the overdetermined system (5.165) by $\mathbf{Q}^T$ reads

$$\mathbf{Q}^T \mathbf{A}\, \mathbf{x} = \mathbf{Q}^T \mathbf{b}, \tag{5.170}$$

which, according to (5.168) and (5.169), can be written as

$$\begin{bmatrix} \widehat{\mathbf{R}} \\ 0 \end{bmatrix} \mathbf{x} = \begin{bmatrix} \widehat{\mathbf{c}} \\ \mathbf{d} \end{bmatrix}, \tag{5.171}$$

where $\widehat{\mathbf{c}} \in \mathbb{R}^m$ and $\mathbf{d} \in \mathbb{R}^{n-m}$ are the first m and last $n - m$ components of the vector $\mathbf{Q}^T \mathbf{b}$, respectively. More specifically, system (5.171) reads

$$(5.172) \qquad \begin{matrix} m\left\{ \vphantom{\begin{bmatrix}1\\1\\1\\1\\1\end{bmatrix}} \right. \\ \\ n-m\left\{ \vphantom{\begin{bmatrix}1\\1\\1\end{bmatrix}} \right. \end{matrix} \begin{bmatrix} r_{11} & r_{12} & \cdots & r_{1m} \\ 0 & r_{22} & \cdots & r_{2m} \\ \vdots & \vdots & \ddots & \vdots \\ 0 & 0 & \cdots & r_{mm} \\ 0 & \cdots & \cdots & 0 \\ \vdots & & & \vdots \\ \vdots & & & \vdots \\ 0 & \cdots & \cdots & 0 \end{bmatrix} \begin{bmatrix} x_1 \\ x_2 \\ \vdots \\ x_m \end{bmatrix} = \begin{bmatrix} \widehat{c}_1 \\ \widehat{c}_2 \\ \vdots \\ \widehat{c}_m \\ d_1 \\ d_2 \\ \vdots \\ d_{n-m} \end{bmatrix},$$

where $r_{ij} \equiv [\widehat{\mathbf{R}}]_{ij}$, $(j \ge i = 1, 2, \ldots, m)$ are the nonzero elements of the upper triangular matrix $\widehat{\mathbf{R}}$, whose rank is m, as can be shown using basic linear algebra. The rank of $\mathbf{A}$ satisfies

$$\text{rank}(\mathbf{A}) = \text{rank}(\mathbf{Q}\mathbf{R}) \le \text{rank}(\mathbf{R}) = \text{rank}(\widehat{\mathbf{R}}) \le m.$$

Since we have assumed that $\text{rank}(\mathbf{A}) \equiv m$, the last inequality implies that $\text{rank}(\widehat{\mathbf{R}}) \equiv m$; therefore $\widehat{\mathbf{R}}$ must be non-singular.

---

18 The remaining $n - m$ columns of $\overline{\mathbf{R}}$ (light gray upper triangular rectangle) are irrelevant.

Summing up, after applying orthogonal transformations, the original system (5.165) has been transformed into system (5.171) or, equivalently, system (5.172). Recall that our original goal was to find the vector $\mathbf{x} \in \mathbb{R}^m$ minimizing the residual in (5.167). However, the residual of the new transformed system (5.171) now reads

$$
s(x_1, x_2, \ldots, x_n) = \left\| \begin{bmatrix} \widehat{\mathbf{R}} \\ \mathbf{0} \end{bmatrix} \mathbf{x} - \begin{bmatrix} \widehat{\mathbf{c}} \\ \mathbf{d} \end{bmatrix} \right\| = \left\{ \sum_{i=1}^m [(\widehat{\mathbf{R}}\,\mathbf{x})_i - \hat{c}_i]^2 + \sum_{j=1}^{n-m} d_j^2 \right\}^{1/2}.
$$

(5.173)

The obvious question is whether the residuals $r(x_1, \ldots, x_m)$ in (5.167) and $s(x_1, \ldots, x_m)$ in (5.173) are the same. The residual of the transformed system is

$$
s = \|\mathbf{R}\mathbf{x} - \mathbf{Q}^{\mathrm{T}}\mathbf{b}\| = \|\mathbf{Q}^{\mathrm{T}}\mathbf{A}\mathbf{x} - \mathbf{Q}^{\mathrm{T}}\mathbf{b}\| = \|\mathbf{Q}^{\mathrm{T}}(\mathbf{A}\mathbf{x} - \mathbf{b})\| = \|\mathbf{Q}^{\mathrm{T}}\,\mathbf{r}\|.
$$

However, since the transformation $\mathbf{Q}$ is orthogonal, it preserves norms and therefore $\|\mathbf{Q}^{\mathrm{T}}\,\mathbf{r}\| = r$ so that $s = r$. As a result, if s attains a minimum at $(x_1, \ldots, x_m)$, so does r. From (5.173), we see that s attains a minimum when

$$
\sum_{i=1}^m [(\widehat{\mathbf{R}}\,\mathbf{x})_i - \hat{c}_i]^2 = 0,
$$

that is, when $\mathbf{x}$ is the solution of the square system

$$
\widehat{\mathbf{R}}\,\mathbf{x} = \widehat{\mathbf{c}}.
$$

(5.174)

Since $\mathbf{R}$ is non-singular, the solution of (5.174) exists and is unique, therefore answering the two initial questions regarding the existence and uniqueness of the original least square problem (5.165).

As a simple example, consider the system

$$
\begin{bmatrix} 1 \\ 1 \end{bmatrix} [x] = \begin{bmatrix} 5 \\ 9 \end{bmatrix}.
$$

The residual in this case is $r(x) = \sqrt{(x-5)^2 + (x-9)^2}$. Its minimum can be easily determined just by imposing $r'(x) = 0$, where prime stands for differentiation with respect to x. The reader may check that the minimum is attained at $x = 7$. This minimizer can also be obtained by left multiplying the original system by the reflector

$$
\mathbf{Q} = \begin{bmatrix} -\sqrt{2}/2 & -\sqrt{2}/2 \\ -\sqrt{2}/2 & \sqrt{2}/2 \end{bmatrix},
$$

leading to

$$
\begin{bmatrix} -\sqrt{2} \\ 0 \end{bmatrix} [x] = \begin{bmatrix} -7\sqrt{2} \\ -2\sqrt{2} \end{bmatrix},
$$

whose least squares solution is obviously $x = 7$, as expected.

Practical 5.2: Least Squares Polynomial Fit

The figure on the right shows the plot of the function

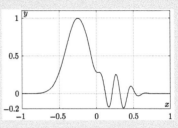

$$f(x) = e^{-20\left(x+\frac{1}{4}\right)^2} + \frac{\sin(30\,x)\,e^{-20\left(x-\frac{1}{4}\right)^2}}{4},$$

within the domain $[-1, 1]$.

In Part I we studied how to approximate functions by means of polynomial interpolation, where the interpolant $\Pi_n f(x) \in \mathbb{R}^n[x]$ adopted the values of the function at $n+1$ given nodes $x_0, x_1, \ldots, x_n$, i.e.

$$\Pi_n f(x_j) = f(x_j), \quad j = 0, 1, \ldots, n.$$

However, in this practical we will see how to approximate $f(x)$ with polynomial expansions of arbitrary degree $m \leq n$.

(a) Consider the approximation $p(x) \in \mathbb{R}_m[x]$:

$$p(x) = \sum_{j=0}^{m} a_j\, x^j = a_0 + a_1 x_1 + a_2 x^2 + \cdots + a_m x^m.$$

Impose the polynomial to take the values $p(x_j) = f(x_j) = f_j$ over a set of given nodes $\{x_0, x_1, \ldots, x_n\}$. Show that this leads to the Vandermonde system (use MATLAB's **vander(x)** command)

$$\begin{bmatrix} 1 & x_0 & x_0^2 & \cdots & x_0^m \\ 1 & x_1 & x_1^2 & \cdots & x_1^m \\ 1 & x_2 & x_2^2 & \cdots & x_2^m \\ \vdots & \vdots & \vdots & & \vdots \\ 1 & x_n & x_n^2 & \cdots & x_n^m \end{bmatrix} \begin{bmatrix} a_0 \\ a_1 \\ a_2 \\ \vdots \\ a_m \end{bmatrix} = \begin{bmatrix} f_0 \\ f_1 \\ f_2 \\ \vdots \\ f_n \end{bmatrix}$$

(b) Approximate $f(x)$ using the nodes

$$x_j = -1 + 2j/n, \quad (j = 0, 1, \ldots, n),$$

and minimize the residual of the system with your QR-code. For the cases $(n, m) = \{(14, 7); (28, 14); (28, 20); (64, 30)\}$, plot the resulting polynomials and check if they capture the oscillations in the middle. Do you observe any anomaly toward the ends of the interval?

(c) Repeat (b) using the Chebyshev nodes

$$x_j = \cos(j\pi/n), \quad (j = 0, 1, \ldots, n).$$

5.6 Matrix Norms and Conditioning

In Part I, we addressed the concept of conditioning within the context of root-finding methods. We saw that an iterative solver for nonlinear scalar equations such as Newton's method could lose its quadratic order of convergence if the sought root was ill-conditioned due to algebraic multiplicity. Within that context, ill-conditioning led to extreme sensitivity of the results with respect to very small inaccuracies, deteriorating the efficiency of the solver.

In this section, we want to study the sensitivity of the solution of an arbitrary linear system $\mathbf{A}\,\mathbf{x} = \mathbf{b}$ with respect to small inaccuracies in $\mathbf{b}$, for example. Consider, for instance, the trivial one-dimensional equation $ax = b$ whose solution is $x = b/a$. Changing b to $b + \Delta b$ changes the solution by an amount $\Delta x = \Delta b/a$, so that the condition number in this toy case is simply $K = |a|^{-1}$. Therefore, the smaller the magnitude of $|a|$, the larger the condition number. From that simple observation, it is a natural assumption to think that what makes the solution of the linear system $\mathbf{A}\,\mathbf{x} = \mathbf{b}$ sensitive to small changes in $\mathbf{b}$ is the magnitude of the *determinant* of the matrix $\mathbf{A}$, that is, $|\det(\mathbf{A})|$. However, as we will see in the following example, the magnitude of the determinant may sometimes be misleading. Consider the linear system $\mathbf{A}\,\mathbf{x} = \mathbf{b}$,

$$\begin{bmatrix} 1000 & 999 \\ 999 & 998 \end{bmatrix} \begin{bmatrix} x_1 \\ x_2 \end{bmatrix} = \begin{bmatrix} 1000 \\ 999 \end{bmatrix},$$

whose solution is $\mathbf{x} = [x_1\ x_2]^{\mathrm{T}} = [1\ 0]^{\mathrm{T}}$. The reader may check that the determinant of the matrix is $\det(A) = -1$. If the right-hand side of the previous system is slightly perturbed to $\tilde{\mathbf{b}} = [1001\ 999]^{\mathrm{T}}$, the new solution becomes $\tilde{\mathbf{x}} = [-997\ 999]^{\mathrm{T}}$. We would have never expected, by simply evaluating the magnitude of the determinant, that such a small relative perturbation $\Delta\mathbf{b} = [1; 0]$ with relative norm $\|\Delta\mathbf{b}\|/\|\mathbf{b}\| \approx 10^{-3}$ could lead to changes in the solution as large as $\Delta\mathbf{x} = [-998\ 999]^{\mathrm{T}}$, with relative norm $\|\Delta\mathbf{x}\|/\|\mathbf{x}\| \approx 10^{3}$.

In order to measure the sensitivity of linear systems, we need new elements from the analysis of matrices and linear operators, such as the concept of *matrix norm*:

Matrix Norm

A matrix norm is any map

$$\| \cdot \| : \mathrm{M}_n(\mathbb{R}) \to \mathbb{R}^{+,0}$$

that assigns to every square matrix $\mathbf{A} \in \mathrm{M}_n(\mathbb{R})$ a positive or zero scalar, denoted by $\|\mathbf{A}\|$, satisfying:

1. $\|\mathbf{A}\| > 0,\ \forall \mathbf{A} \neq 0\ (\|\mathbf{A}\| = 0\ \Leftrightarrow\ \mathbf{A} = 0)$.
2. $\|\alpha\mathbf{A}\| = |\alpha|\|\mathbf{A}\|,\ \forall \alpha \in \mathbb{R}$.
3. $\|\mathbf{A} + \mathbf{B}\| \leq \|\mathbf{A}\| + \|\mathbf{B}\|$.
4. $\|\mathbf{AB}\| \leq \|\mathbf{A}\|\|\mathbf{B}\|$.

The concept of norm can also be extended to non-square matrices. We refer the reader to the recommended monographs on matrix analysis at the end of the chapter for more details. The theory of matrices provides a wide variety of matrix norms, some of them explicitly written in terms of the matrix elements a_{ij} of **A**. Such is the case of the *Frobenius norm*,

$$\|\mathbf{A}\|_F \doteq \left[\sum_{i,j=1}^{n} a_{ij}^2 \right]^{1/2}, \tag{5.175}$$

the 1-norm,

$$\|\mathbf{A}\|_1 \doteq \max_{1 \le j \le n} \sum_{i=1}^{n} |a_{ij}|, \tag{5.176}$$

or the so-called ∞-norm,

$$\|\mathbf{A}\|_\infty \doteq \max_{1 \le i \le n} \sum_{j=1}^{n} |a_{ij}|. \tag{5.177}$$

Other matrix norms are implicitly defined using the vector norm used in $\mathbb{R}^n$. These norms are usually called *induced*, *subordinate*, or *operator* norms. They are defined as follows:

Induced Vector or Operator Norm of a Matrix

Let $\| \cdot \|_\mathbf{v}$ be a vector norm in $\mathbb{R}^n$. The *operator norm* of a matrix $\mathbf{A} \in M_n(\mathbb{R})$ *induced* by the vector norm $\| \cdot \|_\mathbf{v}$ is

$$\|\mathbf{A}\|_\mathbf{v} \doteq \max_{\mathbf{x} \ne 0} \frac{\|\mathbf{A}\,\mathbf{x}\|_\mathbf{v}}{\|\mathbf{x}\|_\mathbf{v}}. \tag{5.178}$$

Since $\mathbf{x} = \|\mathbf{x}\|_\mathbf{v}\, \hat{\mathbf{x}}$, where $\hat{\mathbf{x}}$ is the normalized vector $\mathbf{x}$ so that $\|\hat{\mathbf{x}}\|_\mathbf{v} = 1$, we may write

$$\frac{\|\mathbf{A}\,\mathbf{x}\|_\mathbf{v}}{\|\mathbf{x}\|_\mathbf{v}} = \frac{\|\mathbf{A}\,\|\mathbf{x}\|_\mathbf{v}\, \hat{\mathbf{x}}\|_\mathbf{v}}{\|\|\mathbf{x}\|_\mathbf{v}\, \hat{\mathbf{x}}\|_\mathbf{v}} = \frac{\|\mathbf{x}\|_\mathbf{v}\|\mathbf{A}\, \hat{\mathbf{x}}\|_\mathbf{v}}{\|\mathbf{x}\|_\mathbf{v}\|\hat{\mathbf{x}}\|_\mathbf{v}} = \|\mathbf{A}\, \hat{\mathbf{x}}\|_\mathbf{v},$$

so that the norm operator (5.178) is also frequently defined as

$$\|\mathbf{A}\|_\mathbf{v} \doteq \max_{\|\mathbf{x}\|_\mathbf{v}=1} \|\mathbf{A}\, \mathbf{x}\|_\mathbf{v}. \tag{5.179}$$

It can be shown (see references in the Complementary Reading section at the end of the chapter) that induced vector norms are matrix norms. In particular, if the matrix norm is that induced by the standard Euclidean 2-norm of a vector $\mathbf{x} = [x_1 \ x_2 \ \cdots \ x_n]^T$ given by

$$\|\mathbf{x}\|_2 = [x_1^2 + x_2^2 + \cdots + x_n^2]^{1/2},$$

then the norm

$$\|\mathbf{A}\|_2 = \max_{\mathbf{x} \ne 0} \frac{\|\mathbf{A}\,\mathbf{x}\|_2}{\|\mathbf{x}\|_2} = \max_{\|\mathbf{x}\|_2=1} \|\mathbf{A}\mathbf{x}\|_2 = \max_{x_1^2+x_2^2+\cdots+x_n^2=1} \left[\sum_{j=1}^{n} (\mathbf{A}\mathbf{x})_j^2 \right]^{1/2} \tag{5.180}$$

(Continued)

is usually known as the *spectral norm* of $\mathbf{A}$. This norm measures the maximum *amplification* that the linear operator $\mathbf{A} \in M_n(\mathbb{R})$ may produce when acting on the set of unit vectors $\mathbf{x} \in \mathbb{R}^n$. This is one of the reasons why this norm is frequently used. At this point, it is important to avoid terminology confusion and to clarify that the spectral 2-norm *is not* the Frobenius norm.

Unless stated otherwise, we will henceforth denote the Euclidean norm of a vector $\|\mathbf{x}\|_2$ and its corresponding induced spectral norm of a matrix $\|\mathbf{A}\|_2$ simply by $\|\mathbf{x}\|$ and $\|\mathbf{A}\|$, respectively.

From the definition (5.178), it follows that for any nonzero vector $\mathbf{x}$,

$$\frac{\|\mathbf{A}\,\mathbf{x}\|}{\|\mathbf{x}\|} \leq \max_{\mathbf{z} \neq 0} \frac{\|\mathbf{A}\,\mathbf{z}\|}{\|\mathbf{z}\|} = \|\mathbf{A}\|, \qquad (5.181)$$

or

$$\|\mathbf{A}\,\mathbf{x}\| \leq \|\mathbf{A}\|\|\mathbf{x}\|. \qquad (5.182)$$

We have seen before that the trivial scalar equation $ax = b$ has a very sensitive solution x when $|a|$ is very small. The matrix norms just described above provide the right measure to generalize the conditioning problem to linear systems of the form $\mathbf{A}\,\mathbf{x} = \mathbf{b}$. The diagram below depicts the essential ingredients of condition analysis of a linear system.

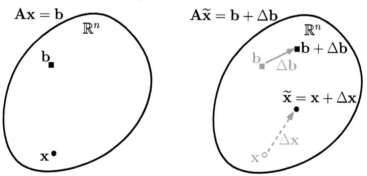

On the left, the diagram shows the solution $\mathbf{x}$ (black bullet) of $\mathbf{A}\,\mathbf{x} = \mathbf{b}$. When the right-hand side $\mathbf{b}$ (black square on the left) is perturbed by an amount $\Delta\mathbf{b}$ (gray vector on the right) so that the right-hand side is now $\mathbf{b} + \Delta\mathbf{b}$ (black square on the right), the solution changes by an amount $\Delta\mathbf{x}$, so that the new solution has moved to $\widetilde{\mathbf{x}} = \mathbf{x} + \Delta\mathbf{x}$ (black bullet on the right). In numerical linear algebra it is said that the system $\mathbf{A}\,\mathbf{x} = \mathbf{b}$ is ill-conditioned or *badly conditioned* if

$$\frac{\|\Delta\mathbf{x}\|}{\|\mathbf{x}\|} \quad \text{is large even when} \quad \frac{\|\Delta\mathbf{b}\|}{\|\mathbf{b}\|} \quad \text{is small.}$$

In other words, the system is ill-conditioned whenever small deviations in the right-hand side lead to large changes in the resulting solution. A relation between Δx and Δb can be obtained by using the concept of matrix norm. The perturbed system reads

$$A(x + \Delta x) = b + \Delta b,$$

but, since $Ax = b$, the previous equation provides a relation between Δx and Δb:

$$A \, \Delta x = \Delta b,$$

or, equivalently,

$$\Delta x = A^{-1} \, \Delta b.$$

The norm of Δx can be bounded by means of (5.182):

$$\|\Delta x\| = \|A^{-1} \, \Delta b\| \leq \|A^{-1}\| \, \|\Delta b\|. \tag{5.183}$$

By the same rule, we can bound $\|b\|$:

$$\|b\| = \|A \, x\| \leq \|A\| \, \|x\|. \tag{5.184}$$

The multiplication of inequalities (5.183) and (5.184) leads to

$$\|\Delta x\| \, \|b\| \leq \|A^{-1}\| \, \|\Delta b\| \, \|A\| \, \|x\|,$$

or

$$\frac{\|\Delta x\|}{\|x\|} \leq \|A\| \|A^{-1}\| \frac{\|\Delta b\|}{\|b\|}. \tag{5.185}$$

We therefore conclude that the magnitude of the relative uncertainties in the solution can potentially be as large as the relative deviations of the right-hand side magnified by the factor $\|A\| \|A^{-1}\|$. This magnifying factor is therefore the condition number of our problem. For this reason, we define

---

Condition Number of a Matrix

$$\kappa(A) \doteq \|A\| \|A^{-1}\|. \tag{5.186}$$

Definition (5.186) assumes that A is non-singular. By convention, the condition number of a singular matrix A is $\kappa(A) = \infty$. The condition number depends on the matrix norm used. Definition (5.186) implicitly makes use of the spectral norm and for this reason some authors refer to (5.186) as $\kappa_2(A)$. Unless stated otherwise, henceforth in this book we will refer to $\kappa(A) = \kappa_2(A)$ as the condition number of the matrix A.

Computing the spectral norm and the corresponding condition number of a 2×2 or even 3×3 matrix by hand may be a tedious task, but feasible. The computation of the norm and the condition number of a large matrix requires numerical algorithms that unfortunately are beyond the limited scope of this textbook. The reader can find suitable references regarding these algorithms in the Complementary Reading section, at the end of the chapter. For our applications, we will rely on Matlab's **norm** and **cond** commands to compute these quantities using the spectral norm by default. For example, consider again the 2×2 matrix of the system introduced at the beginning of this section:

$$\mathbf{A} = \begin{bmatrix} 1000 & 999 \\ 999 & 998 \end{bmatrix}.$$

Using Matlab's command **cond**, the reader may check that the condition number of $\mathbf{A}$ is approximately $\kappa(\mathbf{A}) = 3.992 \times 10^6$. This result is consistent with our initial experiment, where the introduction of uncertainties of relative size $\|\Delta \mathbf{b}\|/\|\mathbf{b}\| \approx 10^{-3}$ led to deviations in the solution of relative size $\|\Delta \mathbf{x}\|/\|\mathbf{x}\| \approx 10^3$, that is, 6 orders of magnitude larger, as predicted by the bound (5.185)

$$\frac{\|\Delta \mathbf{x}\|}{\|\mathbf{x}\|} \leq \kappa(\mathbf{A}) \frac{\|\Delta \mathbf{b}\|}{\|\mathbf{b}\|}.$$

Our conditioning analysis has so far only considered small variations or uncertainties in the right-hand side of the linear system $\mathbf{A} \mathbf{x} = \mathbf{b}$. In actuality, the matrix $\mathbf{A}$ is also affected by this type of inaccuracies. In general we must consider deviations $\Delta \mathbf{x}$ in the solution $\mathbf{x}$ caused not only by inaccuracies $\Delta \mathbf{b}$ in each component of the right-hand side $\mathbf{b}$, but also within the entries a_{ij} of $\mathbf{A}$, namely $\Delta \mathbf{A}$. Overall, if the perturbed linear system reads

$$(\mathbf{A} + \Delta \mathbf{A})\hat{\mathbf{x}} = \mathbf{b} + \Delta \mathbf{b},$$

with resulting perturbed solutions $\hat{\mathbf{x}} = \mathbf{x} + \Delta \mathbf{x}$, then it can be shown that

$$\frac{\|\Delta \mathbf{x}\|}{\|\mathbf{x}\|} \leq \kappa(\mathbf{A}) \left\{ \frac{\|\Delta \mathbf{A}\|}{\|\mathbf{A}\|} + \frac{\|\Delta \mathbf{b}\|}{\|\mathbf{b}\|} + \frac{\|\Delta \mathbf{A}\|}{\|\mathbf{A}\|} \frac{\|\Delta \mathbf{b}\|}{\|\mathbf{b}\|} \right\}. \tag{5.187}$$

The third term on the right-hand side of the previous inequality is quadratic and therefore negligible in front of the other two.

Numerical computations involving matrices with large condition numbers (ill-conditioned matrices) may lead to highly inaccurate results. In particular, within the context of direct solvers for linear systems, the following rule of thumb is commonly accepted:

Accuracy of GEM-LU Solutions with Partial Pivoting

Suppose the system $\mathbf{A} \mathbf{x} = \mathbf{b}$ is solved using LU-factorization with partial pivoting. Assuming that

1. the entries a_{ij} and b_i of $\mathbf{A}$ and $\mathbf{b}$, respectively, are accurate to n_1 decimal places,
2. the condition number of $\mathbf{A}$ satisfies

$$\log_{10}\kappa(\mathbf{A}) \approx n_2 \quad (\text{with} \quad n_2 < n_1),$$

then, the entries x_i of the solution $\mathbf{x}$ are accurate to about $n_1 - n_2$ decimal places.

We end this section by revisiting Practical 4.1 of Part I. The Vandermonde matrix appearing in that practical was used for the computation of quadrature weights using equispaced nodes $x_j = j/n$ $(j = 0, 1, \ldots, n)$ within the unit interval $[0, 1]$:

$$
\mathbf{V}_n \equiv \begin{bmatrix}
1 & 1 & 1 & \cdots & 1 \\
0 & 1/n & 2/n & \cdots & 1 \\
0 & 1/n^2 & 4/n^2 & \cdots & 1 \\
\vdots & \vdots & \vdots & & \vdots \\
0 & 1/n^n & (2/n)^n & \cdots & 1
\end{bmatrix}.
$$

For $n \geq 1$, $\mathbf{V}_n$ can be built by typing: `x=[0:n]/n; Vn = flipud(vander(x)')`. The reader may check that the condition number of this matrix is approximately $\kappa(\mathbf{V}_4) \approx 686.4$. However, for $n = 16$, $\kappa(\mathbf{V}_{16})$ is already of order 10^{13}, therefore deteriorating the accuracy of the computed weights when solving the system using Matlab's backslash \ command. By contrast, the matrix appearing later in (4.64),

$$
\begin{bmatrix}
1 & 1 & 1 & \cdots & 1 \\
1 & \cos(\pi/n) & \cos(2\pi/n) & \cdots & -1 \\
1 & \cos(2\pi/n) & \cos(4\pi/n) & \cdots & 1 \\
1 & \cos(3\pi/n) & \cos(6\pi/n) & \cdots & -1 \\
\vdots & \vdots & \vdots & & \vdots \\
1 & -1 & 1 & \cdots &
\end{bmatrix},
\tag{5.188}
$$

used to compute the Clenshaw–Curtis quadrature weights for the Chebyshev nodes $x_j = \cos(j\pi/n)$, $(j = 0, 1, \ldots, n)$, is extremely well conditioned. The reader may also check that the condition number of the matrix above is of order 1 even for n within the hundreds.

5.7 Gram–Schmidt Orthonormalization

The reader is probably familiar with the classical *Gram–Schmidt* orthonormalization algorithm, henceforth referred to as CGS. Consider an arbitrary basis of an m-dimensional subspace $S \subset \mathbb{R}^n$ with $m \leq n$, that is,

$$
S \equiv \text{span}\{\mathbf{v}_1, \mathbf{v}_2, \ldots, \mathbf{v}_m\}.
\tag{5.189}
$$

The CGS algorithm provides a sequence of embedded subspaces

$$
S_j = \text{span}\{\mathbf{q}_1, \mathbf{q}_2, \ldots, \mathbf{q}_j\}, \quad (j = 1, 2, \ldots, m),
\tag{5.190}
$$

of dimension $\dim(S_j) = j$, satisfying $S_1 \subset S_2 \subset S_3 \subset \cdots \subset S_{m-1} \subset S_m$ and

$$
\begin{aligned}
S_1 &= \mathrm{span}\{\mathbf{q}_1\} = \mathrm{span}\{\mathbf{v}_1\}\\
S_2 &= \mathrm{span}\{\mathbf{q}_1, \mathbf{q}_2\} = \mathrm{span}\{\mathbf{v}_1, \mathbf{v}_2\}\\
S_3 &= \mathrm{span}\{\mathbf{q}_1, \mathbf{q}_2, \mathbf{q}_3\} = \mathrm{span}\{\mathbf{v}_1, \mathbf{v}_2, \mathbf{v}_3\}
\end{aligned}
$$

$$\vdots$$

$$
S_m = \mathrm{span}\{\mathbf{q}_1, \ldots, \mathbf{q}_m\} = \mathrm{span}\{\mathbf{v}_1, \ldots, \mathbf{v}_m\} = S, \tag{5.191}
$$

where the resulting vectors $\{\mathbf{q}_1, \ldots, \mathbf{q}_m\}$ have all norm 1 and are pairwise orthogonal (orthonormal), that is,

$$
\langle \mathbf{q}_i, \mathbf{q}_j \rangle = \delta_{ij}, \quad (i, j = 1, 2 \ldots, m). \tag{5.192}
$$

Figure 5.5 outlines schematically the first two stages of the CGS algorithm. The first stage is simply a normalization of $\mathbf{v}_1$, that is,

$$
\mathbf{q}_1 = \frac{\mathbf{v}_1}{r_{11}}, \tag{5.193}
$$

where we have defined $r_{11} = \|\mathbf{v}_1\|$ (this notation will be justified later) so that $S_1 = \mathrm{span}\{\mathbf{q}_1\}$. In the second stage we create an orthogonal vector, namely $\widetilde{\mathbf{q}}_2$, by subtracting from $\mathbf{v}_2$ its own projection $\langle \mathbf{v}_2, \mathbf{q}_1 \rangle \mathbf{q}_1$ onto the subspace S_1 (see Stage 2 in Figure 5.5),

$$
\widetilde{\mathbf{q}}_2 = \mathbf{v}_2 - \langle \mathbf{v}_2, \mathbf{q}_1 \rangle \mathbf{q}_1 = \mathbf{v}_2 - r_{12}\mathbf{q}_1, \tag{5.194}
$$

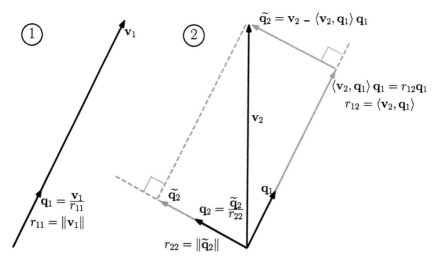

Figure 5.5 Classical Gram–Schmidt (CGS) algorithm.

where $r_{12} = \langle \mathbf{v}_2, \mathbf{q}_1 \rangle$. Finally, we define $r_{22} = \|\widetilde{\mathbf{q}}_2\|$ so that the normalized vector is

$$\mathbf{q}_2 = \frac{\widetilde{\mathbf{q}}_2}{r_{22}}. \tag{5.195}$$

A third vector $\widetilde{\mathbf{q}}_3$ orthogonal to $\mathbf{q}_1$ and $\mathbf{q}_2$ can similarly be created by subtracting from $\mathbf{v}_3$ its components parallel to S_2, generated by these two previous unit vectors,

$$\begin{aligned}
\widetilde{\mathbf{q}}_3 &= \mathbf{v}_3 - \langle \mathbf{v}_3, \mathbf{q}_1 \rangle \, \mathbf{q}_1 - \langle \mathbf{v}_3, \mathbf{q}_2 \rangle \, \mathbf{q}_2 \\
&= \mathbf{v}_3 - r_{13}\mathbf{q}_1 - r_{23}\mathbf{q}_2,
\end{aligned} \tag{5.196}$$

where $r_{13} = \langle \mathbf{v}_3, \mathbf{q}_1 \rangle$ and $r_{23} = \langle \mathbf{v}_3, \mathbf{q}_2 \rangle$. Finally we normalize $\widetilde{\mathbf{q}}_3$, so that

$$\mathbf{q}_3 = \frac{\widetilde{\mathbf{q}}_3}{\|\widetilde{\mathbf{q}}_3\|} = \frac{\widetilde{\mathbf{q}}_3}{r_{33}}. \tag{5.197}$$

In our analysis of CGS, we have systematically defined the projections by

$$r_{ik} \doteq \langle \mathbf{v}_k, \mathbf{q}_i \rangle, \tag{5.198}$$

for $i = 1, 2, \ldots, k-1$, and the normalizing factors by

$$r_{kk} \doteq \|\widetilde{\mathbf{q}}_k\|, \tag{5.199}$$

for $k = 1, 2, \ldots, m$. In general, once the subspace $S_{k-1} = \mathrm{span}\{\mathbf{q}_1, \mathbf{q}_2, \ldots, \mathbf{q}_{k-1}\}$ has been built, the next subspace S_k is generated by first creating the vector

$$\widetilde{\mathbf{q}}_k = \mathbf{v}_k - \sum_{j=1}^{k-1} r_{jk}\mathbf{q}_j = \mathbf{v}_k - r_{1k}\mathbf{q}_1 - r_{2k}\mathbf{q}_2 - r_{3k}\mathbf{q}_3 - \cdots - r_{k-1,k}\mathbf{q}_{k-1}, \tag{5.200}$$

that is orthogonal to S_{k-1}, as it satisfies

$$\langle \mathbf{q}_i, \widetilde{\mathbf{q}}_k \rangle = \langle \mathbf{q}_i, \mathbf{v}_k \rangle - \sum_{j=1}^{k-1} r_{jk}\langle \mathbf{q}_i, \mathbf{q}_j \rangle = r_{ik} - \sum_{j=1}^{k-1} r_{jk}\delta_{ij} = 0, \tag{5.201}$$

for $i = 1, 2, \ldots, k-1$. Then the vector $\mathbf{q}_k$ is obtained by normalizing $\widetilde{\mathbf{q}}_k$,

$$\mathbf{q}_k = \frac{\widetilde{\mathbf{q}}_k}{r_{kk}}, \tag{5.202}$$

and appended to S_{k-1} in order to span S_k. One of the most important features of the CGS algorithm is that it performs a QR-factorization in disguise. To make this last statement more clear, first notice that Eqs. (5.193), (5.194), (5.196), and (5.200) can be written as

$$\mathbf{v}_1 = r_{11}\mathbf{q}_1, \tag{5.203}$$

$$\mathbf{v}_2 = r_{12}\mathbf{q}_1 + \widetilde{\mathbf{q}}_2 = r_{12}\mathbf{q}_1 + r_{22}\mathbf{q}_2, \tag{5.204}$$

$$\mathbf{v}_3 = r_{13}\mathbf{q}_1 + r_{23}\mathbf{q}_2 + \widetilde{\mathbf{q}}_3 = r_{13}\mathbf{q}_1 + r_{23}\mathbf{q}_2 + r_{33}\mathbf{q}_3, \tag{5.205}$$

and

$$\mathbf{v}_k = \sum_{j=1}^{k=1} r_{jk}\mathbf{q}_j + \tilde{\mathbf{q}}_k = \sum_{j=1}^{k=1} r_{jk}\mathbf{q}_j + r_{kk}\mathbf{q}_k = \sum_{j=1}^{k} r_{jk}\mathbf{q}_j, \tag{5.206}$$

respectively. For $k \equiv m$, (5.206) becomes

$$\mathbf{v}_m = r_{1m}\mathbf{q}_1 + r_{2m}\mathbf{q}_2 + r_{3m}\mathbf{q}_3 + \cdots r_{mm}\mathbf{q}_m. \tag{5.207}$$

Arranging the bases $\{\mathbf{v}_1, \mathbf{v}_2, \ldots, \mathbf{v}_m\}$ and $\{\mathbf{q}_1, \mathbf{q}_2, \ldots, \mathbf{q}_m\}$ columnwise as matrices in $M_{nm}(\mathbb{R})$, Eqs. (5.203), (5.204), (5.205), and (5.207) now read

$$\begin{bmatrix} \vert & \vert & & \vert \\ \mathbf{v}_1 & \mathbf{v}_2 & \cdots & \mathbf{v}_m \\ \vert & \vert & & \vert \end{bmatrix} = \begin{bmatrix} \vert & \vert & & \vert \\ \mathbf{q}_1 & \mathbf{q}_2 & \cdots & \mathbf{q}_m \\ \vert & \vert & & \vert \end{bmatrix} \begin{bmatrix} r_{11} & r_{12} & r_{13} & \cdots & r_{1m} \\ 0 & r_{22} & r_{23} & \cdots & r_{2m} \\ 0 & 0 & r_{33} & \cdots & r_{3m} \\ \vdots & & & \ddots & \vdots \\ 0 & \cdots & & 0 & r_{mm} \end{bmatrix}. \tag{5.208}$$

Recall the QR-factorization of a non-square matrix, outlined in Figure 5.4. From those schematics we can clearly identify that $\mathbf{A}$ is the result of the product of the first m columns of the matrix $\mathbf{Q}$ (depicted again as $\hat{\mathbf{Q}}$ on the left of Figure 5.6) times the $m \times m$ upper triangular matrix $\hat{\mathbf{R}}$. This product is represented in condensed form on the right of Figure 5.6, where $\mathbf{A}$ can be expressed as the product $\hat{\mathbf{Q}} \, \hat{\mathbf{R}}$, that is,

$$\begin{bmatrix} a_{11} & a_{12} & \cdots & a_{1m} \\ a_{21} & a_{22} & \cdots & a_{2m} \\ \vdots & \vdots & & \vdots \\ a_{n1} & a_{n2} & \cdots & a_{nm} \end{bmatrix} = \begin{bmatrix} q_{11} & q_{12} & \cdots & q_{1m} \\ q_{21} & q_{22} & \cdots & q_{2m} \\ \vdots & \vdots & & \vdots \\ q_{n1} & q_{n2} & \cdots & q_{nm} \end{bmatrix} \begin{bmatrix} r_{11} & r_{12} & \cdots & r_{1m} \\ 0 & r_{22} & \cdots & r_{2m} \\ \vdots & 0 & \ddots & \vdots \\ 0 & \cdots & & r_{mm} \end{bmatrix}. \tag{5.209}$$

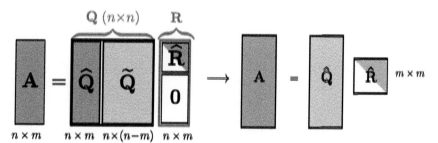

Figure 5.6 CGS and QR-factorization equivalence.

The QR-CGS equivalence becomes apparent just by identifying matrices $\mathbf{A}$ and $\widehat{\mathbf{Q}}$ of (5.209) with matrices $[\mathbf{v}_1|\mathbf{v}_2|\cdots|\mathbf{v}_m]$ and $[\mathbf{q}_1|\mathbf{q}_2|\cdots|\mathbf{q}_m]$ of (5.208), respectively. Matrix $\widehat{\mathbf{Q}} \in \mathrm{M}_{nm}(\mathbb{R})$ satisfies

$$\widehat{\mathbf{Q}}^{\mathrm{T}}\widehat{\mathbf{Q}} = \mathbf{I}_m, \tag{5.210}$$

but, if $m < n$, it is not orthogonal (that is, $\widehat{\mathbf{Q}}^{-1}$ is not defined and $\widehat{\mathbf{Q}}\widehat{\mathbf{Q}}^{\mathrm{T}} \neq \mathbf{I}$). In this case, $\widehat{\mathbf{Q}}$ is formally known as an *isometry* since it preserves lengths and angles, that is,

$$\|\widehat{\mathbf{Q}}\mathbf{x}\|^2 = \langle\widehat{\mathbf{Q}}\mathbf{x}, \widehat{\mathbf{Q}}\mathbf{x}\rangle = \mathbf{x}^{\mathrm{T}}\widehat{\mathbf{Q}}^{\mathrm{T}}\widehat{\mathbf{Q}}\mathbf{x} = \|\mathbf{x}\|^2$$

and

$$\langle\widehat{\mathbf{Q}}\mathbf{x}, \widehat{\mathbf{Q}}\mathbf{y}\rangle = \mathbf{x}^{\mathrm{T}}\widehat{\mathbf{Q}}^{\mathrm{T}}\widehat{\mathbf{Q}}\mathbf{y} = \langle\mathbf{x}, \mathbf{y}\rangle,$$

for any two vectors $\mathbf{x}$ and $\mathbf{y}$ in $\mathbb{R}^m$.

5.7.1 Instability of CGS: Reorthogonalization

The CGS orthonormalization just seen provides deep theoretical insight but is rarely used in practice. The main reason is that, as we will see in a moment, CGS is numerically *unstable*. The next code is a simple implementation of the CGS algorithm that orthonormalizes the columns of an arbitrary full-rank matrix $\mathbf{A} \in \mathrm{M}_{nm}(\mathbb{R})$ whose columns are the original set of m vectors $[\mathbf{v}_1|\mathbf{v}_2|\cdots|\mathbf{v}_m]$. The code provides the resulting orthonormalized vectors $[\mathbf{q}_1|\mathbf{q}_2|\cdots|\mathbf{q}_m]$ as columns of the output matrix $\mathbf{Q} \in \mathrm{M}_{nm}(\mathbb{R})$, as well as the matrix $\mathbf{R} \in \mathrm{M}_m(\mathbb{R})$, whose upper triangular part contains the factors r_{ij} appearing on the right-hand side of (5.209) and defined in (5.198) and (5.199):

```
% Classical Gram--Schmidt (CGS) unstable algorithm
function [Q,R,S] = mycgs(A)
[n,m] = size(A); Q = 0*A; R = zeros(m); S = R;
R(1,1)= norm(A(:,1)); Q(:,1) = A(:,1)/R(1,1);
for k = 2:m
  R(1:k-1,k) = Q(:,1:k-1)'*A(:,k);
  Q(:,k) = A(:,k) - Q(:,1:k-1)*R(1:k-1,k);
  S(1:k-1,k) = Q(:,1:k-1)'*Q(:,k);
  R(k,k) = norm(Q(:,k)); Q(:,k) = Q(:,k)/R(k,k);
end
```

The code also provides a matrix $\mathbf{S}$ whose contents will be described later in this section but which are not relevant at the moment. As an example, let us revisit Practical 5.2, where we had to perform a QR-factorization of the Vandermonde matrix

$$
\begin{bmatrix}
1 & x_0 & x_0^2 & \cdots & x_0^m \\
1 & x_1 & x_1^2 & \cdots & x_1^m \\
1 & x_2 & x_2^2 & \cdots & x_2^m \\
\vdots & \vdots & \vdots & & \vdots \\
1 & x_n & x_n^2 & \cdots & x_n^m
\end{bmatrix}.
\tag{5.211}
$$

Consider the equispaced nodes in the unit interval $[0,1]$ given by $x_j = j/n$, $(j = 0, 1, 2, \ldots, n)$. For $(n, m) = (14, 7)$, the Vandermonde matrix associated with these nodes can be easily generated and its columns orthonormalized by typing:

```
n=14;m=7; x = [0:n]'/n;
V = fliplr(vander(x)); V = V(:,1:m+1);
[Q,R,S] = mycgs(V);
```

The reader may check that, as expected, typing `Q'*Q` provides the matrix of inner products $\langle \mathbf{q}_i, \mathbf{q}_j \rangle$, that is, the identity matrix $\mathbf{I}_8$ (to some extent). A careful inspection however reveals a progressive deterioration of the orthogonality of the vectors. For instance, whereas the top left matrix elements of `Q'*Q` such as $\langle \mathbf{q}_2, \mathbf{q}_1 \rangle$ or $\langle \mathbf{q}_3, \mathbf{q}_2 \rangle$ are of order 10^{-16} or 10^{-15}, the bottom right last inner products $\langle \mathbf{q}_7, \mathbf{q}_6 \rangle$ or $\langle \mathbf{q}_8, \mathbf{q}_7 \rangle$ are nearly of order 10^{-8} or 10^{-6}, respectively. A reliable numerical test to validate the orthogonality of the vectors is to measure the magnitude $\|\mathbf{Q}^T\mathbf{Q} - \mathbf{I}_m\|$; this, in principle, should be zero. Typing in Matlab's prompt `norm(Q'*Q-eye(8))` provides an approximate value $\|\mathbf{Q}^T\mathbf{Q} - \mathbf{I}_8\|$ close to 10^{-6}. Increasing n or m just worsens the accuracy of CGS.[19] In this example, we have applied CGS on the Vandermonde matrix (5.211), which is particularly ill-conditioned. Applying CGS on well-conditioned matrices leads to better results, of course, but in practice ill-conditioned matrices are frequently the rule, not the exception.

We can obtain $\mathbf{Q}$ more accurately using Matlab's `qr` command by typing `[Q,R]=qr(V,0)` and checking that $\|\mathbf{Q}^T\mathbf{Q} - \mathbf{I}_8\| \approx 7.1 \times 10^{-16}$. In this sense, QR-orthonormalization is numerically very robust. The reader may logically wonder why we address an orthonormalization algorithm such as CGS that is numerically unstable and whose results can be alternatively (and accurately) obtained using the QR-factorization method. The main reason is that certain

---

19 For $(n, m) = (28, 14)$, $\|\mathbf{Q}^T\mathbf{Q} - \mathbf{I}_{15}\|$ is of order 1, which is by all means unacceptable.

types of orthonormalizations are not always feasible by means of QR. Instead, the Gram–Schmidt algorithm (suitably improved) provides accurate orthonormalizations and lies at the very heart of modern iterative methods to solve *matrix-free* linear systems of equations (as we will see in Section 5.8) and to compute eigenvalues of very large matrices. But before improving the accuracy of CGS, we need to understand the origin of the problem.

Let us take a closer look at the diagonal terms r_{jj} of the matrix $\mathbf{R}$ provided by the CGS algorithm. These terms are the norms $r_{jj} = \|\tilde{\mathbf{q}}_j\|$, and their magnitude may become very small, in this case, $\|\tilde{\mathbf{q}}_8\| \approx 3.4 \times 10^{-4}$. The reason for the presence of such small norms is that the column vectors that constitute the Vandermonde matrix (5.211) are, to describe it in a bit informal way, *far from being orthogonal* or *nearly parallel*. The angle $\theta(\mathbf{v}_i, \mathbf{v}_j)$ between two arbitrary vectors $\mathbf{v}_i$ and $\mathbf{v}_j$ is related to their scalar product and the corresponding norms by means of the expression

$$\cos\theta(\mathbf{v}_i, \mathbf{v}_j) = \frac{\langle \mathbf{v}_i, \mathbf{v}_j \rangle}{\|\mathbf{v}_i\| \, \|\mathbf{v}_j\|}. \tag{5.212}$$

If we compute the matrix $\mathbf{C}$ with elements $c_{ij} = \cos\theta(\mathbf{v}_i, \mathbf{v}_j)$, obtained from the column vectors $\{\mathbf{v}_1, \mathbf{v}_2, \ldots, \mathbf{v}_8\}$ extracted from the Vandermonde matrix (5.211), we observe that many of these matrix elements are close to unity, clearly indicating that the vectors $\mathbf{v}_j$ are far from being orthogonal. As an exercise, we recommend the reader to compute the symmetric matrix $\mathbf{C}$, whose approximate values rounded to two decimal places are given below:

$$\mathbf{C} \approx \begin{bmatrix} 1 & 0.85 & 0.73 & 0.65 & 0.59 & 0.55 & 0.51 & 0.48 \\ & 1 & 0.97 & 0.92 & 0.87 & 0.82 & 0.78 & 0.75 \\ & & 1 & 0.99 & 0.96 & 0.93 & 0.9 & 0.87 \\ & & & 1 & 0.99 & 0.98 & 0.96 & 0.93 \\ & & & & 1 & 1 & 0.98 & 0.97 \\ & & & & & 1 & 1 & 0.99 \\ & & & & & & 1 & 1 \\ & & & & & & & 1 \end{bmatrix}.$$

In this case, we can see that $\mathbf{v}_8$ is remarkably aligned with $\mathbf{v}_2$ ($c_{2,8} \approx 0.75$), but also almost parallel to $\mathbf{v}_6$ and $\mathbf{v}_7$ ($c_{6,8} \approx 0.99$ and $c_{7,8} \approx 1$). From this example, we can easily see that two column vectors $\mathbf{v}_i$ and $\mathbf{v}_j$ arbitrarily extracted from a large *dense*[20] matrix such as the Vandermonde matrix will typically lead to a nonzero scalar product $\langle \mathbf{v}_i, \mathbf{v}_j \rangle$.

---

20 In numerical mathematics, a matrix is considered dense if most of its elements are nonzero

Figure 5.7 Near parallelism between $\mathbf{v}_1$ and $\mathbf{v}_2$ results in a vector $\widetilde{\widetilde{\mathbf{q}}}_2$ with very small norm r_{22}.

Let us now analyze geometrically the effects of near parallelism of vectors in the CGS algorithm. Figure 5.7 reproduces again Figure 5.5 in a particular case of two nearly parallel vectors. After normalizing $\mathbf{v}_1$ to obtain $\mathbf{q}_1$, CGS provides the vector $\widetilde{\widetilde{\mathbf{q}}}_2 \equiv \mathbf{v}_2 - \mathbf{q}_1\langle\mathbf{q}_1, \mathbf{v}_2\rangle$ that has a very small norm $r_{22} \equiv \|\widetilde{\widetilde{\mathbf{q}}}_2\|$. The problem comes when this vector is normalized to obtain $\mathbf{q}_2 \equiv \widetilde{\widetilde{\mathbf{q}}}_2/r_{22}$: any small inaccuracy in the computation of $\widetilde{\widetilde{\mathbf{q}}}_2$ will potentially be amplified (and propagated afterwards) when dividing by the small quantity r_{22}. We have already experienced the effects of division by small quantities in the solution of linear systems when using GEM. Partial pivoting was precisely devised to minimize those effects. The question lies now in whether $\widetilde{\widetilde{\mathbf{q}}}_2$ or, in general, the kth vector arising from the CGS algorithm

$$\widetilde{\widetilde{\mathbf{q}}}_k \equiv \mathbf{v}_k - r_{1k}\mathbf{q}_1 - r_{2k}\mathbf{q}_2 - r_{3k}\mathbf{q}_3 - \cdots - r_{k-1,k}\mathbf{q}_{k-1} \tag{5.213}$$

may contain such inaccuracies. A simple test can be performed by computing the inner products

$$s_j \equiv \langle\mathbf{q}_j, \widetilde{\widetilde{\mathbf{q}}}_k\rangle, \tag{5.214}$$

for $j \equiv 1, 2, \ldots, k - 1$, right after the computation of $\widetilde{\widetilde{\mathbf{q}}}_k$. According to (5.201), the inner products s_j (5.214) should be zero by construction. For this purpose, our CGS code computes these inner products[21] and stores them in matrix $\mathbf{S}$ so that

$$\mathbf{S}_{jk} \equiv \langle\mathbf{q}_j, \widetilde{\widetilde{\mathbf{q}}}_k\rangle, \tag{5.215}$$

for $j \equiv 1, 2, \ldots, k - 1$. A simple inspection of the elements $\mathbf{S}_{jk}$ for the Vandermonde test carried out before for $n \equiv 14$ and $m \equiv 7$ reveals a remarkable deterioration of the resulting inner products $\mathbf{S}_{jk}$ when increasing j. For example, the reader may check that while $\mathbf{S}_{1,8} \approx 3.5 \times 10^{-16}$, the last two products are $\mathbf{S}_{6,8} \approx 2.1 \times 10^{-11}$ and $\mathbf{S}_{7,8} \approx 3.1 \times 10^{-10}$. In order to understand this lack of accurate orthogonality, let us first assume that the scalar product $\langle\mathbf{q}_j, \mathbf{v}_k\rangle$ is known to

---

21 See command line: $S(1:k-1,k) = Q(:,1:k-1)'*Q(:,k)$.

infinite precision. The origin of the problem is that we have assumed that the quantity r_{jk} computed in (5.198) is *exactly* $\langle \mathbf{q}_j, \mathbf{v}_k \rangle$, but in actuality it is

$$r_{jk} \equiv \langle \mathbf{q}_j, \mathbf{v}_k \rangle \pm \Delta, \tag{5.216}$$

where Δ represents a small deviation from the actual value due to round-off errors, for example. These deviations may have negligible effects during the first stages of the CGS algorithm, but they will eventually become larger afterwards, since every normalization will amplify (and propagate) them. For example, the second vector provided by the CGS iteration is actually

$$\widetilde{\mathbf{q}}_2 \equiv \mathbf{v}_2 - r_{12}\mathbf{q}_1 \equiv \mathbf{v}_2 - (\langle \mathbf{q}_1, \mathbf{v}_2 \rangle \pm \Delta)\mathbf{q}_1, \tag{5.217}$$

which, after normalization, becomes

$$\mathbf{q}_2 \equiv \frac{\mathbf{v}_2 - \langle \mathbf{q}_1, \mathbf{v}_2 \rangle \mathbf{q}_1}{r_{22}} \pm \frac{\Delta}{r_{22}} \mathbf{q}_1. \tag{5.218}$$

As a result, those components of $\mathbf{v}_2$ parallel to $\mathbf{q}_1$ have not been completely removed, but they are still present in $\mathbf{q}_2$ with size Δ/r_{22}. At the kth stage of CGS, these inaccuracies are amplified by the factor r_{kk}^{-1}, which may be potentially large ($r_{88}^{-1} \approx 3 \times 10^9$ in our previous Vandermonde example, for instance). In CGS, these operations of subtracting inaccurate projections from $\mathbf{v}_k$ and dividing by potentially small norms are carried out over and over, sometimes with catastrophic results.

One technique that helps to remove inaccuracies in r_{jk} is called *reorthogonalization*. Let $\widetilde{\widetilde{\mathbf{q}}}_k^{(1)}$ be the vector obtained at the kth stage of CGS:

$$\widetilde{\widetilde{\mathbf{q}}}_k^{(1)} \equiv \mathbf{v}_k = \sum_{j=1}^{k-1} r_{jk}^{(1)} \mathbf{q}_j, \tag{5.219}$$

where $r_{jk}^{(1)} \equiv \langle \mathbf{v}_k, \mathbf{q}_j \rangle$, $(j = 1, 2, \ldots, k-1)$ are the *inaccurate* orthogonal projections of $\mathbf{v}_k$ onto S_k. The remnants of $\widetilde{\widetilde{\mathbf{q}}}_k^{(1)}$ parallel to S_k, given by the quantities

$$s_j \doteq \langle \mathbf{q}_j, \widetilde{\widetilde{\mathbf{q}}}_k^{(1)} \rangle, \quad (j = 1, 2, \ldots, k-1), \tag{5.220}$$

can be removed from $\widetilde{\widetilde{\mathbf{q}}}_k^{(1)}$ by simply subtracting, leading to the vector

$$\widetilde{\widetilde{\mathbf{q}}}_k^{(2)} \equiv \widetilde{\widetilde{\mathbf{q}}}_k^{(1)} - \sum_{j=1}^{k-1} s_j \mathbf{q}_j \equiv \mathbf{v}_k - \sum_{j=1}^{k-1} r_{jk}^{(1)} \mathbf{q}_j - \sum_{j=1}^{k-1} s_j \mathbf{q}_j \equiv \mathbf{v}_k - \sum_{j=1}^{k-1} r_{jk}^{(2)} \mathbf{q}_j, \tag{5.221}$$

where

$$r_{jk}^{(2)} \equiv r_{jk}^{(1)} + s_j, \quad (j = 1, 2, \ldots, k-1). \tag{5.222}$$

The unit vector resulting from this reorthogonalization is therefore

$$\mathbf{q}_k = \frac{\widetilde{\mathbf{q}}_k^{(2)}}{r_{kk}}, \tag{5.223}$$

where $r_{kk} = \|\widetilde{\mathbf{q}}_k^{(2)}\|$. Code 17 implements reorthogonalization in the classical Gram–Schmidt algorithm, henceforth referred to as GSR:

```
% Code 17: GSR (Gram--Schmidt-Reorthogonalized)
% Input:  1) A (n x m matrix - n >= m)
% Output: 2) Q (n x m isometric matrix)
%           R (m x m upper triangular matrix)
function [Q,R] = gsr(A)
[n,m] = size(A); Q = 0*A; R = zeros(m);
R(1,1)= norm(A(:,1)); Q(:,1) = A(:,1)/R(1,1);
for k = 2:m
 R(1:k-1,k) = Q(:,1:k-1)'*A(:,k);
 Q(:,k) = A(:,k) - Q(:,1:k-1)*R(1:k-1,k);
 S = Q(:,1:k-1)'*Q(:,k);
 Q(:,k) = Q(:,k) - Q(:,1:k-1)*S;
 R(k,k) = norm(Q(:,k));
 if R(k,k) > 1e-16
     Q(:,k) = Q(:,k)/R(k,k);
 else
     ['Lin. Dep.']
 end
end
```

The reader may easily check that applying Code 17 to orthonormalize the Vandermonde matrix (5.211) for $n = 14$ and $m = 7$ provides an outstandingly accurate residual norm $\|\mathbf{Q}^T\mathbf{Q} - \mathbf{I}_8\|$ close to 10^{-16}. We also recommend the reader to check the order of magnitude of the inner products $\langle \mathbf{q}_i, \mathbf{q}_j \rangle$ and confirm that the vectors $\mathbf{q}_j$ are, after reorthogonalization, pairwise accurately orthonormal.

The bad news about reorthogonalization is that it makes CGS computationally more expensive. However, the good news is that this technique needs to be applied just once to get accurate results in practice. More details about the performance of this and other techniques to improve CGS, such as *Modified Gram–Schmidt*, can be found in the references included at the end of the chapter. In Section 5.8 we will see how this improved Gram–Schmidt algorithm can be adapted for the solution of matrix-free systems of linear equations.

Practical 5.3: Gram–Schmidt and Reorthogonalization

In this practical you have to compare three different methods of orthogonalization: CGS, Gram–Schmidt with reorthogonalization (GSR), and QR. This comparison will be carried out by orthogonalizing *Hilbert matrices* $H(n) \in \mathbb{M}_n(\mathbb{R})$ whose explicit elements are

$$H_{ij} = \frac{1}{i+j-1}, \quad (i,j = 1,2,\ldots,n).$$

In Matlab, $H(n)$ is generated typing `hilb(n)`. You need three external codes to orthogonalize H so they provide the **Q** orthogonalized columns of the matrix. To measure the accuracy of the resulting orthonormal basis, compute $\|\mathbf{I}_n - \mathbf{Q}^\mathrm{T}\mathbf{Q}\|$ using Matlab's **norm** command (that by default provides the 2-norm). If the orthogonalization is accurate, this norm should be very small.

(a) CGS: Generate $H(n)$ (with $2 \leq n \leq 16$, for example) and monitor the inner products $\langle q_j, \widetilde{q}_k \rangle$ for $j \leq k-1$ of the resulting normalized basis. At every stage k, check the norm of $\widetilde{q}_k$ and observe how it is decreasing. Do you observe an increase of $\|\mathbf{I}_n - \mathbf{Q}^\mathrm{T}\mathbf{Q}\|$ when you increase n?

(b) GSR: Same as in (a) but now monitor the non-orthogonal roundoffs $s_j = \langle q_j, \widetilde{q}_k \rangle$, $j \leq k$. Subtract these projections and recompute the $q_k^{(2)}$ vectors. Recheck orthogonality. Do you observe an improvement when compared with CGS?

(c) QR: Repeat (a) and (b) with MATLAB's `qr`.

On a semi-logarithmic plot, represent $\|\mathbf{I}_n - \mathbf{Q}^\mathrm{T}\mathbf{Q}\|$ versus n for the three methods.

5.8 Matrix-Free Krylov Solvers

This is the last section devoted to numerical linear algebra. By no means are its contents less important than the previous ones. Quite to the contrary, the methodologies that are about to be presented here open the door to a new perspective of solving linear systems. As computers have become more powerful during the last three or four decades, numerical algorithms arising in physics and engineering have been capable of addressing larger scale problems with more unknowns, frequently requiring the solution of linear systems of large dimension. One aspect that we have deliberately overlooked so far is that related to the computational cost of building and storing the matrix that later needs to be factorized to solve the problem. These matrices have been always presumed to be known in advance, ignoring that their elements are typically obtained at a computational cost that sometimes may be quite high. When a linear system involves a very large number of degrees of freedom, building its corresponding matrix may be computationally unfeasible. This section is an introduction to the concept of linear solvers where, as we will see, the matrices are not explicitly built nor required.

We begin this section in a rather unconventional manner by experimenting in Matlab with a linear system of equations that we formulate in a slightly different way. Consider a linear map $\mathbf{f}(\mathbf{x}) : \mathbb{R}^5 \to \mathbb{R}^5$ explicitly given by

$$\mathbf{x} \equiv \begin{bmatrix} x_1 \\ x_2 \\ x_3 \\ x_4 \\ x_5 \end{bmatrix} \xrightarrow{\mathbf{f}(\mathbf{x})} \begin{bmatrix} 2x_1 - x_2 \\ -x_1 + 2x_2 - x_3 \\ -x_2 + 2x_3 - x_4 \\ -x_3 + 2x_4 - x_5 \\ -x_4 + 2x_5 \end{bmatrix} . \tag{5.224}$$

This map is of course trivially represented (in the canonical basis of $\mathbb{R}^5$) by the matrix

$$\mathbf{A} \equiv \begin{bmatrix} 2 & -1 & 0 & 0 & 0 \\ -1 & 2 & -1 & 0 & 0 \\ 0 & -1 & 2 & -1 & 0 \\ 0 & 0 & -1 & 2 & -1 \\ 0 & 0 & 0 & -1 & 2 \end{bmatrix} . \tag{5.225}$$

However, we aim to avoid this matrix in our analysis. Instead, we implement the map (5.224) in Matlab by means of the simple function **Afun**:

```
function Y = Afun(X)
Y = [2*X(1)-X(2); -X(1)+2*X(2)-X(3);...
     -X(2)+2*X(3)-X(4); -X(3)+2*X(4)-X(5); -X(4)+2*X(5)];
```

For any vector $\mathbf{x} \in \mathbb{R}^5$, the function above provides the image

$$\mathbf{y} \equiv \mathbf{f}(\mathbf{x}) \equiv \mathbf{A}\,\mathbf{x}, \tag{5.226}$$

where $\mathbf{A}$ is given in (5.225). As a matter of fact, the function **Afun**, evaluated at $\mathbf{x}$, performs *de facto* the product $\mathbf{A}\,\mathbf{x}$, that is, it provides the *action* of the matrix $\mathbf{A}$ on any given vector $\mathbf{x} \in \mathbb{R}^5$.

Imagine that we are given the function **Afun.m** but that we *do not* have access to its contents, that is, we do not know the inner functional structure of the map (5.224). In that sense, we are just given a *black box*

$$\mathbf{x} \Mapsto \boxed{\mathbf{f}} \Mapsto \mathbf{f}(\mathbf{x}) \equiv \mathbf{A}\mathbf{x},$$

that only allows us to evaluate the image of an arbitrary vector $\mathbf{x} \in \mathbb{R}^5$ under the action of $\mathbf{f}$, which, in turn, is the product $\mathbf{A}\,\mathbf{x}$, where $\mathbf{A}$ is also *unknown*. In general, $\mathbf{f}$ will map vectors of $\mathbb{R}^n$ – with n typically being very large – so that computing the explicit matrix elements of $\mathbf{A}$ by mapping the n canonical vectors of $\mathbb{R}^n$ is out of the question. Under these constraints, suppose that we are asked to find $\mathbf{x} \equiv [x_1\ x_2\ x_3\ x_4\ x_5]^{\mathrm{T}}$ satisfying

$$\mathbf{f}(\mathbf{x}) = [1\ 1\ 1\ 1\ 1]^{\mathrm{T}}. \tag{5.227}$$

If we knew the matrix $\mathbf{A}$ representing $\mathbf{f}$, we could easily find the solution of (5.227) simply by solving the linear system

$$\begin{bmatrix} 2 & -1 & 0 & 0 & 0 \\ -1 & 2 & -1 & 0 & 0 \\ 0 & -1 & 2 & -1 & 0 \\ 0 & 0 & -1 & 2 & -1 \\ 0 & 0 & 0 & -1 & 2 \end{bmatrix} \begin{bmatrix} x_1 \\ x_2 \\ x_3 \\ x_4 \\ x_5 \end{bmatrix} = \begin{bmatrix} 1 \\ 1 \\ 1 \\ 1 \\ 1 \end{bmatrix} \qquad (5.228)$$

with LU or QR direct methods seen previously. Unfortunately, we do not have access to $\mathbf{A}$. Equation (5.227) is the *matrix-free* version of Eq. (5.228).

Summing up, we are asked to solve $\mathbf{A}\mathbf{x} = \mathbf{b}$ with $\mathbf{b} \equiv [1\ 1\ 1\ 1\ 1]^{\mathrm{T}}$ ignoring $\mathbf{A}$, but only knowing the action $\mathbf{A}\mathbf{x}$ on an arbitrary vector $\mathbf{x}$ by means of a black-box linear map $\mathbf{f}(\mathbf{x})$ provided in `Afun.m`. As paradoxical as it may initially seem, solving linear systems just by knowing the action of the linear operator is a practice that is more common than one may think, particularly if the dimension of the problem is large. The main question is, how do we proceed to solve (5.227)?

Since we can use `Afun.m` to evaluate the linear map at our convenience, let us explore its effects on vectors of $\mathbb{R}^5$. The only reference vector we have is precisely the right-hand side $\mathbf{b} \equiv [1\ 1\ 1\ 1\ 1]^{\mathrm{T}}$, so let us start by evaluating[22] $\mathbf{A}\mathbf{b}$, $\mathbf{A}^2\mathbf{b}$, $\mathbf{A}^3\mathbf{b}$, etc., in Matlab's prompt by typing:

```
b = [1;1;1;1;1]; Ab = Afun(b);
A2b = Afun(Ab); A3b = Afun(A2b); A4b = Afun(A3b);
```

The reader may easily check that

$$\left[\, \mathbf{b} \,\middle|\, \mathbf{A}\mathbf{b} \,\middle|\, \mathbf{A}^2\mathbf{b} \,\middle|\, \mathbf{A}^3\mathbf{b} \,\middle|\, \mathbf{A}^4\mathbf{b} \,\right] = \begin{bmatrix} 1 & 1 & 2 & 5 & 14 \\ 1 & 0 & -1 & -4 & -15 \\ 1 & 0 & 0 & 2 & 12 \\ 1 & 0 & -1 & -4 & -15 \\ 1 & 1 & 2 & 5 & 14 \end{bmatrix} . \qquad (5.229)$$

It may seem apparently pointless to evaluate the actions $\mathbf{A}^j\mathbf{b}$ in order to find the solution of (5.227). However, a careful inspection of the *rank*[23] of the system of vectors generated by this procedure reveals a very interesting phenomenon:

$$\dim\{\mathbf{b}\} = 1,$$
$$\dim\{\mathbf{b},\ \mathbf{A}\mathbf{b}\} = 2,$$
$$\dim\{\mathbf{b},\ \mathbf{A}\mathbf{b},\ \mathbf{A}^2\mathbf{b}\} = 3,$$
$$\dim\{\mathbf{b},\ \mathbf{A}\mathbf{b},\ \mathbf{A}^2\mathbf{b},\ \mathbf{A}^3\mathbf{b}\} = 3. \qquad (5.230)$$

---

22 Henceforth, $\mathbf{A}\mathbf{x}$ will stand for the action $\mathbf{x} \mapsto \mathbf{f}(\mathbf{x})$, i.e. $\mathbf{A}^k\mathbf{x} = (\mathbf{f} \circ \mathbf{f} \circ \overset{(k)}{\cdots} \circ \mathbf{f})\mathbf{x}$.
23 We recommend the reader to use Matlab's `rank` command.

In other words, the vectors $\mathbf{b}$, $\mathbf{A}\,\mathbf{b}$, $\mathbf{A}^2\mathbf{b}$, and $\mathbf{A}^3\mathbf{b}$ are linearly dependent. The reader can check that, in this particular case, $\mathbf{A}^3\mathbf{b}$ can be expressed as

$$\mathbf{A}^3\mathbf{b} = 2\mathbf{b} - 9\mathbf{A}\mathbf{b} + 6\mathbf{A}^2\mathbf{b}. \tag{5.231}$$

The last expression can be written as

$$\mathbf{A}^3\mathbf{b} - 6\mathbf{A}^2\mathbf{b} + 9\mathbf{A}\mathbf{b} = 2\mathbf{b}, \tag{5.232}$$

or, equivalently,

$$\mathbf{A}\left(\frac{1}{2}\mathbf{A}^2\mathbf{b} - 3\mathbf{A}\mathbf{b} + \frac{9}{2}\mathbf{b}\right) = \mathbf{b}. \tag{5.233}$$

Probably the reader has already noticed that we have just obtained the solution of $\mathbf{A}\mathbf{x} = \mathbf{b}$, which in this case reads

$$\mathbf{x} = \frac{1}{2}\mathbf{A}^2\mathbf{b} - 3\mathbf{A}\mathbf{b} + \frac{9}{2}\mathbf{b} = \begin{bmatrix} \frac{5}{2} & 4 & \frac{9}{2} & 4 & \frac{5}{2} \end{bmatrix}^{\mathrm{T}}. \tag{5.234}$$

This example clearly illustrates that, even though the system is five-dimensional, this particular solution lies in a subspace of smaller dimension (three-dimensional, to be exact) spanned by the basis $\{\mathbf{b}, \mathbf{A}\mathbf{b}, \mathbf{A}^2\mathbf{b}\}$ obtained by computing $\mathbf{f}(\mathbf{b})$ and $(\mathbf{f} \circ \mathbf{f})\mathbf{b}$. Proceeding similarly with other right-hand side vectors $\mathbf{b}$ may require higher dimensional spaces. For example, if $\mathbf{b} = [1\ 2\ 3\ 4\ 5]^{\mathrm{T}}$, we need the five-dimensional space $\mathrm{span}\{\mathbf{b}, \mathbf{A}\,\mathbf{b}, \mathbf{A}^2\mathbf{b}, \mathbf{A}^3\mathbf{b}, \mathbf{A}^4\mathbf{b}\}$. The solvers that we are about to introduce aim to provide, if possible, low dimensional approximate solutions of large linear systems of equations. We begin by introducing a few new concepts.

Krylov[24] Spaces and Krylov Matrices

Let $\mathbf{A} \in \mathbb{M}_n(\mathbb{R})$ and $\mathbf{b} \in \mathbb{R}^n$. We define the *Krylov space* $\mathbb{K}_m(\mathbf{A},\,\mathbf{b})$ as

$$\mathbb{K}_m(\mathbf{A},\,\mathbf{b}) \doteq \mathrm{span}\{\mathbf{b}, \mathbf{A}\mathbf{b}, \mathbf{A}^2\mathbf{b}, \ldots, \mathbf{A}^{m-1}\mathbf{b}\}, \tag{5.235}$$

and its associated *Krylov matrix* as

$$\mathbf{K}_m \doteq \begin{bmatrix} \mathbf{b} \,\big|\, \mathbf{A}\mathbf{b} \,\big|\, \mathbf{A}^2\mathbf{b} \,\big|\, \cdots \,\big|\, \mathbf{A}^{m-1}\mathbf{b} \end{bmatrix}. \tag{5.236}$$

Since $\mathbf{A} \in \mathbb{M}_n(\mathbb{R})$, $\dim \mathbb{K}_m \leq n$. Any vector $\mathbf{x}$ in the Krylov subspace $\mathbb{K}_m(\mathbf{A},\,\mathbf{b})$, henceforth referred to simply as $\mathbb{K}_m$, can be expressed as the linear combination

$$\mathbf{x} = c_0\mathbf{b} + c_1\mathbf{A}\mathbf{b} + c_2\mathbf{A}^2\mathbf{b} + \cdots + c_{m-1}\mathbf{A}^{m-1}\mathbf{b}, \tag{5.237}$$

---

24 A. N. Krylov (1863–1945), Russian naval engineer and applied mathematician who made remarkable contributions within the field of hydrodynamics.

or, in terms of the matrix $\mathbf{K}_m$ and the vector $\mathbf{c} = [c_0 \; c_1 \; c_2 \; \cdots \; c_{m-1}]^{\mathrm{T}} \in \mathbb{R}^m$,

$$
\mathbf{x} = \left[\; \mathbf{b} \,\middle|\, \mathbf{Ab} \,\middle|\, \mathbf{A}^2\mathbf{b} \,\middle|\, \cdots \,\middle|\, \mathbf{A}^{m-1}\mathbf{b} \; \right]
\begin{bmatrix} c_0 \\ c_1 \\ c_2 \\ \vdots \\ c_{m-1} \end{bmatrix} = \mathbf{K}_m \, \mathbf{c}. \tag{5.238}
$$

Suppose we have to solve a large linear system $\mathbf{Ax} = \mathbf{b}$, with $\mathbf{A} \in \mathrm{M}_n(\mathbb{R})$ and $\mathbf{b} \in \mathbb{R}^n$. In a first attempt, we may proceed as we did at the beginning of this section by generating the corresponding Krylov subspaces $\mathbb{K}_1$, $\mathbb{K}_2$, $\mathbb{K}_3$, etc., hoping for a lucky stagnation of $\dim(\mathbb{K}_m)$ for some iteration m onward. Unfortunately, this is in general a very unlikely situation. Instead, we may search within each Krylov subspace $\mathbb{K}_m$ for the vector $\mathbf{x}$ that *minimizes* the residual $\|\mathbf{Ax} - \mathbf{b}\|$, that is,

$$
\min_{\mathbf{x} \,\in\, \mathbb{K}_m} \|\mathbf{Ax} - \mathbf{b}\| = \min_{\mathbf{c} \,\in\, \mathbb{R}^m} \|\mathbf{AK}_m\mathbf{c} - \mathbf{b}\|. \tag{5.239}
$$

At first glance, this is a least squares problem that should be easily solved by the QR algorithm. However, the matrix $\mathbf{AK}_m$ may be numerically intractable. To explain the reason, consider the eigenvalues λ_j and associated eigenvectors $\mathbf{v}_j$ of $\mathbf{A}$ satisfying

$$
\mathbf{Av}_j = \lambda_j \mathbf{v}_j \quad (j = 1, 2, \ldots, n), \tag{5.240}
$$

Let the eigenvalues be sorted in modulus so that $|\lambda_1| \geq |\lambda_2| \geq \cdots \geq |\lambda_n|$. Since $\mathrm{span}\{\mathbf{v}_1, \mathbf{v}_2, \ldots, \mathbf{v}_n\}$ is a basis of $\mathbb{R}^n$, the right-hand side $\mathbf{b}$ of (5.239) can be expanded as a linear combination of the form

$$
\mathbf{v} = \beta_1 \, \mathbf{v}_1 + \beta_2 \mathbf{v}_2 + \cdots + \beta_n \mathbf{v}_n. \tag{5.241}
$$

The iterated actions of $\mathbf{A}$ on $\mathbf{b}$ are therefore

$$
\mathbf{Ab} = \beta_1 \lambda_1 \, \mathbf{v}_1 + \beta_2 \lambda_2 \mathbf{v}_2 + \cdots + \beta_n \lambda_n \mathbf{v}_n,
$$
$$
\mathbf{A}^2\mathbf{b} = \beta_1 \lambda_1^2 \, \mathbf{v}_1 + \beta_2 \lambda_2^2 \mathbf{v}_2 + \cdots + \beta_n \lambda_n^2 \mathbf{v}_n,
$$

$$
\vdots
$$

$$
\mathbf{A}^m\mathbf{b} = \beta_1 \lambda_1^m \, \mathbf{v}_1 + \beta_2 \lambda_2^m \mathbf{v}_2 + \cdots + \beta_n \lambda_n^m \mathbf{v}_n. \tag{5.242}
$$

The last iterate can be expressed as

$$
\mathbf{A}^m\mathbf{b} = \lambda_1^m \left[\beta_1 \, \mathbf{v}_1 + \beta_2 \left(\frac{\lambda_2}{\lambda_1}\right)^m \mathbf{v}_2 + \cdots + \beta_n \left(\frac{\lambda_n}{\lambda_1}\right)^m \mathbf{v}_n \right]. \tag{5.243}
$$

Since $|\lambda_j| \leq |\lambda_1|$ for $j = 2, 3, \ldots, n$, the ratios $(\lambda_2/\lambda_1)^m$, $(\lambda_3/\lambda_1)^m$, etc. will typically tend to zero as m increases. As a result, the vectors $\{\mathbf{Ab}, \mathbf{A}^2\mathbf{b}, \mathbf{A}^3\mathbf{b}, \ldots, \mathbf{A}^m\mathbf{b}\}$ become nearly parallel to $\mathbf{v}_1$. Consequently, the matrix $\mathbf{AK}_m$ appearing in (5.239) turns out to be constituted by nearly parallel vectors and its QR factorization requires new strategies capable of generating well-conditioned bases of Krylov subspaces. One of these strategies consists in a wise Gram-Schmidt orthonormalization of the Krylov vectors $\mathbf{A}^j\mathbf{b}$ *on the fly*, that is, as soon as they are being generated. This technique, better known as the *Arnoldi iteration*[25] is outlined below:

$$\mathbb{K}_1 \equiv \text{span}\{\mathbf{b}\} \equiv \text{span}\{\mathbf{q}_1\},$$

$$\mathbb{K}_2 \equiv \text{span}\{\mathbf{b}, \ \mathbf{Ab}\} \equiv \text{span}\{\mathbf{q}_1, \ \mathbf{Aq}_1\} \equiv \text{span}\{\mathbf{q}_1, \ \mathbf{q}_2\},$$

$$\mathbb{K}_3 \equiv \text{span}\{\mathbf{b}, \ \mathbf{Ab}, \mathbf{A}^2\mathbf{b}\} \equiv \text{span}\{\mathbf{q}_1, \mathbf{q}_2, \mathbf{Aq}_2\} \equiv \text{span}\{\mathbf{q}_1, \mathbf{q}_2, \mathbf{q}_3\},$$

$$\vdots$$

$$\mathbb{K}_m \equiv \text{span}\{\mathbf{b}, \mathbf{Ab}, \dots, \mathbf{A}^{m-1}\mathbf{b}\} \equiv \text{span}\{\mathbf{q}_1, \mathbf{q}_2, \dots, \mathbf{q}_{m-1}, \mathbf{Aq}_{m-1}\}$$

$$\equiv \text{span}\{\mathbf{q}_1, \mathbf{q}_2, \dots, \mathbf{q}_m\}. \tag{5.244}$$

As in the Gram-Schmidt algorithm, we start the Arnoldi iteration by normalizing $\mathbf{b}$ to obtain

$$\mathbf{q}_1 = \frac{\mathbf{b}}{\|\mathbf{b}\|}. \tag{5.245}$$

However, in order to generate $\mathbb{K}_2$, we compute $\mathbf{Aq}_1$ *instead* of $\mathbf{Ab}$. A vector orthogonal to $\mathbf{q}_1$ can be obtained by subtracting from $\mathbf{Aq}_1$ its corresponding projection on $\mathbb{K}_1$, that is,

$$\tilde{\mathbf{q}}_2 \equiv \mathbf{Aq}_1 - h_{11}\mathbf{q}_1, \tag{5.246}$$

where

$$h_{11} \equiv \langle \mathbf{q}_1, \mathbf{Aq}_1 \rangle. \tag{5.247}$$

Finally we normalize $\tilde{\mathbf{q}}_2$:

$$\mathbf{q}_2 = \frac{\tilde{\mathbf{q}}_2}{h_{21}}, \tag{5.248}$$

where we have defined the normalization factor $h_{21} \doteq \|\tilde{\mathbf{q}}_2\|$. Similarly, to generate an orthonormal basis of $\mathbb{K}_3$, we first compute $\mathbf{A}\,\mathbf{q}_2$ (and not $\mathbf{A}^2\mathbf{b}$) and devise a third vector

$$\tilde{\mathbf{q}}_3 \equiv \mathbf{Aq}_2 - h_{12}\mathbf{q}_1 - h_{22}\mathbf{q}_2, \tag{5.249}$$

which is orthogonal to $\mathbb{K}_2$ if

$$h_{12} \equiv \langle \mathbf{q}_1, \mathbf{Aq}_2 \rangle \quad \text{and} \quad h_{22} \equiv \langle \mathbf{q}_2, \mathbf{Aq}_2 \rangle. \tag{5.250}$$

This third vector is normalized accordingly:

$$\mathbf{q}_3 = \frac{\tilde{\mathbf{q}}_3}{h_{32}}, \tag{5.251}$$

---

[25] W. E. Arnoldi (1917-1995), American engineer who first devised this algorithm within the context of eigenvalue computations.

where $h_{32} \equiv \|\tilde{\mathbf{q}}_3\|$. In this algorithm, the coefficients h_{ij} are being defined as

$$h_{ij} \doteq \begin{cases} \langle \mathbf{q}_i, \mathbf{A}\mathbf{q}_j \rangle & (i \le j), \\ \|\tilde{\mathbf{q}}_{j+1}\| & (i = j + 1). \end{cases} \tag{5.252}$$

Using (5.248), we can write (5.246) as

$$\mathbf{A}\mathbf{q}_1 = h_{11}\mathbf{q}_1 + h_{21}\mathbf{q}_2. \tag{5.253}$$

Similarly, in virtue of (5.251), Eq. (5.249) can be expressed as

$$\mathbf{A}\mathbf{q}_2 = h_{12}\mathbf{q}_1 + h_{22}\mathbf{q}_2 + h_{32}\mathbf{q}_3. \tag{5.254}$$

The reader has probably noticed that these last two equations are very similar to (5.204) and (5.205) used previously to show the CGS-QR equivalence. As a matter of fact, Eqs. (5.253) and (5.254) can be understood as the matrix-matrix products

$$\mathbf{A} \begin{bmatrix} \mathbf{q}_1 & \mathbf{q}_2 \end{bmatrix} = \begin{bmatrix} \mathbf{q}_1 & \mathbf{q}_2 & \mathbf{q}_3 \end{bmatrix} \begin{bmatrix} h_{11} & h_{12} \\ h_{21} & h_{22} \\ 0 & h_{32} \end{bmatrix}. \tag{5.255}$$

At stage m, the Arnoldi iteration reads

$$\mathbf{A} \underbrace{\begin{bmatrix} \mathbf{q}_1 & \mathbf{q}_2 & \cdots & \mathbf{q}_m \end{bmatrix}}_{\mathbf{Q}_m} = \underbrace{\begin{bmatrix} \mathbf{q}_1 & \mathbf{q}_2 & \cdots & \mathbf{q}_{m+1} \end{bmatrix}}_{\mathbf{Q}_{m+1}} \overbrace{\begin{bmatrix} h_{11} & h_{12} & \cdots & & h_{1m} \\ h_{21} & h_{22} & \cdots & & h_{2m} \\ 0 & h_{32} & \ddots & & h_{3m} \\ \vdots & & \ddots & \ddots & \vdots \\ 0 & 0 & & \ddots & h_{mm} \\ 0 & 0 & & & h_{m+1,m} \end{bmatrix}}^{\tilde{\mathbf{H}}_m}, \tag{5.256}$$

The last column on the left-hand and right-hand sides of (5.256) is

$$\mathbf{A}\mathbf{q}_m = h_{1m}\mathbf{q}_1 + h_{2m}\mathbf{q}_2 + \cdots + h_{mm}\mathbf{q}_m + h_{m+1,m}\mathbf{q}_{m+1}. \tag{5.257}$$

In particular, if the last coefficient $h_{m+1,m} \equiv \|\tilde{\mathbf{q}}_{m+1}\|$ is zero, then $\mathbf{A}\mathbf{q}_m \in \mathbb{K}_m$, that is, $\mathbb{K}_{m+1} \equiv \mathbb{K}_m$, and the Arnoldi iteration must be stopped. For brevity, we write Eq. (5.256) simply as

$$\mathbf{A}\mathbf{Q}_m \equiv \mathbf{Q}_{m+1}\tilde{\mathbf{H}}_m, \tag{5.258}$$

where matrices $\mathbf{Q}_m$ and $\mathbf{Q}_{m+1}$ are explicitly indicated in (5.256), and where $\tilde{\mathbf{H}}_m \in \mathbb{M}_{m+1,m}(\mathbb{R})$ is an *upper Hessenberg*[26] matrix.

---

26 An upper Hessenberg matrix is any matrix with zero entries below its first subdiagonal. K. A. Hessenberg (1904–1959), German mathematician and engineer who first exploited the structure of these matrices in the numerical computation of eigenvalues.

Summing up, recall that the main motivation for the Arnoldi iteration lies in the fact that the least squares problem

$$\min_{\mathbf{c} \in \mathbb{R}^m} \|\mathbf{A}\mathbf{K}_m\mathbf{c} - \mathbf{b}\| \tag{5.259}$$

was ill-conditioned because the desired minimizing vector $\mathbf{c} \in \mathbb{R}^m$ was expressed in the nearly parallel Krylov basis $\mathbb{K}_m$. The good news is that now we have an orthonormal basis for $\mathbb{K}_m$, so that the vector $\mathbf{K}_m\mathbf{c}$ can also be expressed as the linear combination $\mathbf{Q}_m\mathbf{y}$ in terms of this new basis, that is,

$$\mathbf{K}_m\mathbf{c} = c_0\mathbf{b} + c_1\mathbf{A}\mathbf{b} + \cdots + c_{m-1}\mathbf{A}^{m-1}\mathbf{b} = y_1\mathbf{q}_1 + \cdots + y_m\mathbf{q}_m = \mathbf{Q}_m\mathbf{y}. \tag{5.260}$$

Therefore, the least squares problem (5.259) now reads

$$\min_{\mathbf{y} \in \mathbb{R}^m} \|\mathbf{A}\mathbf{Q}_m\mathbf{y} - \mathbf{b}\|. \tag{5.261}$$

By virtue of (5.258), we can write (5.261) as

$$\min_{\mathbf{y} \in \mathbb{R}^m} \|\mathbf{Q}_{m+1}\widetilde{\mathbf{H}}_m\mathbf{y} - \mathbf{b}\|. \tag{5.262}$$

In addition, recall that $\mathbf{b}$ was the first vector to be normalized, so that $\mathbf{b} = \|\mathbf{b}\|\mathbf{q}_1$ and, expressed in terms of the new basis, reads

$$\mathbf{b} = \|\mathbf{b}\|\mathbf{Q}_{m+1}\mathbf{e}_1, \tag{5.263}$$

where $\mathbf{e}_1 = [1\ 0\ 0\ \cdots\ 0]^{\mathrm{T}}$ is the first canonical vector in $\mathbb{R}^{m+1}$. Therefore (5.262) now reads

$$\min_{\mathbf{y} \in \mathbb{R}^m} \|\mathbf{Q}_{m+1}\widetilde{\mathbf{H}}_m\mathbf{y} - \|\mathbf{b}\|\mathbf{Q}_{m+1}\mathbf{e}_1\| = \min_{\mathbf{y} \in \mathbb{R}^m} \|\mathbf{Q}_{m+1}(\widetilde{\mathbf{H}}_m\mathbf{y} - \|\mathbf{b}\|\mathbf{e}_1)\|. \tag{5.264}$$

Since $\mathbf{Q}_{m+1}$ preserves norms, that is, for all $\mathbf{z} \in \mathbb{R}^{m+1}$ we have

$$\|\mathbf{Q}_{m+1}\mathbf{z}\|^2 = \mathbf{z}^{\mathrm{T}}\mathbf{Q}_{m+1}^{\mathrm{T}}\mathbf{Q}_{m+1}\mathbf{z} = \mathbf{z}^{\mathrm{T}}\mathbf{I}_{m+1}\mathbf{z} = \|\mathbf{z}\|^2. \tag{5.265}$$

the least squares problem (5.264) is equivalent to

$$\boxed{\min_{\mathbf{y} \in \mathbb{R}^m} \left\|\widetilde{\mathbf{H}}_m\mathbf{y} - \|\mathbf{b}\|\mathbf{e}_1\right\|.} \tag{5.266}$$

While the original problem (5.259) is ill-conditioned and of large dimension $n \times m$, problem (5.266) is formulated on a well-conditioned basis and also has a much smaller dimension $(m+1) \times m$. Besides, solving (5.266) involves an upper Hessenberg matrix whose QR-factorization is numerically stable, but also computationally cheaper, since all the entries below the subdiagonal are zero.

Using the Arnoldi iteration to generate a proper orthonormal Krylov basis that reduces the least squares problem (5.266) into upper triangular Hessenberg form is commonly known as the *generalized minimum residual method*, or GMRES for short. This method is just one of the many matrix-free iterative solvers that are used nowadays to solve matrix-free systems of linear equations.

Code 18 implements GMRES performing reorthogonalization during the Arnoldi iteration and makes use of Householder's QR-algorithm to minimize the residual in (5.266) by invoking `myqr.m` and `qrsolve.m` (see Codes 15 and 16).

```
% Code 18: GMRES (Arnoldi iteration - Reorthogonalized)
function [x,k] = mygmres(Afun,b,tol,dimkryl)
% Requires: myqr.m and qrsolve.m (Codes 15 & 16)
% Input:   Afun (x --> Ax function) | b  (r.h.s)
%          tol (iteration stops if |A*x_m-b| < tol)
%          dimkryl: max. dimension of krylov space.
% Output:  x: approximation found
%          k: no. of iterations used
hkp1 = 1; resd = 1; k = 1;
Q(:,1) = b/norm(b); H = [];          % q_1 = b/|b|
while resd > tol & hkp1 > eps & k < dimkryl
  Q(:,k+1) = feval(Afun,Q(:,k));     % q_k+1 = Aq_k
  h = Q(:,1:k)'*Q(:,k+1);            % h_{ij} - coeffs.
  Q(:,k+1)  = Q(:,k+1) - Q(:,1:k)*h; % See (5.255)
  S = Q(:,1:k)'*Q(:,k+1); Q(:,k+1) = Q(:,k+1) - Q(:,1:k)*S;
  h = h + S;                         % Reorth.
  hkp1 = norm(Q(:,k+1)); H(1:k+1,k) = [h; hkp1];
  Q(:,k+1) = Q(:,k+1)/hkp1;
  UR = myqr(H(1:k+1,1:k)); e1b = [norm(b); zeros(k,1)];
  y = qrsolve(UR,e1b); x = Q(:,1:k)*y; k = k + 1;
  resd = norm(feval(Afun,x)-b);
end
```

There are more efficient techniques that perform the QR-reduction of (5.266) in just $O(m^2)$ operations. These methodologies make use of a different type of orthogonal transformations known as *Givens rotations* (see references at the end of the chapter, in the Complementary Reading section) that allow for a selective annihilation of the subdiagonal elements of the Hessenberg matrix $\widetilde{\mathbf{H}}_m$. However, within the academic context of this introductory text, we will simply use the standard QR-algorithm using reflectors. One of the major advantages of this iterative approach is that we have control of the residual variable **resd** so that the algorithm stops when we consider that the quantity is sufficiently small.

What we have seen in this last section is just an *oversimplified* introduction to a vast area of numerical analysis, namely, *iterative linear solvers*. The study of different features of these methods, such as their convergence rates and computational cost or suitable methodologies to improve their performance such as *preconditioning* is far beyond this introductory course. The reader will find these and many other concepts addressed in detail in the monographs on numerical linear algebra recommended at the end of the chapter.

Practical 5.4: Matrix-Free Krylov Solvers (GMRES)

The figure below shows the same circuit as in Practical 5.1: identical resistors ($R = 3\ \Omega$) connected to a battery $V = 1$ V. We aim to compute the intensities I_k up to a fixed tolerance and *without explicitly* building the matrix by looking for $O(m)$-*low dimensional* approximations of an $O(n)$ dimensional problem.

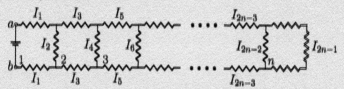

(a) Kirchhoff: the solution vector of intensities $[I_1\ I_2\ \cdots\ I_{2n-1}]^T$ must satisfy the $2n-1$ equations

$$
\begin{aligned}
2R\,I_1 + R\,I_2 &= V, \\
I_1 - I_2 - I_3 &= 0, \\
2R\,I_{2j-3} + R\,I_{2j-2} - R\,I_{2j-4} &= 0, \quad (j = 3,\ldots,n), \\
I_{2j-3} - I_{2j-2} - I_{2j-1} &= 0, \quad (j = 3,\ldots,n), \\
3R\,I_{2n-1} - R I_{2n-2} &= 0.
\end{aligned}
$$

In this case, we only need the linear action on the vector $[I_1\ I_2\ \cdots\ I_{2n-1}]^T$, such as the function below for $n = 30$:

```
function Ax = Afun_prac14(x)
R = 3; V = 1;  n = 30; Ax = [];
Ax = [Ax;  2*R*x(1) + R*x(2)];
Ax = [Ax;  x(1)-x(2)-x(3)];
for j = 3:n
 Ax = [Ax;  2*R*x(2*j-3) + R*x(2*j-2) - R*x(2*j=4)];
 Ax = [Ax;  x(2*j-3) - x(2*j-2) - x(2*j-1)];
end
Ax = [Ax;  3*R*x(2*n-1) - R*x(2*n-2)];
```

Check that the output coincides with the product **Ax** in Practical 5.1.

(b) Write your own GMRES algorithm allowing for an arbitrary tolerance ε and maximum dimension of the Krylov subspace m (tol and dimkryl in Code 18, respectively). Find approximate solutions satisfying $\|\mathbf{Ax} - \mathbf{b}\| \le \varepsilon$, for $\varepsilon = 10^{-2}, 10^{-3},\ldots,10^{-8}$. For these cases, compare the solutions with the one obtained in Practical 5.1 and plot the required dimension $m(\varepsilon)$ of the Krylov subspace as a function of the tolerance ε.

Complementary Reading

For a quite modern and comprehensive theoretical review of linear algebra, Meyer's *Matrix Analysis and Applied Linear Algebra* is a good choice. In this chapter, we have only addressed a *few* topics. Numerical linear algebra is a huge and active research area and it should deserve a whole book. Fortunately, there are dozens of outstanding monographs specialized on this topic. For a nice and friendly introduction, I particularly recommend Watkins' *Fundamentals of Matrix Computations*, where the reader will find many topics not covered here such as *Cholesky decomposition* for positive definite systems, *backward stability*, *Givens rotations*, *Modified Gram–Schmidt* orthogonalization, and the *Francis*-QR step method for the computation of eigenvalues of matrices. Watkins' book also covers in detail formal aspects regarding uniqueness of LU and QR factorizations, or the induced matrix norms, but also includes more recent elements such as reorthogonalization.

Another reference that has become a classic in the topic is Trefethen and Bau's *Numerical Linear Algebra*. This book provides deep insight into many topics but also manages to give the subject a global view, which sometimes is lost in other treatises. This monograph addresses very important concepts of numerical analysis not seen in this chapter such as *stability* and *backward stability* of an algorithm. For a deep analysis of the stability of GEM and the concept of *growing factors*, this is the reference to look at. This book also addresses beautifully the *singular value decomposition* (SVD) and its useful applications in the efficient computation of norms and condition numbers. I also strongly recommend this book to anyone interested in Krylov–Arnoldi methods for the computation of eigenvalues and the solution of linear systems.

For over three decades, the classical reference has been Golub and Van Loan's *Matrix Computations* (already in its 4th edition). For those readers interested in the theoretical foundations of this topic as well as in its software implementation, this is a must-read reference. For a comprehensive treatment of iterative methods for the solution of linear systems, Saad's book *Iterative Methods for Sparse Linear Systems* is a very good choice. This applied book addresses classical topics not covered here such as *Jacobi* or *Gauss–Seidel* methods as well as the modern Krylov matrix-free approach. Saad's text also addresses crucial aspects such as *preconditioning*, or the origin of sparse matrices after the discretization of partial differential equations (that we have not covered). Other two excellent references on Krylov methods are van der Vorst's *Iterative Krylov Methods for Large Linear Systems* and Greenbaum's *Iterative Methods for Solving Linear Systems*.

Finally, I recommend the reader to explore Matlab functions such as **qr**, **lu**, and **gmres** and the references included in the documentation.

Problems and Exercises

The symbol (A) means that the problem has to be solved analytically, whereas (N) means to use numerical Matlab codes. Problems with the symbol * are slightly more difficult.

1. (A) Prove that if a matrix $\mathbf{A}$ has an inverse $\mathbf{A}^{-1}$, then there can be no nonzero $\mathbf{y}$ for which $\mathbf{A}\mathbf{y} = 0$.

2. (A) Prove that if a matrix $\mathbf{A}$ has an inverse $\mathbf{A}^{-1}$ then $\det \mathbf{A} \neq 0$.

3. (A) Show that the computational cost of solving an arbitrary n-dimensional triangular system with BS (backward substitution) or FS (forward substitution) is $O(n^2)$.

4. (A*) Show that the computational cost of solving an arbitrary n-dimensional system with GEM is $O(n^3)$.

   Hint: First show that $\sum_{k=0}^{n} k^2 = n(n+1)(2n+1)/6$.

5. (A) Let $\mathbf{L}, \mathbf{M} \in \mathbb{M}_n(\mathbb{R})$ be two lower triangular matrices. Prove that $\mathbf{LM}$ is also lower triangular.

6. (A*) Let $\mathbf{L} \in \mathbb{M}_n(\mathbb{R})$ be a non-singular, lower triangular matrix. Prove that $\mathbf{L}^{-1}$ is also lower triangular.

   Hint: Use induction on n, define $\mathbf{M} = \mathbf{L}^{-1}$, and consider the partitions:

   $$\mathbf{L} = \begin{bmatrix} \mathbf{L}_{11} & \mathbf{0} \\ \mathbf{L}_{21} & \mathbf{L}_{22} \end{bmatrix}, \quad \mathbf{M} = \begin{bmatrix} \mathbf{M}_{11} & \mathbf{M}_{12} \\ \mathbf{M}_{21} & \mathbf{M}_{22} \end{bmatrix},$$

   where $\mathbf{L}_{11}$ and $\mathbf{M}_{11}$ are square matrices of the same size, and $\mathbf{L}_{11}$ and $\mathbf{L}_{22}$ are lower triangular and non-singular.

7. (N) Develop two codes for LU factorization with and without pivoting of an arbitrary non-singular square matrix $\mathbf{A}$. Compute the LU factorization of the matrix

   $$\mathbb{A} = \begin{bmatrix} -8 & -9 & 7 & 19 & 3 \\ 9 & 10 & 9 & 1 & -16 \\ 7 & 2 & -8 & -2 & 3 \\ 19 & 8 & -18 & -8 & -3 \\ 15 & 10 & 16 & -16 & -18 \end{bmatrix}$$

using both codes. Solve the linear system $\mathbf{Ax} = \mathbf{b}$, with $\mathbf{b}$ an arbitrary nonzero vector of $\mathbb{R}^5$. Let $\mathbf{x}_1$ and $\mathbf{x}_2$ be the obtained solutions without and with pivoting, respectively. Since $\mathbf{Ax} - \mathbf{b} = \mathbf{0}$, one way of measuring the accuracy of the numerical solutions $\mathbf{x}_1$ and $\mathbf{x}_2$ consists of evaluating their corresponding relative *residuals* $r_1 = \|\mathbf{Ax}_1 - \mathbf{b}\|/\|\mathbf{x}_1\|$ and $r_2 = \|\mathbf{Ax}_2 - \mathbf{b}\|/\|\mathbf{x}_2\|$, respectively using Matlab's `norm`.[27] Which of them is smaller? Why?

8. (A) Consider three blocks attached to springs of stiffness k_1, k_2, k_3, and k_4 as depicted in the figure below. Assume that initially the springs are not compressed nor stretched and neglect frictional forces between the blocks and the ground.

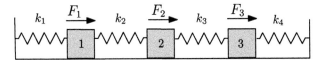

Suppose that three forces F_1, F_2, and F_3 are applied on each one of the blocks 1, 2, and 3, displacing the blocks to new equilibrium coordinates x_1, x_2, and x_3 with respect their initial positions, respectively. Write the equations of equilibrium. Obtain the new equilibrium coordinates if the forces are $F_1 = 1$ N, $F_2 = 2$ N, and $F_3 = 3$ N. Assume that the three springs have the same stiffness $k_1 = k_2 = k_3 = k_4 = 4$ N m^{-1}.

9. (A) Using the recurrences (5.104) and (5.105), along with the properties of exchange matrices, obtain the explicit expressions for $\mathbf{L}_3$ and $\mathbf{L}_4$ seen in (5.110) and (5.111), respectively. Generalize your calculations to obtain the last term of the LU-factorization with partial pivoting $\mathbf{L}_{n-1}$ shown in (5.112).

10. Show that any orthogonal matrix $\mathbf{Q} \in \mathbb{M}_n(\mathbb{R})$ satisfies $\det \mathbf{Q} = \pm 1$.

11. (A) Show that the reflector defined in (5.135) satisfies $\mathbf{Qu} = -\mathbf{u}$ and $\mathbf{Qv} = \mathbf{v}$, for any $\mathbf{v} \perp \mathbf{u}$.

12. (A) Show that two proportional Householder vectors $\mathbf{u}$ and $\alpha \mathbf{u}$ ($\alpha \neq 0$) generate the same reflector.

---

27 $\|\mathbf{x}\| \doteq [x_1^2 + x_2^2 + \cdots + x_n^2]^{1/2}$ is the *Euclidean* or 2-norm.

13. (A) Computational cost of the Householder QR-factorization via reflectors:

    (a) Show that if $\mathbf{A}_j \in M_{nm}(\mathbb{R})$ and $\mathbf{u}_j$, $\mathbf{v}_j \in \mathbb{R}^n$, the total computation of the term $\mathbf{u}_j \mathbf{v}_j^T \mathbf{A}_j$ appearing in (5.160) can be reduced to $3\,nm$ by pre-computing the auxiliary vector $\mathbf{v}_{\text{aux.}} = \mathbf{v}_j^T \mathbf{A}_j \in \mathbb{R}^m$ and then the product $\mathbf{u}_j \mathbf{v}_{\text{aux.}}$.

    (b) For $m = n$, show that the total computational cost of the $n-1$ transformations (5.160) for $j = 1, 2, \ldots, n-1$ is $4n^3/3$ (that is, the computational cost of the QR-factorization is twice that of an LU decomposition).

14. (A) Find an orthogonal matrix $\mathbf{Q} \in M_3(\mathbb{R})$ corresponding to the reflector of $\mathbb{R}^3$ through the plane $x - z = 0$.

15. (A) Consider an arbitrary reflector in $\mathbb{R}^n$:

$$\mathbf{F} \equiv \mathbf{I} - 2\mathbf{w}\mathbf{w}^T, \quad \text{with } \mathbf{w} \in \mathbb{R}^n, \text{ and } \|\mathbf{w}\| = 1.$$

Find the eigenvalues λ of $\mathbf{F}$ and their corresponding eigenspace dimension $\dim\{\ker(\mathbf{F} - \lambda\mathbf{I})\}$.

16. (A) Consider the QR-factorization: $\mathbf{A} = \mathbf{QR}$, with $\mathbf{A}$, $\mathbf{Q} \in M_4(\mathbb{R})$, $\mathbf{QQ}^T = \mathbf{I}$, and

$$\mathbf{R} \equiv \begin{bmatrix} 1 & 2 & 3 & 4 \\ & 5 & 6 & 7 \\ & & 8 & 9 \\ & & & 1 \end{bmatrix}.$$

Let $\mathbf{R} = \mathbf{Q}_3\, \mathbf{Q}_2\, \mathbf{Q}_1\, \mathbf{A}$, with $\mathbf{Q}_k = \mathbf{I} - \gamma_k\, \mathbf{u}_k \mathbf{u}_k^T$, $\gamma_k = 2/\|\mathbf{u}_k\|^2$ and

$$\mathbf{u}_1 = \begin{bmatrix} 1 & 1 & 1 & 0 \end{bmatrix}^T, \quad \mathbf{u}_2 = \begin{bmatrix} 0 & 1 & 1 & 1 \end{bmatrix}^T, \quad \mathbf{u}_3 = \begin{bmatrix} 0 & 0 & 1 & 1 \end{bmatrix}^T.$$

Determine the matrices $\mathbf{Q}$ and $\mathbf{A}$ *exactly*, not numerically.

17. (N) Find the quadratic polynomial $p_2(x) \equiv a_0 + a_1 x + a_2 x^2$ that better fits the data below:

x_i	-1	-0.75	-0.5	0	0.25	0.5	0.75
y_i	1.0	0.8125	0.75	1.00	1.3125	1.75	2.3125

18. (N) The figure given on the right shows an electrical circuit consisting of a resistor $R_1 = 10\ \Omega$ in parallel with another device of unknown electrical resistance R_2.
The experimental setting only allows to connect different batteries (whose internal resistance can be neglected) of known voltages V_j to the whole system via the connectors a and b so that we can measure the intensity I_j with an ammeter A. The registered data are included in the following table:

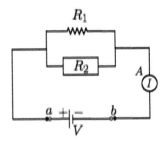

V (V)	12.0	24.0	30.0	36.0	40.0	52.0	60.0
I (A)	1.35	2.66	3.41	4.16	4.53	5.97	6.90

Fitting the data above, determine the value of R_2.

19. (N)[28] An astronomer in some city in the Northern Hemisphere decides to keep track of the time of sunrise during a 365-day year. The recorded data is summarized in the following table:

Date	1 Jan.	29 Jan.	26 Feb.	26 Mar.	23 Apr.	21 May	18 Jun.	16 Jul.	13 Aug.	10 Sept.	8 Oct.	5 Nov.	3 Dec.
Time	8:14	7:49	7:02	6:06	5:13	4:36	4:29	4:51	5:27	6:03	6:42	7:24	8:02
j	1	29	57	85	113	141	169	197	225	253	281	309	337

We want to fit the data with suitable functions capable of reproducing the *periodicity* of the phenomenon (certainly polynomials will not work). Let $T(\tau)$ be the time of sunrise, where $\tau \equiv j\,K$ is a variable proportional to the j-th day of the year (third row in the table). We decide to fit the data with the three periodic functions $(1,\ \cos\tau,\ \text{and}\ \sin\tau)$:

$$T(\tau) = a_0 + a_1 \cos\tau + a_2 \sin\tau.$$

Determine the coefficients a_0, a_1, and a_2 that best fit the data above. Represent the data and the curve resulting from your fitting on a suitable (τ, T) graph. According to your fitting, what are the dates of the *longest* and *shortest* day of the year?

20. (A) Show that for any two invertible matrices $\mathbf{A},\ \mathbf{B} \in M_n(\mathbb{R})$ and $\lambda \in \mathbb{R}$ $(\lambda \neq 0)$: (a) $\kappa(\mathbf{A}) \geq 1$, (b) $\kappa(\lambda \mathbf{A}) = \kappa(\mathbf{A})$, and (c) $\kappa(\mathbf{A}\,\mathbf{B}) \leq \kappa(\mathbf{A})\,\kappa(\mathbf{B})$.

---

28 Courtesy of André Weideman, based on an original problem by Peter Henrici.

21. (N) Consider the linear system

$$\begin{bmatrix} 8 & 6 & 4 & 1 \\ 1 & 4 & 5 & 1 \\ 8 & 4 & 1 & 1 \\ 1 & 4 & 3 & 6 \end{bmatrix} \begin{bmatrix} x_1 \\ x_2 \\ x_3 \\ x_4 \end{bmatrix} = \begin{bmatrix} 19 \\ 11 \\ 14 \\ 14 \end{bmatrix}.$$

Assume that the elements of the 4×4 matrix are *exact*. The right-hand side is slightly modified to $[19.01\ 11.05\ 14.07\ 14.05]^{\mathrm{T}}$. Let the new solution be $\mathbf{x} + \Delta\mathbf{x}$. Compute $\|\Delta\mathbf{x}\|$ and its corresponding upper bound, i.e. find $M > 0$ such that $\|\Delta\mathbf{x}\| \leq M$.

6

Systems of Nonlinear Equations

In Chapter 1, we introduced the concept of transcendental or nonlinear equation, that is, an algebraic equation that cannot be solved by means of analytical methods. The equations seen in that chapter are *scalar* or *univariate* equations, involving just *one* unknown and occasionally including some fixed parameters. However, physical problems typically involve many variables or unknowns, as well as equations (not necessarily linear, as in Chapter 5). In order to find solutions of systems of nonlinear equations, we need to generalize the univariate root-finding methodologies addressed in the first part of the book. Henceforth, we will assume that the reader is already familiar with fundamental concepts of multivariate calculus such as the *inverse* and *implicit* function theorems, Taylor series of vector fields, or Jacobian matrices.

In this chapter, we aim to solve systems of n nonlinear equations with n unknowns of the form

$$f_1(x_1, x_2, \ldots, x_n) = 0,$$
$$f_2(x_1, x_2, \ldots, x_n) = 0,$$

$$\vdots$$

$$f_n(x_1, x_2, \ldots, x_n) = 0, \tag{6.1}$$

or

$$\mathbf{f}(\mathbf{x}) = \mathbf{0}, \tag{6.2}$$

where $\mathbf{x} = [x_1 \, x_2 \, \cdots \, x_n]^{\mathrm{T}}$ and $\mathbf{f} : \mathbb{R}^n \to \mathbb{R}^n$ is supposed to be a nonlinear map with continuous partial derivatives[1]

$$\partial_{x_j} f_i \doteq \frac{\partial f_i}{\partial x_j}, \tag{6.3}$$

---

1 Continuity of higher order partial derivatives of $\mathbf{f}$ may also be assumed, if required.

Fundamentals of Numerical Mathematics for Physicists and Engineers, First Edition. Alvaro Meseguer.
© 2020 John Wiley & Sons, Inc. Published 2020 by John Wiley & Sons, Inc.

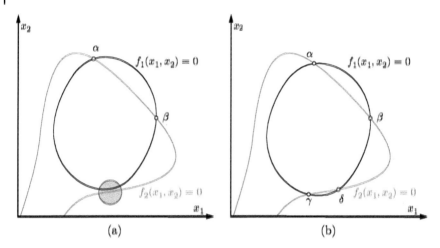

Figure 6.1 Geometrical interpretation of a system of two nonlinear equations.

for $i, j = 1, 2, \ldots, n$. In general, it is difficult to provide a geometrical description of Eq. (6.1) or (6.2). Figure 6.1a shows a particular case for $n = 2$, where the system of equations is

$$f_1(x_1, x_2) = 0,$$
$$f_2(x_1, x_2) = 0. \qquad (6.4)$$

Each equation above implicitly defines a curve in $\mathbb{R}^2$ (black and gray curves in Figure 6.1a, respectively). The sought solution (or solutions) must therefore lie on both curves simultaneously: points $\boldsymbol{\alpha} = [\alpha_1 \ \alpha_2]^{\mathrm{T}}$ and $\boldsymbol{\beta} = [\beta_1 \ \beta_2]^{\mathrm{T}}$. In this simple case we already observe that we may encounter certain problems when searching for possible solutions of (6.4). In particular, Figure 6.1a shows a shadowed encircled region where both curves approach, but never intersect. By contrast, Figure 6.1b shows how the slightest change in f_1 may lead to the emergence of two new solutions in the neighborhood of that region, namely, $\boldsymbol{\gamma} = [\gamma_1 \ \gamma_2]^{\mathrm{T}}$ and $\boldsymbol{\delta} = [\delta_1 \ \delta_2]^{\mathrm{T}}$. Nonlinear systems involving hundreds or thousands of nonlinear equations may be even more sensitive to small changes. From this simple example, we already anticipate that the numerical solution of nonlinear systems may be a challenging task.

6.1 Newton's Method for Nonlinear Systems

In order to solve $\mathbf{f}(\mathbf{x}) = \mathbf{0}$, we need to have clear indications of the existence of a solution nearby a point, namely an *initial guess* $\mathbf{x}^{(0)} \in \mathbb{R}^n$. Consider the

Taylor series of $\mathbf{f}(\mathbf{x})$ at $\mathbf{x}^{(0)}$

$$\mathbf{f}(\mathbf{x}) = \mathbf{f}(\mathbf{x}^{(0)}) + \mathbf{J}(\mathbf{x}^{(0)})\,(\mathbf{x} - \mathbf{x}^{(0)}) + O\left(\|\mathbf{x} - \mathbf{x}^{(0)}\|^2\right), \tag{6.5}$$

where $\mathbf{J}(\mathbf{x}^{(0)}) \in \mathrm{M}_n(\mathbb{R})$ is the *Jacobian* of $\mathbf{f}$ evaluated at $\mathbf{x}^{(0)}$ given by the matrix

$$\mathbf{J}(\mathbf{x}^{(0)}) \equiv \begin{bmatrix} \partial_{x_1} f_1 & \partial_{x_2} f_1 & \cdots & \partial_{x_n} f_1 \\ \partial_{x_1} f_2 & \partial_{x_2} f_2 & \cdots & \partial_{x_n} f_2 \\ \vdots & \vdots & & \vdots \\ \partial_{x_1} f_n & \partial_{x_2} f_n & \cdots & \partial_{x_n} f_n \end{bmatrix}_{\mathbf{x} \,=\, \mathbf{x}^{(0)}}, \tag{6.6}$$

If $\mathbf{x}^{(0)}$ is reasonably close to the solution of $\mathbf{f}(\mathbf{x}) = \mathbf{0}$, then the first order approximation

$$\mathbf{f}(\mathbf{x}) \approx \mathbf{f}(\mathbf{x}^{(0)}) + \mathbf{J}(\mathbf{x}^{(0)})\,(\mathbf{x} - \mathbf{x}^{(0)}) \tag{6.7}$$

should reasonably reproduce the function. As a result, instead of trying to solve $\mathbf{f}(\mathbf{x}) = \mathbf{0}$, we may try to solve its linear approximation:

$$\mathbf{f}(\mathbf{x}^{(0)}) + \mathbf{J}(\mathbf{x}^{(0)})\,(\mathbf{x} - \mathbf{x}^{(0)}) = \mathbf{0}. \tag{6.8}$$

The crucial difference between (6.2) and (6.8) is that the second is a system of *linear* equations, which, if $\mathbf{J}(\mathbf{x}^{(0)})$ is non-singular, can successfully be solved using any of the methods seen in Chapter 5. The solution of (6.8) is

$$\mathbf{x}^{(1)} = \mathbf{x}^{(0)} - \mathbf{J}^{-1}(\mathbf{x}^{(0)})\,\mathbf{f}(\mathbf{x}^{(0)}), \tag{6.9}$$

where $\mathbf{J}^{-1}(\mathbf{x}^{(0)})$ stands for the inverse matrix of $\mathbf{J}(\mathbf{x}^{(0)})$ and where we have denoted this solution as $\mathbf{x}^{(1)}$ since it is the first improvement on the initial guess $\mathbf{x}^{(0)}$. We should emphasize that (6.9) is just one way of expressing the solution of (6.8) since, in general, the inverse of $\mathbf{J}(\mathbf{x}^{(0)})$ is seldom computed in practice. We can improve this approximation by taking $\mathbf{x}^{(1)}$ as a new initial guess, and repeat the process to obtain the second iterate

$$\mathbf{x}^{(2)} = \mathbf{x}^{(1)} - \mathbf{J}^{-1}(\mathbf{x}^{(1)})\,\mathbf{f}(\mathbf{x}^{(1)}), \tag{6.10}$$

so that the general iteration is obviously

$$\mathbf{x}^{(k+1)} = \mathbf{x}^{(k)} - \mathbf{J}^{-1}(\mathbf{x}^{(k)})\,\mathbf{f}(\mathbf{x}^{(k)}), \quad k = 0, 1, 2, \ldots \tag{6.11}$$

Iterative formula (6.11) is the multidimensional version of Newton's method (1.9) seen in Part I for scalar transcendental equations. For convenience, let us rewrite (6.11) as

$$\mathbf{J}(\mathbf{x}^{(k)})(\mathbf{x}^{(k+1)} - \mathbf{x}^{(k)}) = -\mathbf{f}(\mathbf{x}^{(k)}). \tag{6.12}$$

If we define

$$\Delta\mathbf{x}^{(k)} \doteq \mathbf{x}^{(k+1)} - \mathbf{x}^{(k)}, \tag{6.13}$$

then iteration (6.12) reads

$$\mathbf{J}^{(k)}\Delta\mathbf{x}^{(k)} = -\mathbf{f}^{(k)}, \tag{6.14}$$

where, to avoid cumbersome notation, we have abbreviated $\mathbf{J}(\mathbf{x}^{(k)})$ and $\mathbf{f}(\mathbf{x}^{(k)})$ by $\mathbf{J}^{(k)}$ and $\mathbf{f}^{(k)}$, respectively.

Newton's Iteration for Nonlinear Systems

Let $\mathbf{f} : \mathbb{R}^n \to \mathbb{R}^n$ be a continuously differentiable function. Consider the system of nonlinear equations

$$\mathbf{f}(\mathbf{x}) = \mathbf{0}. \tag{6.15}$$

Starting from an initial guess $\mathbf{x}^{(0)}$, the Newton's iteration to find approximate solutions of (6.15) consists in first solving the linear system

$$\mathbf{J}^{(k)}\Delta\mathbf{x}^{(k)} = -\mathbf{f}^{(k)}, \tag{6.16}$$

for $\Delta\mathbf{x}^{(k)}$, where $\mathbf{f}^{(k)} = \mathbf{f}(\mathbf{x}^{(k)})$ and $\mathbf{J}^{(k)}$ is the Jacobian of $\mathbf{f}$ evaluated at $\mathbf{x}^{(k)}$ with matrix elements

$$\left[\mathbf{J}^{(k)}\right]_{ij} = \left(\frac{\partial f_i}{\partial x_j}\right)_{\mathbf{x}\,=\,\mathbf{x}^{(k)}}. \tag{6.17}$$

Once (6.16) is solved, the next iterate is

$$\mathbf{x}^{(k+1)} = \mathbf{x}^{(k)} + \Delta\mathbf{x}^{(k)}. \tag{6.18}$$

For low-dimensional nonlinear systems (that is, $n = 2$ or $n = 3$, for example), we can calculate by hand the matrix elements of the Jacobian matrix (6.17). Even for $n = 3$, the required nine partial differentiations (along with their eventual codification in a program) may be a tedious and error-prone task. The common practice is to *approximate* the Jacobian elements by means of *partial finite differences*, that is,

$$\frac{\partial f_i}{\partial x_j} \approx \frac{f_i(x_1, \ldots, x_j + h, \ldots, x_n) - f_i(x_1, \ldots, x_j, \ldots, x_n)}{h}. \tag{6.19}$$

This is simply the forward finite difference FD1 formula (3.3) seen in Part I, but generalized to a multivariate function, where all variables but x_j remain *frozen*. As in the univariate case, the finite increment h appearing in (6.19) needs to be small in order to provide a reasonably accurate approximation of the partial derivative. However, we must contemplate the fact that not all variables will in general have the same order of magnitude. For example, while x_1 can be related to the distance between two neighboring satellites, which typically could be (in meters) of order 10^5, x_j may represent a latitude angle (in radians) between 0 and π. Therefore, while a unitary increment of $h = 1.0$ m in x_1 would in general provide a good approximation of the partial derivative with respect to x_1, an increment of $h = 1.0$ rad in x_j to approximate ∂_{x_j} would be unacceptable. For this reason, it is common practice to use increments $\{h_1, h_2, \ldots, h_n\}$, which are small according to the order of magnitude of each one of the variables $\{x_1, x_2, \ldots, x_n\}$, respectively. A simple choice for the jth increment is

$$h_j = \begin{cases} \varepsilon^{1/2} |x_j| & (x_j \neq 0), \\ \varepsilon^{1/2} & (x_j = 0), \end{cases} \tag{6.20}$$

for $j = 1, 2, \ldots, n$, where ε is the machine epsilon. More accurate formulas for the increments can be found in the references at the end of this book. Code 19 is a simple implementation of the partial forward finite difference approximation (6.19) of the Jacobian of an arbitrary function $\mathbf{f} : \mathbb{R}^m \to \mathbb{R}^n$, using the *same* increment $h = \varepsilon^{1/2}$ for the m independent variables.

```
% Code 19: Computation of the Jacobian J
% Input: F(x):R^m ---> R^n
%     x: (m x 1)-vector ; F: (n x 1)-vector
% Output: DF(x) (n x m) Jacobian matrix at x
function DF = jac(F,x)
f1 = feval(F,x); n = length(f1); m = length(x);
DF = zeros(n,m); H = sqrt(eps)*eye(m);
for j = 1:m
    f2 = feval(F,x+H(:,j)); DF(:,j) = (f2 - f1)/H(j,j);
end
```

Every Newton step implicitly involves three main tasks. On the one hand, we need the evaluation of the function $\mathbf{f}$, as well as the approximation of its Jacobian $\mathbf{J}$, at $\mathbf{x}^{(k)}$. On the other hand, we need to solve the system of n linear equations (6.16). Assuming that the computational cost of evaluating $\mathbf{f}$ is, at least, of order $O(n)$, the approximation of its Jacobian is accordingly of order $O(n^2)$. However, the main cost of every Newton step may actually come from the solution of (6.16), requiring $O(n^3)$ operations using LU or QR factorizations. Code 20 is a simple implementation of Newton's method for systems of nonlinear equations that requires function `jac` to approximate the

Jacobian, as well as functions `pplu` and `plusolve` to solve the linear system (6.16) at every Newton step:

```
% Code 20: Newton's method for n-dimensional systems
% Input: x0 - initial guess (column vector)
%        tol - tolerance so that ||x_{k+1} - x_{k} || < tol
%        itmax - max number of iterations
%        fun - function's name
% Output: XK - iterated
% resd: resulting residuals of iteration: ||F_k||
% it:   number of required iterations to satisfy tolerance
function [XK,resd,it] = newtonn(x0,tol,itmax,fun)
xk = [x0]; resd = [norm(feval(fun,xk))]; XK = [x0]; it = 1;
tolk = 1.0; n = length(x0);
while it < itmax & tolk > tol
  Fk = feval(fun,xk);
  DFk = jac(fun,xk); [P,L,U] = pplu(DFk);
  dxk = plusolve(L,U,P,-Fk);
  xk = xk + dxk; XK = [XK xk]; resd = [resd norm(Fk)];
  tolk = norm(XK(:,end)-XK(:,end-1)); it = it + 1;
end
```

As an illustration, let us solve the system of nonlinear equations

$$x_1^2 + x_2^2 - 1 \equiv 0,$$

$$x_2 - \left(x_1 - \frac{1}{2}\right)^2 \equiv 0. \tag{6.21}$$

Geometrically, the solutions to these equations are the coordinates (x_1, x_2) corresponding to any of the possible intersections of the parabola $x_2 \equiv (x_1 - 1/2)^2$ with the unit circle centered at the origin $x_2 \equiv \pm(1 - x_1^2)^{1/2}$: points α and β depicted in Figure 6.2a. In this example, we initiate the Newton's iteration from two different initial guesses. Table 6.1 shows the resulting iterates when starting from $\mathbf{x}^{(0)} \equiv [-1.2 \ 0.25]^T$ (the first three being depicted as white bullets in Figure 6.2a). The last column of Table 6.1 also shows the absolute error $\varepsilon_k \equiv \|\mathbf{x}^{(k)} - \alpha\|$, clearly exhibiting quadratic convergence from $k \equiv 3$ onward (ε_k nearly squares from one iterate to the next). The reader may also confirm that starting the iteration from $\mathbf{x}^{(0)} \equiv [0.75 \ 1.1]^T$ also leads to quadratic convergence to the other solution β (indexed gray bullets in Figure 6.2a).

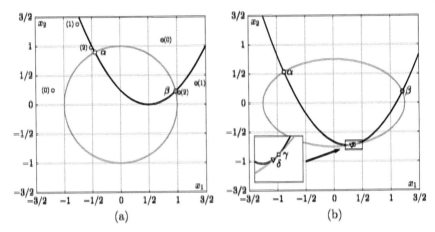

Figure 6.2 (a) Solutions of system (6.21) showing the first three iterates of Newton's method starting from $\mathbf{x}^{(0)} \equiv [-1.2 \; 0.25]^{\mathrm{T}}$ (small indexed white bullets) and the resulting converged solution $\alpha \approx [-0.446 \; 0.895]^{\mathrm{T}}$ (large white bullet). Starting from $\mathbf{x}^{(0)} \equiv [0.75 \; 1.1]^{\mathrm{T}}$ (small gray bullets) the iteration converges to the other solution $\beta \approx [0.974 \; 0.225]^{\mathrm{T}}$ (large gray circle). (b) Roots of system (6.22). The inset figure zooms in the region containing the two neighboring solutions γ and δ depicted with a square and a triangle, respectively.

Table 6.1 Iterates of Newton's methods applied to system (6.21).

k	$x_1^{(k)}$	$x_2^{(k)}$	$\varepsilon_k \equiv \|\mathbf{x}^{(k)} - \alpha\|$
0	-1.2	0.25	9.9×10^{-1}
1	$-0.755\ 487\ 803\ 693\ 313$	$1.378\ 658\ 538\ 739\ 43$	5.7×10^{-1}
2	$-0.516\ 436\ 894\ 022\ 781$	$0.975\ 998\ 611\ 917\ 637$	1.1×10^{-1}
3	$-0.450\ 285\ 040\ 746\ 61$	$0.898\ 665\ 592\ 102\ 37$	5.6×10^{-3}
4	$-0.446\ 063\ 656\ 676\ 88$	$0.895\ 018\ 622\ 492\ 74$	1.8×10^{-5}
5	$-0.446\ 048\ 929\ 565\ 43$	$0.895\ 008\ 576\ 915\ 01$	1.9×10^{-10}
6	$-0.446\ 048\ 929\ 400\ 408$	$0.895\ 008\ 576\ 819\ 658$	1.1×10^{-16}

As a second example, let us try to find the solutions of a similar system given by the equations

$$\frac{x_1^2}{25} + \frac{x_2^2}{9} - \frac{1}{16} \equiv 0,$$

$$x_2 - \left(x_1 - \frac{1}{4}\right)^2 + 0.73 \equiv 0. \tag{6.22}$$

Table 6.2 Solutions and conditioning of system (6.22).

Solution	x_1	x_2	$\kappa(\mathbf{J})$
α	$-0.874\ 993\ 704\ 304\ 283$	$0.535\ 610\ 834\ 724\ 273$	17.9
β	$1.209\ 197\ 601\ 168\ 76$	$0.190\ 060\ 038\ 087\ 907$	26.4
γ	0.35	-0.72	266.6
δ	$0.315\ 796\ 103\ 135\ 521$	$-0.725\ 670\ 872\ 812\ 18$	258.2

In this case, we have slightly deformed the unit circle into an ellipse of semi-major and semi-minor axes 5/4 and 3/4, respectively, and we have shifted the parabola 0.25 and 0.73 length units leftwards and downwards, respectively. In this second example, the solutions of (6.22) are the coordinates of the four possible intersections, namely $\{\alpha, \beta, \gamma, \delta\}$, between the ellipse and the parabola depicted in Figure 6.2b, where the two new solutions γ and δ (denoted by a triangle and a square, respectively) are so close to each other that we have included an inset to better visualize their location. Table 6.2 outlines the coordinates of these four roots. For reasons that will be justified later, the last column of Table 6.2 also includes the condition number $\kappa(\mathbf{J})$ of the Jacobian matrix evaluated at each one of the four solutions.

As an exercise, the reader may confirm that starting Newton's iteration from initial guesses close to α or β usually leads to a successful quadratic convergence to any of these two solutions, so they can be approximated to full accuracy in just six or seven iterations. However, starting from an initial guess close to roots γ or δ leads to remarkably worst convergence rates or even divergent iterates. For example, Table 6.3 shows the resulting iterates when Newton's method is started from $\mathbf{x}^{(0)} = [0.2 - 0.2]^{\mathrm{T}}$. In this particular case, the iteration converges to γ, but not quadratically, as can be clearly concluded from the very slow decay rate shown by the absolute errors $\varepsilon_k = \|\mathbf{x}^{(k)} - \gamma\|$ of the last column reported in Table 6.3. We could have anticipated the loss of quadratic convergence by looking at the condition number of the Jacobian of system (6.22) at root γ included in Table 6.2, which is roughly 1 order of magnitude larger than its corresponding value at α or β. This is essentially the same phenomenon already observed in Part I when using Newton's univariate method to search double roots. The reader may check that replacing 0.73 by 0.7295 (less than 0.1%) in the second equation of (6.22) results in dramatic changes, since both roots γ and δ disappear. In this case, initiating Newton's method from $\mathbf{x}^{(0)} = [0.2 - 0.2]^{\mathrm{T}}$ leads to endless iterations that never converge, leading to a wandering sequence $\mathbf{x}^{(k)}$ searching for a *ghost* solution. Decreasing 0.73 by a suitable accurate amount makes solutions γ and δ merge, resulting in one

Table 6.3 Iterates of Newton's methods applied to system (6.22).

k	$x_1^{(k)}$	$x_2^{(k)}$	$\varepsilon_k = \|\mathbf{x}^{(k)} - \boldsymbol{\gamma}\|$
0	0.2	−0.2	5.4×10^{-1}
1	1.814 673 909 628 69	−0.888 967 432 829 059	1.5
2	1.057 811 390 130 83	−0.650 281 664 090 034	7.1×10^{-1}
3	0.698 133 986 215 266	−0.658 543 774 314 817	3.5×10^{-1}
4	0.515 562 684 077 448	−0.692 808 742 361 594	1.7×10^{-1}
5	0.425 052 035 528 963	−0.707 548 963 135 772	7.6×10^{-2}
6	0.380 561 335 226 263	−0.714 933 160 432 333	3.1×10^{-2}
7	0.359 797 852 499 263	−0.718 375 553 789 262	1.0×10^{-2}
8	0.351 784 361 550 707	−0.719 704 159 750 727	1.8×10^{-3}
9	0.350 084 292 190 131	−0.719 986 024 682 207	8.5×10^{-5}
10	0.350 000 206 773 81	−0.719 999 965 717 362	2.1×10^{-7}
11	0.350 000 000 001 471	−0.719 999 999 999 753	1.5×10^{-12}
12	0.350 000 000 000 003	−0.719 999 999 999 999	3.6×10^{-15}

multiple root where the Jacobian becomes singular and its condition number diverges to infinity.

In general, multivariate Newton's method converges quadratically if certain conditions, summarized below, are satisfied:

Quadratic Convergence (Newton–Kantorovich Conditions)

Let $\mathbf{f} : \mathbb{R}^n \to \mathbb{R}^n$ be a continuously differentiable function, and let $\boldsymbol{\alpha} \in \mathbb{R}^n$ be a solution of the equation

$$\mathbf{f}(\mathbf{x}) = \mathbf{0},$$

such that $\mathbf{J}^{-1}(\boldsymbol{\alpha})$ is non-singular. If for any two points $\mathbf{x}$ and $\mathbf{y}$ nearby $\boldsymbol{\alpha}$ there are two positive constants γ and β such that

$$C_1 : \|\mathbf{J}(\mathbf{x}) - \mathbf{J}(\mathbf{y})\| \leq \gamma \|\mathbf{x} - \mathbf{y}\| \quad \text{and} \quad C_2 : \|\mathbf{J}^{-1}(\mathbf{x})\| \leq \beta, \qquad (6.23)$$

then the Newton's iteration sequence $\mathbf{x}^{(k)}$ in (6.16–6.18) satisfies

$$\lim_{k \to \infty} \mathbf{x}^{(k)} = \boldsymbol{\alpha} \quad \text{with} \quad \|\mathbf{x}^{(k+1)} - \boldsymbol{\alpha}\| \leq \gamma \beta \|\mathbf{x}^{(k)} - \boldsymbol{\alpha}\|^2. \qquad (6.24)$$

These two last convergence properties are valid only *locally*, assuming that the initial guess $\mathbf{x}^{(0)}$, from which Newton's iteration is initiated, is very close to the root $\boldsymbol{\alpha}$. Condition C_1 in (6.23) is usually known as *Lipschitz continuity* and it is satisfied by continuously differentiable fields. Condition C_2, however, is more subtle, since it imposes that the norm of $\mathbf{J}^{-1}(\mathbf{x})$ – and consequently the condition number of the Jacobian $\kappa(\mathbf{J}) \equiv \|\mathbf{J}(\mathbf{x})\| \, \|\mathbf{J}^{-1}(\mathbf{x})\|$ – must be bounded by a suitable constant β that also appears on the right-hand side of the inequality of (6.24) for the quadratic convergence to be satisfied.

Whenever quadratic convergence is at stake because of a clearly diagnosed ill-conditioning of the system of equations, computing the Jacobian at every Newton step may not be completely justified. In this case, it is normal practice to update and factorize the Jacobian less frequently. To some extent, this practice is similar to the univariate chord method seen in Part I, where we used the same estimation of the function's slope for the whole iteration. As a result, the lack of quadratic convergence can, in turn, be an advantage in terms of computational cost. Since we just need to approximate and factorize the Jacobian from to time to time, we can advance a few steps by just updating the right-hand side $\mathbf{f}^{(k)}$ at a $O(n^2)$ cost (FS and BS algorithms) per step.

In general, systems involving a large number of nonlinear equations and unknowns lead to very complex vector fields $\mathbf{f}$ with Jacobian matrices $\mathbf{J}$ containing elements of quite diverse orders of magnitude. Trying to anticipate how Newton's method is going to perform at every step is almost impossible. Typically, if $\mathbf{f}$ is continuously differentiable, Newton's linear system (6.16) tends to provide the right direction $\Delta\mathbf{x}^{(k)}$ but not always the right modulus. In practice, advanced multivariate Newton solvers incorporate what are called *trust region* algorithms that explore around the vicinity of each iterate $\mathbf{x}^{(k)}$ for local minimizers of $\|\mathbf{f}\|$, usually improving the convergence. The technical details of these algorithms are out of the scope of this book and we refer the interested reader to look at the recommended bibliography at the end of this chapter, in the Complementary Reading section.

Practical 6.1: Newton's Method in $\mathbb{R}^n$

The *van der Waals*[2] equation in reduced thermodynamic coordinates is

$$p(v,T) = \frac{8T}{3v-1} - \frac{3}{v^2}, \quad \left(v > \frac{1}{3}, \ T > \frac{27}{32} \right).$$

Mathematically, the liquid–vapor coexistence curve shown in gray in the figure on the right is the boundary that satisfies what is usually called *Maxwell construction*[3]: the shadowed areas I and II enclosed by the isotherm $T < T_e$ and ordinate $p_\ell = p_g$ shown in the figure on the left must be equal.

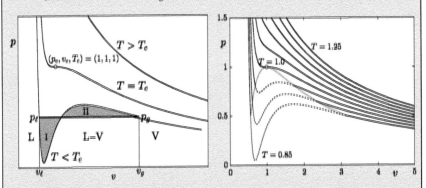

For a fixed value of T, the equal areas can be imposed by integration, reducing the problem to find v_ℓ and v_g satisfying the equations

$$\log\left(\frac{3v_g - 1}{3v_\ell - 1} \right) + \frac{9}{4T}\left(\frac{1}{v_g} - \frac{1}{v_\ell} \right) - \frac{1}{3v_g - 1} + \frac{1}{3v_\ell - 1} = 0$$

$$\frac{8T}{3}\left(\frac{1}{3v_g - 1} - \frac{1}{3v_\ell - 1} \right) - \frac{1}{v_g^2} + \frac{1}{v_\ell^2} = 0.$$

(a) For the temperatures $T = \{0.99, 0.98, 0.97, \ldots, 0.85\}$, use Newton's method to compute the coordinates $v_\ell(T)$ and $v_g(T)$, along with the corresponding pressure. When changing T, use the previous result as initial guess. For $T = 0.99$, use $v_\ell^{(0)} = 0.8$ and $v_g^{(0)} = 1.2$.

(b) Difficult and optional: repeat (a) for the Dieterici's equation:

$$p(v,T) = \frac{T}{2v-1} \exp\left(2 - \frac{2}{vT} \right), \quad \left(v > \frac{1}{2} \right).$$

---

2 J. D. van der Waals (1837–1923), Dutch physicist and thermodynamicist.
3 J. C. Maxwell (1831–1879), Scottish mathematical physicist better known for his theory of electromagnetism.

6.2 Nonlinear Systems with Parameters

In Practical 6.1 we have solved the system of nonlinear equations

$$f_1(x_1, x_2, \alpha) = \log\left(\frac{3x_2 - 1}{3x_1 - 1}\right) + \frac{9}{4\alpha}\left(\frac{1}{x_2} - \frac{1}{x_1}\right) - \frac{1}{3x_2 - 1} + \frac{1}{3x_1 - 1} = 0$$

$$f_2(x_1, x_2, \alpha) = \frac{8\alpha}{3}\left(\frac{1}{3x_2 - 1} - \frac{1}{3x_1 - 1}\right) - \frac{1}{x_2^2} + \frac{1}{x_1^2} = 0, \tag{6.25}$$

where we have replaced v_ℓ, v_g, and T by x_1, x_2, and α, respectively. In particular, we had to compute x_1 and x_2 for different values of α, i.e. for different temperatures. In system (6.25), α plays the role of a parameter, implicitly changing the location of the solution when this parameter is changed. In other words, the system of nonlinear equations (6.25) potentially defines $x_1(\alpha)$ and $x_2(\alpha)$ as implicit functions of this parameter. In that practical it was recommended to start Newton's method from the last solution obtained, that is, to use the solution (x_1, x_2) obtained for $\alpha = 0.99$ as the initial guess for $\alpha = 0.98$, for example. The reason is simple: if α changes by a small amount, we naturally expect the new solution to be close to the previous one. This strategy, usually known as *natural continuation*, helps Newton's method to converge more quickly to the new solution and to continue it as a function of the parameter.

In general, nonlinear systems of algebraic equations appearing in physics or engineering frequently include parameters. Exploration of the solutions of these nonlinear systems for different values of their parameters is common practice in nonlinear science. If the parameters change by small amounts, we expect accordingly small changes in the solutions, and we must take advantage of that when starting Newton's method. Occasionally, however, the variation of one parameter may lead to serious changes in the structure of the solutions. For example, in system (6.25) there are no physical solutions for $\alpha > 1$, that is, above the isotherm $T = 1$ in the original formulation of Practical 6.1. This is clearly shown in Figure 6.3, where we have depicted the solutions of (6.25) $x_1(\alpha)$ and $x_2(\alpha)$. Anticipating the existence (or absence) of solutions of nonlinear systems when changing parameters requires introducing new mathematical conditions. These new conditions will also pave the way to formulate parameter-dependent Newton's methods to continue solutions in a reliable way.

We start by formulating the problem mathematically. In what follows, we will consider systems of n nonlinear equations involving n variables and just

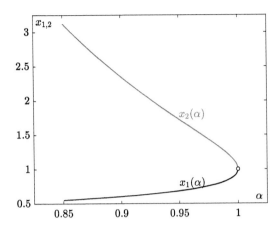

Figure 6.3 Parameter-dependent solution branches of (6.25).

one parameter α:

$$f_1(x_1, x_2, \ldots, x_n, \alpha) = 0,$$
$$f_2(x_1, x_2, \ldots, x_n, \alpha) = 0,$$
$$\vdots$$
$$f_n(x_1, x_2, \ldots, x_n, \alpha) = 0, \tag{6.26}$$

or simply

$$\mathbf{f}(\mathbf{x}, \alpha) = \mathbf{0}, \tag{6.27}$$

where $\mathbf{x} = [x_1 \ x_2 \ \cdots \ x_n]^{\mathrm{T}}$, $\mathbf{f} = [f_1 \ f_2 \ \cdots \ f_n]^{\mathrm{T}}$, $\alpha \in \mathbb{R}$, and $\mathbf{f} : \mathbb{R}^{n+1} \to \mathbb{R}^n$. Suppose $\mathbf{x} = \mathbf{x}^{(0)} = [x_1^{(0)} \ x_2^{(0)} \ \cdots \ x_n^{(0)}]^{\mathrm{T}}$ is a solution of (6.27), for $\alpha = \alpha^{(0)}$, that is

$$\mathbf{f}(\mathbf{x}^{(0)}, \alpha^{(0)}) = \mathbf{0}. \tag{6.28}$$

According to the *Implicit Function Theorem* (IMFT), system (6.26) uniquely defines the n functions

$$x_1 = g_1(\alpha),$$
$$x_2 = g_2(\alpha),$$
$$\vdots$$
$$x_n = g_n(\alpha), \tag{6.29}$$

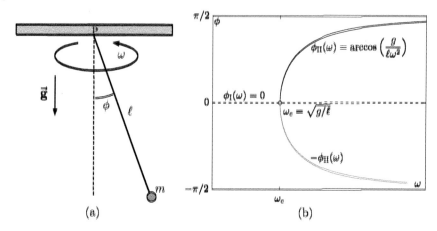

Figure 6.4 (a) Rotating pendulum. (b) Trivial solution branch $\phi_I \equiv 0$ and new solution branches $\phi_{II} = \pm\arccos(g/\ell\omega^2)$.

satisfying $g_j(\alpha^{(0)}) = x_j^{(0)}$, for $j = 1, 2, \ldots, n$, if the Jacobian matrix

$$\mathbf{J}(\mathbf{x}^{(0)}, \alpha^{(0)}) \equiv \begin{bmatrix} \partial_{x_1} f_1 & \partial_{x_2} f_1 & \cdots & \partial_{x_n} f_1 \\ \partial_{x_1} f_2 & \partial_{x_2} f_2 & \cdots & \partial_{x_n} f_2 \\ \vdots & \vdots & & \vdots \\ \partial_{x_1} f_n & \partial_{x_2} f_n & \cdots & \partial_{x_n} f_n \end{bmatrix}_{(\mathbf{x}^{(0)}, \alpha^{(0)})} \tag{6.30}$$

is non-singular, i.e. if $\text{rank}\{\mathbf{J}(\mathbf{x}^{(0)}, \alpha^{(0)})\} \equiv n$. That is, in a neighborhood of $(\mathbf{x}^{(0)}, \alpha^{(0)})$, there exists a unique map $\mathbf{x} = \mathbf{g}(\alpha)$, where $\mathbf{g} = [g_1 \; g_2 \; \cdots \; g_n]^T$, satisfying $\mathbf{g}(\alpha^{(0)}) = \mathbf{x}^{(0)}$. Details regarding the regularity properties of the implicitly defined function $\mathbf{g}$ can be found in any text on classical analysis.

We illustrate some of the concepts previously seen by a very simple problem that can be found in any undergraduate physics course. Consider a pendulum consisting of a small marble of mass m attached to a massless rigid rod of length ℓ. The rod-marble system is suspended under the effects of gravity $g \equiv \|\mathbf{g}\|$ from a fixed point P and is forced to rotate at a constant angular speed $\omega > 0$ around a vertical axis passing through P, as shown in Figure 6.4a. Our purpose is to find the equilibrium angle (or angles) ϕ such that the pendulum remains stationary in a frame rotating with angular speed ω. The equilibrium conditions are reduced to the system of nonlinear equations involving T (the rod's tension) and ϕ (the equilibrium angle):

$$\left.\begin{array}{r} T\cos\phi = mg, \\ T\sin\phi = m\ell\omega^2\sin\phi \end{array}\right\} . \tag{6.31}$$

Therefore, the original system (6.31) is formally written as

$$f_1(T, \phi, \omega) \equiv T \cos \phi - mg \equiv 0,$$
$$f_2(T, \phi, \omega) \equiv \sin \phi (T - m\ell\omega^2) \equiv 0. \tag{6.32}$$

System (6.32) can be understood as a particular case of (6.26), where the parameter α is the external rotating forcing ω, and the variables x_1 and x_2 are the tension T and angle ϕ, respectively. The question is whether system (6.32) defines T and ϕ as unique functions of ω. In this particular case, this question can be answered analytically. A simple calculation shows that for $\omega < \sqrt{g/\ell}$, the only available equilibrium solution (I) is

$$[\phi_\mathrm{I} \ T_\mathrm{I}] = [0 \ mg], \tag{6.33}$$

that is, the pendulum rotates vertically. However, for angular speeds beyond this critical value $\omega_e \equiv \sqrt{g/\ell}$, a new solution

$$[\phi_\mathrm{II} \ T_\mathrm{II}] = [m\ell\omega^2 \ \arccos(g/\ell\omega^2)], \tag{6.34}$$

is born, along with its symmetric counterpart, $-\phi_\mathrm{II}(\omega)$, due to the reflection symmetry of the problem. These new solutions, along with the trivial equilibrium ϕ_I, have been depicted in Figure 6.4b. For $\omega > \omega_e$, the uniqueness of solutions is lost and the two new branches $\pm\phi_\mathrm{II}$ coexist with the trivial solution $\phi_\mathrm{I} \equiv 0$. The loss of uniqueness could have been anticipated just by checking the full-rank condition of the Jacobian of system (6.32)

$$\mathbf{J} \equiv \begin{bmatrix} \partial_T f_1 & \partial_\phi f_1 \\ \partial_T f_2 & \partial_\phi f_2 \end{bmatrix} \equiv \begin{bmatrix} \cos \phi & -T \sin \phi \\ \sin \phi & (T - m\ell\omega^2) \cos \phi \end{bmatrix}, \tag{6.35}$$

with determinant

$$\det\{\mathbf{J}(T, \phi, \omega)\} \equiv T - m\ell\omega^2 \cos^2 \phi,$$

adopting the values

$$\det \mathbf{J} \equiv \begin{cases} m(g - \ell\omega^2), & \text{along branch } (I) \text{ and } \omega \geq 0, \\ m\ell\omega^2 \left(1 - \dfrac{g^2}{\ell^2\omega^4}\right), & \text{along branch } (II) \text{ and } \omega \geq \omega_e \equiv \sqrt{g/\ell}. \end{cases} \tag{6.36}$$

In both cases, the determinants above become zero for $\omega \equiv \omega_e$, and the uniqueness of solutions in a neighborhood of that critical value is lost, as expected. For $\omega > \omega_e$ both determinants never cancel and therefore all branches are locally unique so no new equilibrium solutions are expected to emerge from them.

6.3   Numerical Continuation (Homotopy)

From what has been seen in Section 6.2, it is clear that we need suitable numerical strategies to compute and monitor branches of solutions, as well as to forecast potential pathologies along them such as *branching* or the presence of *turning points* or *folds*.

In this section, we will only address the most fundamental techniques to track or *continuate* branches of solutions of parameter-dependent nonlinear systems of the form given in (6.26). These techniques must address, among other issues, the problem of continuating the solutions in the presence of turning points. Figure 6.5a shows the process of natural continuation that was performed in Practical 6.1, where we plot any of the n variables of the problem, namely x_j $(j = 1, 2, \ldots, n)$, as a function of the parameter α. Starting from a point along the curve (white bullet ① in Figure 6.5a), we increase the parameter by a small amount $\Delta\alpha$ and initiate Newton's iteration from that previous solution (gray bullet ②). The iteration will converge to the real solution for the new value of the parameter (white bullet ③) provided the initial guess is close to it. However, if there is a turning point, this strategy will eventually fail (gray bullet ④), no matter how small $\Delta\alpha$ is.

In order to deal with turning or fold points, it is sometimes useful to think of the problem as involving $n + 1$ independent degrees of freedom, instead of n variables and one parameter, that is, we define the vector

$$
\mathbf{y} \doteq \begin{bmatrix} y_1 \\ y_2 \\ \vdots \\ y_n \\ y_{n+1} \end{bmatrix} = \begin{bmatrix} x_1 \\ x_2 \\ \vdots \\ x_n \\ \alpha \end{bmatrix}. \tag{6.37}
$$

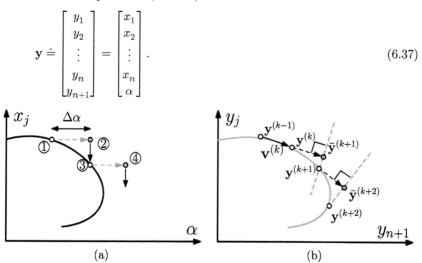

(a)   (b)

Figure 6.5   Continuation methods. (a) Natural continuation fails in the presence of turning points, even using very small increments of $\Delta\alpha$. (b) *Pseudo-arclength* solves the problem by searching in the orthogonal secant direction.

Figure 6.5b shows the same curve as in Figure 6.5a, but in the new system of variables. One way to start the continuation is to assume that we have two neighboring points of the curve, namely $\mathbf{y}^{(k-1)}$ and $\mathbf{y}^{(k)}$, previously obtained from, for example, natural continuation. From these two points, we can generate a *secant* vector to the curve

$$\mathbf{v}^{(k)} = \mathbf{y}^{(k)} - \mathbf{y}^{(k-1)}. \tag{6.38}$$

A reasonable estimate or *predictor* $\widetilde{\mathbf{y}}^{(k+1)}$ of the next point along the secant direction of the curve can be obtained by a simple linear extrapolation

$$\widetilde{\mathbf{y}}^{(k+1)} = \mathbf{y}^{(k)} + s\mathbf{v}^{(k)}, \tag{6.39}$$

where s is a quantity that nearly parameterizes the local *arclength* of the curve in the vicinity of $\mathbf{y}^{(k)}$. For $s = 1$, for example, $\widetilde{\mathbf{y}}^{(k+1)}$ and $\mathbf{y}^{(k-1)}$ are at the same distance from $\mathbf{y}^{(k)}$. For this reason, this prediction method is usually known as *pseudo-arclength* continuation. That predictor point is *not*, however, a point belonging to the curve. In order to *correct* its deviation, we look for a point $\mathbf{y} \in \mathbb{R}^{n+1}$ satisfying

$$\begin{aligned} f_1(y_1, y_2, \ldots, y_n, y_{n+1}) &= 0, \\ f_2(y_1, y_2, \ldots, y_n, y_{n+1}) &= 0, \\ &\vdots \\ f_n(y_1, y_2, \ldots, y_n, y_{n+1}) &= 0 \\ \langle \mathbf{v}^{(k)}, \mathbf{y} - \widetilde{\mathbf{y}}^{(k+1)} \rangle &= 0, \end{aligned} \tag{6.40}$$

or

$$\begin{aligned} \mathbf{f}(\mathbf{y}) &= \mathbf{0}, \\ \langle \mathbf{v}^{(k)}, \mathbf{y} - \widetilde{\mathbf{y}}^{(k+1)} \rangle &= 0, \end{aligned} \tag{6.41}$$

for short. The first of the previous two equations is simply the condition of belonging to the curve. The second condition imposes that the sought point $\mathbf{y}$ also belongs to the hyperplane orthogonal to the secant vector $\mathbf{v}^{(k)}$ passing through the predictor point $\widetilde{\mathbf{y}}^{(k+1)}$ (depicted as gray dashed straight lines in Figure 6.5b). System (6.41) has now $n + 1$ unknowns, but it also involves $n + 1$ equations. This system is solved, as usual, by Newton's method, hopefully converging to the desired next point $\mathbf{y}^{(k+1)}$ on the curve (to some prescribed accuracy). To avoid cumbersome notation, we will denote $\mathbf{v}^{(k)}$ and $\widetilde{\mathbf{y}}^{(k+1)}$ simply by the $(n + 1)$-dimensional vectors $\mathbf{v} = [v_1 \ v_2 \ \cdots \ v_{n+1}]^\mathrm{T}$ and $\widetilde{\mathbf{y}} = [\widetilde{y}_1 \ \widetilde{y}_2 \ \cdots \ \widetilde{y}_{n+1}]^\mathrm{T}$, respectively, bearing in mind that these two vectors will be updated later on, in the search for the next continuation point along the curve. Starting from the initial guess

$$\mathbf{y}^{[0]} \equiv [\tilde{y}_1 \ \tilde{y}_2 \ \cdots \ \tilde{y}_{n+1}]^{\mathrm{T}}, \tag{6.42}$$

the Newton's iteration to solve system (6.41) reads

$$\begin{bmatrix} \partial_{y_1} f_1 & \partial_{y_2} f_1 & \cdots & \partial_{y_{n+1}} f_1 \\ \partial_{y_1} f_2 & \partial_{y_2} f_2 & \cdots & \partial_{y_{n+1}} f_2 \\ \vdots & \vdots & & \vdots \\ \partial_{y_1} f_n & \partial_{y_2} f_n & \cdots & \partial_{y_{n+1}} f_n \\ v_1 & v_2 & \cdots & v_{n+1} \end{bmatrix}_{\mathbf{y}^{[j]}} \begin{bmatrix} \Delta y_1^{[j]} \\ \Delta y_2^{[j]} \\ \vdots \\ \Delta y_n^{[j]} \\ \Delta y_{n+1}^{[j]} \end{bmatrix} = - \begin{bmatrix} f_1(\mathbf{y}^{[j]}) \\ f_2(\mathbf{y}^{[j]}) \\ \vdots \\ f_n(\mathbf{y}^{[j]}) \\ \langle \mathbf{v}, \mathbf{y}^{[j]} - \tilde{\mathbf{y}} \rangle \end{bmatrix}, \tag{6.43}$$

for $j \equiv 0, 1, 2, \ldots$. The next iterate is given by

$$\mathbf{y}^{[j+1]} \equiv \mathbf{y}^{[j]} + \Delta\mathbf{y}^{[j]}, \tag{6.44}$$

where $\Delta\mathbf{y}^{[j]}$ is the vector resulting from solving (6.43):

$$\Delta\mathbf{y}^{[j]} \equiv [\Delta y_1^{[j]} \ \Delta y_2^{[j]} \ \cdots \ \Delta y_n^{[j]} \ \Delta y_{n+1}^{[j]}]^{\mathrm{T}}. \tag{6.45}$$

In this particular case, we have indexed the Newton's iterates using square brackets [ ] to avoid confusion with the indexing of the points $\mathbf{y}^{(k)}$ along the curve. Notice that the first n rows of the matrix appearing on the left-hand side of (6.43) constitute the $n \times (n+1)$ Jacobian of $\mathbf{f}(y_1, \cdots, y_{n+1})$ evaluated at $\mathbf{y}^{[j]}$, whereas its last row is simply the gradient of the orthogonal plane equation, that is, the vector $\mathbf{v}$, which remains constant along the whole iteration.

In general, we cannot guarantee the convergence of Newton's iteration (6.43). A necessary, but not sufficient, condition for convergence is for system (6.41) to admit a solution. For example, Figure 6.5b also shows the next point of the curve $\mathbf{y}^{(k+2)}$, obtained from the predictor $\tilde{\mathbf{y}}^{(k+2)}$ based on the secant vector $\mathbf{v}^{(k+1)} \equiv \mathbf{y}^{(k+1)} - \mathbf{y}^{(k)}$ (not depicted). The reader may have noticed that the predictor $\tilde{\mathbf{y}}^{(k+2)}$ leads to an orthogonal plane that is nearly tangent to the curve. We have deliberately illustrated that phenomenon in order to emphasize that an unsuitable value of s may lead to an ill-conditioning of system (6.41) and to a poor convergence of Newton's iteration. Therefore, if the curve is convoluted, s must be chosen carefully; otherwise, the orthogonal plane and the curve may never intersect, as it almost occurs by a nose in the last point $\mathbf{y}^{(k+2)}$ of Figure 6.5b.

```
% Code 21: secant continuation step
% Input: y0 and y1 (two close column vectors)
% s: pseudo-arclength parameter
% tol - Newton's tolerance: ||y_[k+1] - y_[k] || < tol
% itmax - max number of iterations
% fun - function's name: f(y_1,y_2,...,y_n,y_n+1)
% Output: y - next point along curve f = 0
%          y belongs to plane orth. to y1-y0
%          passing through secant predictor y1 + s(y1-y0)
%          iconv (0 if y is convergenced to desired tol.)
function [y,iconv] = contstep(fun,y0,y1,s,tol,itmax)
tolk = 1.0; it = 0; n = length(y0)-1;
v = y1-y0; yp = y1+s*v; xk = yp;
while tolk > tol & it < itmax
 Fk = [feval(fun,xk); v'*(xk-yp)]; DFk = [jac(fun,xk); v'];
 [P,L,U] = pplu(DFk); dxk = plusolve(L,U,P,-Fk);
 xk = xk + dxk ; tolk = norm(dxk); it = it + 1;
end
y = xk;
if it <= itmax & tolk < tol
    iconv = 0;
else
    iconv = 1;
end
```

Code 21 is a very rudimentary continuation step that uses a secant extrapolated predictor based on two neighboring known points along the solution branch $\mathbf{f}(\mathbf{y}) \equiv 0$ that aims to be continued. These points can be previously obtained by natural continuation, for example. As a simple illustration, we apply Code 21 to compute contour levels of the so-called *Himmelblau's function*

$$f(x,y) \equiv (x^2 + y - 11)^2 + (x + y^2 - 7)^2. \tag{6.46}$$

For example, the reader may easily check that $f(-4,4) \equiv 106$. By slightly incrementing the abscissa $x^{(0)} \equiv -4$ to $x^{(1)} \equiv -3.975$, we can find the corresponding ordinate $y \equiv y^{(1)}$ such that $f(-3.975, y) \equiv 106$. In this case, a simple univariate Newton iteration in y, starting from the initial guess $y^{(0)} \equiv 4$, provides the sought solution to be $y^{(1)} = 4.032\,825\,501\,100\,59$. We can therefore continuate the level curve $f(x,y) = 106$ by repeatedly invoking secant steps using Code 21 starting from the points

$$\mathbf{y}^{(0)} = [x^{(0)}\ y^{(0)}]^{\mathrm{T}} = [-4\ 4]^{\mathrm{T}}, \tag{6.47}$$

and

$$\mathbf{y}^{(1)} = [x^{(1)}\ y^{(1)}]^{\mathrm{T}} = [-3.975\ 4.032\ 825\ 501\ 100\ 59]^{\mathrm{T}}, \tag{6.48}$$

as performed by the code below:

```
y0 = [-4 ; 4]; y1 = [-3.975;4.03282550110059];
xv = [y0(1) y1(1)]; yv = [y0(2) y1(2)]; s = 1;
for ii = 1:1000
  [y,iconv] = contstep(@fun,y0,y1,s,1e-12,10);
  xv = [xv y(1)] ; yv = [yv y(2)] ; y0 = y1 ; y1 = y ;
end
plot(xv,yv,'-k'); axis equal ; axis([-5.5 5.5 -5.5 5.5])
```

In this case, the code applies the secant step function `contstep` on the function $F(x,y) = f(x,y) - 106$ (whose value is 0 along the level curve):

```
function F = fun(X)
x = X(1); y = X(2); F = (x.^2+y-11).^2+(x+y.^2-7).^2-106;
```

After `constep` makes use of $\mathbf{y}^{(0)}$ and $\mathbf{y}^{(1)}$ to provide the new point $\mathbf{y}$, these two vectors $\mathbf{y}^{(0)}$ and $\mathbf{y}^{(1)}$ are replaced by $\mathbf{y}^{(1)}$ and $\mathbf{y}$, respectively, and the process is repeated 1000 times. Figure 6.6a shows the result. The reader may have noticed that in this continuation we used $s = 1$, so that the size of the secant steps is controlled only by the spacing between the two starting points. Starting the iteration from points further apart may lead to misleading or even incorrect results. For example, the reader may easily check that starting the iteration from the points

$$\mathbf{y}^{(0)} = [x^{(0)}\ y^{(0)}]^{\mathrm{T}} = [-4\ 4]^{\mathrm{T}}, \tag{6.49}$$

and

$$\mathbf{y}^{(1)} = [x^{(1)}\ y^{(1)}]^{\mathrm{T}} = [-3.5\ 4.373\ 145\ 082\ 662\ 86]^{\mathrm{T}}, \tag{6.50}$$

fails in tracking the level curve properly. Figure 6.6b clearly shows that, although the continuation points $\{\mathbf{y}^{(2)},\ \mathbf{y}^{(3)},\ \ldots,\mathbf{y}^{(41)}\}$ are correct, the point

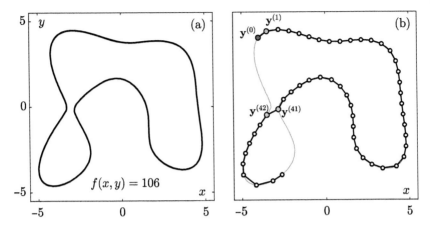

Figure 6.6 Level curve $f(x,y) = 106$ of Himmelblau's function (6.46). (a) Starting the secant continuation from points $\mathbf{y}^{(0,1)}$ given in (6.47) and (6.48). (b) Failure of the continuation when starting from points (6.49) and (6.50). The correct contour is the underlying gray solid curve.

$\mathbf{y}^{(42)}$ is wrong. The reason is that the secant predictor is by far too large and, consequently, the `contstep` function provides a point that *does* belong to the level curve $f = 106$ but not to the local branch that is being tracked. The computation of other isocontour levels of Himmelblau's function such as $f(x,y) = 103$ can be even more challenging. The reader may check that, in this particular case, the lower lobe of the curve lying on the third quadrant splits apart near the coordinates $(x,y) = (-3,0)$, becoming an independent island that may be easily overlooked during the continuation process, sometimes aggravating the convergence to the local branch. The underlying mathematical reason for such potential failure is of course the local presence of a saddle point that makes the problem ill-conditioned.

Robust continuation algorithms include self-adaptive smart criteria to choose and control optimal values of s, sometimes involving predictor techniques based on quadratic or even cubic extrapolations in this variable, using the last three or four computed points along the curve, respectively. These algorithms also incorporate nonplanar conjugate (orthogonality) conditions that improve the conditioning of the problem, and they even monitor the Jacobian in order to anticipate the emergence or nearby presence of new branches. Unfortunately, the study of these techniques is far beyond the scope of this introductory chapter. The reader will find suitable references addressing these topics in the Complementary Reading section.

Practical 6.2: Numerical Continuation

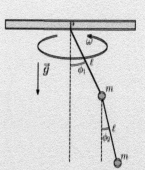

The figure on the right shows a double pendulum consisting of two small spheres of mass m connected by massless rigid rods of equal length ℓ. The system is forced to rotate with a constant angular speed ω around a vertical axis passing through the pivot P.

For a given angular speed ω, we want to find the possible equilibrium solutions, i.e. configurations where the angles ϕ_1 and ϕ_2 remain constant in time. By applying Newton's laws and imposing equilibrium in a co-rotating reference frame, it can be shown that the rod tensions can be eliminated from the problem, leading to the equations

$$f_1(\phi_1, \phi_2, \alpha) \equiv \tan(\phi_1) - \alpha(2\sin\phi_1 + \sin\phi_2) = 0$$
$$f_2(\phi_1, \phi_2, \alpha) \equiv \tan(\phi_2) - 2\alpha(\sin\phi_1 + \sin\phi_2) = 0,$$

where $\alpha = \ell\omega^2/2g$ is a *dimensionless* parameter. For simplicity, we will assume that $\phi_1 \in [0, \pi/2)$ and $\phi_2 \in (-\pi/2, \pi/2)$.

(a) The vertical configuration $(\phi_1, \phi_2) = (0,0)$ is always an equilibrium solution for any α. Use function **jac** in Code 19 to compute the determinant of the Jacobian

$$\det \mathbf{J}(\phi_1, \phi_2) \equiv \begin{vmatrix} \partial_{\phi_1} f_1 & \partial_{\phi_2} f_1 \\ \partial_{\phi_1} f_2 & \partial_{\phi_2} f_2 \end{vmatrix}$$

at $[\phi_1 \ \phi_2]^T = [0\ 0]^T$. Plot $\det \mathbf{J}(0,0)$ as a function of α within the range $\alpha \in [0,3]$. Do you find any indication that new branches of equilibrium solutions may emerge for some value(s) of α? Confirm your result analytically.

(b) **Newton Exploration:** For fixed values of $\alpha \in [0,2]$, start a Newton iteration to solve $\mathbf{f} = \mathbf{0}$ from initial random seeds in the domain $(\phi_1^{(0)}, \phi_2^{(0)}) \in [0, \pi/2) \times (-\pi/2, \pi/2)$ and check if the iteration converges to other solutions. Store *only* those solutions that have converged *within* that domain. Plot their corresponding angles ϕ_1 and ϕ_2 as a function of α.

(c) **Continuation:** Starting from any of the new solutions found, track them and plot their angles $\phi_1(\alpha)$ and $\phi_2(\alpha)$ for $\alpha \in [0,2]$. Optionally, represent physically what these new solutions look like for $\alpha \approx 2.14$. At what values of α are these solutions born? Are more equilibrium solutions expected to appear from these new branches? Why?

Complementary Reading

For a quite comprehensive and detailed analysis of nonlinear solvers, I strongly recommend Dennis and Schnabel's *Numerical Methods for Unconstrained Optimization and Nonlinear Equations*. In this classical monograph, the reader will find a mathematically rigorous approach to the Newton–Kantorovich convergence theorem and a detailed discussion regarding the accurate approximation of the Jacobian using finite differences formulas. The reader can also find many other extremely important topics not covered here, such as updated *trust regions* theory, *hook* and *double dogleg* step strategies for global convergence, or nonlinear least squares problems, to mention just a few.

Another important aspect not addressed in this chapter is the solution of *large scale* nonlinear systems, where the computation of the Jacobian and the complete solution of the Newton step are in general unfeasible tasks. In practice, large nonlinear systems are typically solved using matrix-free Newton's methods, usually known as *Newton–Krylov* methods. For a very nice and succinct exposition of these type of methods I recommend Kelley's *Solving Nonlinear Equations with Newton's Method*. This short monograph includes updated state of the art methodologies in nonlinear solvers, from Newton's or Broyden's direct methods, to matrix-free Newton–Krylov techniques using GMRES, for example. The text also includes Matlab solvers ready to be used. Kelley's monograph also addresses the numerical problem of finite difference Jacobian approximation.

The numerical solution of parameter-dependent nonlinear systems is a very active research field that has a tremendous impact in the development of nonlinear sciences. For many years, the classical reference has been Allgower and Georg's *Numerical Continuation Methods*. A more modern and applied treatment can be found in Govaert's *Numerical Methods for Bifurcations of Dynamical Equilibria*. In particular, I recommend the interested reader to have a look at software packages such as MATCONT and their underlying algorithms for a comprehensive and computationally efficient treatment of this type of problems.

Problems and Exercises

The symbol (A) means that the problem has to be solved analytically, whereas (N) means to use numerical Matlab codes. Problems with the symbol * are slightly more difficult.

1. (AN) Himmelblau's function,

$$V(x, y) = (x^2 + y - 11)^2 + (x + y^2 - 7)^2,$$

has many critical points of different type (minima, maxima, or saddles) within the domain $(x, y) \in [-6, 6] \times [-6, 6]$. Find two minima, one maximum, and two saddles. Provide the coordinates of these points (with at least six exact digits) and classify their type according to the classical Hessian criteria.[4] Let **H** be the Hessian matrix

$$\mathbf{H} = \begin{bmatrix} V_{xx} & V_{xy} \\ V_{yx} & V_{yy} \end{bmatrix};$$

Then, the critical point is a *saddle* if $\det(\mathbf{H}) < 0$. The point is a relative *maximum* if $\det(\mathbf{H}) > 0$ (with $V_{xx} < 0$ and $V_{yy} < 0$). Finally, the point is a relative *minimum* if $\det(\mathbf{H}) > 0$ (with $V_{xx} > 0$ and $V_{yy} > 0$). Optionally, confirm your results by computing isocontours of $V(x, y)$ using Matlab's `contour` function.

2. (AN) A flat wooden board (black rectangle in the figure on the right) is lying on a curved surface with equation

$$f(x) = e^{-x^2} + \frac{1}{2} e^{-(x-2)^2}.$$

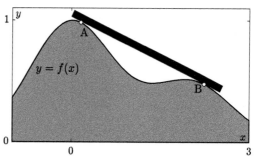

The contact points between the board and the surface are A and B. Find the geometrical conditions satisfied by the coordinates of the two contact points and provide such coordinates with at least six exact figures.

3. (AN) *Chebyshev Quadrature*: In 1874, P. L. Chebyshev proposed to approximate the integral $I(f) = \int_{-1}^{1} f(x) \, dx$ using a quadrature rule with equal

---

4 For brevity, $V_{x_1 x_2} = \dfrac{\partial^2 V}{\partial x_1 \partial x_2}$.

weights $w_1 = w_2 = \cdots = w_n = C$:

$$I(f) \approx I_n(f) = \sum_{k=1}^{n} w_k f(x_k) = C \sum_{k=1}^{n} f(x_k),$$

where the weight factor C and the nodes

$$-1 \le x_1 < x_2 < \cdots < x_{n-1} < x_n \le 1$$

have to be determined by imposing the quadrature formula to be exact for polynomials in $\mathbb{R}_n[x]$, i.e. $I_n(x^j) = I(x^j)$, for $j = 0, 1, 2, \ldots, n$.

(a) Find C as a function of n.

(b) For $n = 4$, find the equations satisfied by the nodes $\{x_1, x_2, x_3, x_4\}$ and compute them with at least six exact figures.

(c) Repeat (b) for $n = 5, 6$, and 7 and provide for at each case the value of the rightmost node x_n.

(d) For $8 \le n \le 12$, is there any value n for which you can compute the nodes? If yes, provide its rightmost x_n.

4. (AN) *Chebyshev Nodes and Electrostatic Equilibrium:* Two identical particles of electrical charge $Q > 0$ are glued to the real axis at the positions $x = -1$ and $x = 1$. You are asked to place n identical particles of charge $q > 0$ (that are free to move along the axis) at suitable coordinates $x_1, x_2, \ldots, x_n$ within the interval $(-1, 1)$ so that they remain at rest in electrostatic *equilibrium* (see figure below, to the left). In this problem, we assume that two charges interact with a force whose modulus is inversely proportional to the distance between them (*not* squared, like in *Coulomb's law*). The figure below, to the right shows the force F_{ij} exerted by an arbitrary charge q_i located at x_i on another charge q_j located at x_j.

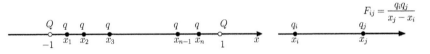

(a) For arbitrary n, write the equilibrium equations for the n particles.

(b) For $Q = 1/4$, $q = 1$, and $n = 4$, find the equilibrium positions using a suitable (equispaced) initial guess in Newton's iteration. Provide at least six exact figures.

(c) Try to repeat (b) for $n = 12$, starting Newton's iteration from an equispaced distribution. Does it converge? If it does not, use a *damping* of the Newton iteration, that is, if $\Delta \mathbf{x}^{(k)}$ is the solution of the Newton step (6.16), update according to

$$\mathbf{x}^{(k+1)} = \mathbf{x}^{(k)} + \gamma \Delta \mathbf{x}^{(k)},$$

where γ is a small damping factor within the range $0 < \gamma < 1$ that controls the size of the step. Find a suitable value of γ that makes your iteration converge and provide the abscissa x_{12}.

(d) Do you find any relation between the equilibrium locations x_k and the Chebyshev polynomial $T_n(x) = \cos(n \arccos x)$?

5. (AN) Consider the matrix $\mathbf{A} = \begin{bmatrix} 7 & 10 \\ 15 & 23 \end{bmatrix}$.

(a) Knowing that $\begin{bmatrix} 1 & 2 \\ 3 & 4 \end{bmatrix}^2 = \begin{bmatrix} 7 & 10 \\ 15 & 22 \end{bmatrix}$, use Newton's method to compute a matrix (or matrices) $\mathbf{X} \in M_2(\mathbb{R})$ such that $\mathbf{X}^2 = \mathbf{A}$. This problem is supposed to be solved without diagonalizing $\mathbf{A}$. Provide at least six correct digits in each element of $\mathbf{X}$.

(b) Find a matrix $\mathbf{X} \in M_2(\mathbb{R})$ such that $\mathbf{X}^3 = \begin{bmatrix} 37 & 54 \\ 81 & 119 \end{bmatrix}$.

6. (AN) The *Death Star* (black dot located at the origin of coordinates) is ready to destroy the *Rebel base* located somewhere in the planet *Alderaan*, whose circular shape is given by the equation

$$(x - 5)^2 + (y - 5)^2 - 9 = 0.$$

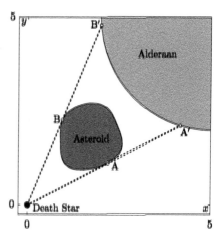

Fortunately, an asteroid with profile given by

$$\left(x + \frac{y}{10} - 2\right)^2 + \left(y - \frac{x}{4} - \frac{3}{2}\right)^4 + \frac{xy^2}{20} - 1 = 0$$

is right in the middle. However, the superlaser weapon of the Death Star has a range (dashed straight lines) capable of destroying any part of the planet beyond points A′ and B′.

(a) Write the equations whose solution lead to the coordinates of the points A and B lying on the boundary of the asteroid.

(b) Compute the coordinates (x, y) of points A and B with at least six exact figures.

(c) Compute the coordinates (x, y) of points A′ and B′.

7

Numerical Fourier Analysis

Spatiotemporal periodic phenomena are ubiquitous in nature. An accurate mathematical characterization of these types of phenomena such as oscillatory motion in mechanical systems or wave propagation requires *Fourier analysis*.[1] In a very few idealized situations, this analysis can be carried out by hand, without the use of numerical mathematics. In practice, however, time signals obtained from registering mechanical vibrations typically include multiplicity of frequencies and noise. Waves (mechanical or electromagnetic) are seldom monochromatic, exhibiting dispersion phenomena, etc. In these situations, numerical Fourier analysis is typically required, and in this chapter we will provide the essential tools to perform such analysis, paying attention to its potential failures such as *aliasing*. Henceforth in this chapter, we will assume that the reader is already familiar with the most basic concepts of classical functional analysis, such as inner-product function spaces, orthogonal functions, and Fourier series. At the end of the chapter in the Complementary Reading section, we suggest some bibliography that may help the reader to review these and many other concepts.

7.1 The Discrete Fourier Transform

We start this section by introducing the space of *square integrable* complex-valued functions in $(0, 2\pi)$

$$L^2(0, 2\pi) \doteq \left\{ f : (0, 2\pi) \to \mathbb{C} \, , \, \int_0^{2\pi} |f(x)|^2 \, \mathrm{d}x \leqslant +\infty \right\}. \tag{7.1}$$

---

1 Jean-Baptiste Joseph Fourier (1768–1830), French mathematician and physicist who formulated the analytic theory of heat and who proposed to model all mathematical functions as trigonometric series.

Fundamentals of Numerical Mathematics for Physicists and Engineers, First Edition. Alvaro Meseguer.
© 2020 John Wiley & Sons, Inc. Published 2020 by John Wiley & Sons, Inc.

The *scalar product* between two arbitrary functions $f(x)$ and $g(x)$ in $L^2(0, 2\pi)$ is defined as

$$(f, g) \doteq \int_0^{2\pi} f(x)\overline{g(x)} \, dx, \tag{7.2}$$

where $\overline{g(x)}$ stands for the complex conjugate of $g(x)$. By definition, two arbitrary functions $f(x)$ and $g(x)$ in $L^2(0, 2\pi)$ are said to be *orthogonal* with respect to the inner product (7.2) if $(f, g) = 0$. The scalar product (7.2) induces the L^2-*norm* of an arbitrary square integrable function $f(x)$ as being

$$\|f\|_{L^2(0,2\pi)} \doteq \sqrt{(f, f)} = \left[\int_0^{2\pi} |f(x)|^2 \, dx \right]^{1/2}. \tag{7.3}$$

We define the infinite set of 2π-periodic functions

$$\varphi_k(x) \doteq e^{ikx} = \cos(kx) + i \sin(kx), \quad k = 0, \pm 1, \pm 2, \ldots, \tag{7.4}$$

satisfying the orthogonality relation

$$(\varphi_j, \varphi_k) = \int_0^{2\pi} e^{ijx} e^{-ikx} \, dx = 2\pi \delta_{jk}, \tag{7.5}$$

so that φ_j and φ_k are orthogonal if $j \neq k$.

The *Fourier series* of an arbitrary function $f(x) \in L^2(0, 2\pi)$ is the infinite sum

$$\sum_{k=-\infty}^{+\infty} \widehat{f}_k \, \varphi_k(x), \tag{7.6}$$

where

$$\widehat{f}_k = \frac{1}{2\pi}(f, \varphi_k) = \frac{1}{2\pi} \int_0^{2\pi} f(x) \, e^{-ikx} \, dx, \tag{7.7}$$

are known as the *Fourier coefficients* of $f(x)$. Unless stated otherwise, we will henceforth assume that $f(x)$ is a smooth 2π-periodic function. In this case, the convergence of the Fourier series is *pointwise*, that is,

$$\lim_{n \to \infty} \sum_{k=-n}^{n} \widehat{f}_k \, \varphi_k(x) = f(x), \tag{7.8}$$

for any $x \in [0, 2\pi]$. However, the convergence of Fourier trigonometric series may be at stake if the function or its derivative has discontinuities. For a comprehensive analysis of the convergence properties of trigonometric Fourier series we refer the reader to the classical texts on the topic included in the Complementary Reading section.

Computationally, it is impossible to evaluate the trigonometric series (7.6) involving infinite terms, and therefore we need to *truncate* this series up to finite order. For this reason, we define the *truncation of order N* of the Fourier series (7.6) by the sum

$$f_N \doteq \sum_{k=-\frac{N}{2}}^{\frac{N}{2}-1} \widehat{f}_k \, e^{ikx}, \tag{7.9}$$

where we assume that N is an *even* nonnegative integer. The reader must have noticed that the truncation formula (7.9) is *asymmetric*, since the term $\widehat{f}_{\frac{N}{2}} \, e^{i\frac{N}{2}x}$ is absent. This (rather unorthodox) expression follows a convention that has long been used by many authors in the vast literature on numerical Fourier methods.[2] Later on in this chapter we will justify this particular truncated form when addressing the numerical approximation of the Fourier coefficients.

The truncation formula (7.9) is actually a linear combination of the set of functions $\varphi_k(x) = e^{ikx}$ previously introduced in (7.4). That is, the sum

$$f_N = \widehat{f}_{-\frac{N}{2}} \, e^{-i\frac{N}{2}x} + \cdots + \widehat{f}_{-1} \, e^{-ix} + \widehat{f}_0 + \widehat{f}_1 e^{ix} + \cdots + \widehat{f}_{\frac{N}{2}-1} \, e^{i(\frac{N}{2}-1)x}, \tag{7.10}$$

can be written as the N-dimensional vector

$$\mathbf{f}_N = \begin{bmatrix} \widehat{f}_{-\frac{N}{2}} & \cdots & \widehat{f}_{-1} & \widehat{f}_0 & \widehat{f}_1 & \cdots & \widehat{f}_{\frac{N}{2}-1} \end{bmatrix}_{\{\varphi\}}, \tag{7.11}$$

where the subscript $\{\varphi\}$ means that the vector $\mathbf{f}_N$ appearing in (7.11) is expressed in the N-dimensional basis of trigonometric functions

$$\begin{aligned}
S_N &= \operatorname{span}\left\{\varphi_{-\frac{N}{2}}, \, \ldots, \, \varphi_{-1}, \, \varphi_0, \, \varphi_1, \, \ldots, \, \varphi_{\frac{N}{2}-1}\right\} \\
&= \operatorname{span}\left\{e^{-i\frac{N}{2}x}, \, \ldots, \, e^{-ix}, \, 1, \, e^{ix}, \, \ldots, \, e^{i(\frac{N}{2}-1)x}\right\}.
\end{aligned} \tag{7.12}$$

To avoid cumbersome indexing in the lower and upper limits of the sum, it is usual practice to rename the basis functions of S_N above as follows:

$$\begin{aligned}
S_N &= \operatorname{span}\left\{\widetilde{\varphi}_0, \, \ldots, \, \widetilde{\varphi}_{\frac{N}{2}-1}, \, \widetilde{\varphi}_{\frac{N}{2}}, \, \widetilde{\varphi}_{\frac{N}{2}+1}, \, \ldots, \, \widetilde{\varphi}_{N-1}\right\} \\
&= \operatorname{span}\left\{e^{-i\frac{N}{2}x}, \, \ldots, \, e^{-ix}, \, 1, \, e^{ix}, \, \ldots, \, e^{i(\frac{N}{2}-1)x}\right\},
\end{aligned} \tag{7.13}$$

or, in general,

$$\widetilde{\varphi}_k(x) \doteq e^{i(k-\frac{N}{2})x}, \quad (k = 0, 1, 2, \ldots, N-1). \tag{7.14}$$

In terms of the *shifted* basis $\widetilde{\varphi}_k(x)$, the truncated series (7.9) now reads

$$f_N(x) \doteq \sum_{k=0}^{N-1} \widehat{f}_k \, \widetilde{\varphi}_k(x) = \sum_{k=0}^{N-1} \widehat{f}_k \, e^{i(k-\frac{N}{2})x}, \tag{7.15}$$

---

2 Although other (equivalent) formulations are also used.

where the coefficients $\widehat{f}_k$ are now redefined as

$$\widehat{f}_k \equiv \frac{1}{2\pi}(f,\widetilde{\varphi}_k) = \frac{1}{2\pi}\int_0^{2\pi} f(x)\,e^{-i(k-\frac{N}{2})x}\,dx, \qquad (7.16)$$

for $k = 0,1,2,\ldots,N-1$. In general, these coefficients cannot be calculated exactly and must be approximated by some suitable quadrature rule. In Part I we studied that for integrals of the form

$$I(g) \equiv \int_0^{2\pi} g(x)\,dx, \qquad (7.17)$$

where $g(x)$ is a 2π-periodic smooth integrand, the composite trapezoidal rule (4.81) provides exponential convergence. Let $h = 2\pi/N$ and the equidistant abscissas $\{0,\ h,\ 2h,\ \ldots,2\pi-h,2\pi\}$. The composite trapezoidal rule (4.81) applied to integral $I(g)$ in (7.17) reads

$$I_{1,N}(g) \equiv h\left(\frac{1}{2}g(0) + g(h) + g(2h) + \cdots + g(2\pi-h) + \frac{1}{2}g(2\pi)\right). \quad (7.18)$$

Since $g(2\pi) \equiv g(0)$, the last sum can be written as

$$I_{1,N}(g) \equiv h[g(0) + g(h) + g(2h) + \cdots + g(2\pi-h)] \equiv h\sum_{j=0}^{N-1} g(x_j), \quad (7.19)$$

where $x_j \equiv jh$, for $j = 0,1,2,\ldots,N-1$.

Formula (7.19) is the ideal quadrature rule to approximate arbitrary inner products of the form

$$(f,g) \doteq \int_0^{2\pi} f(x)\overline{g(x)}\,dx, \qquad (7.20)$$

assuming that $f(x)$ and $g(x)$ are smooth and 2π-periodic functions. Therefore, we introduce the *discrete scalar product* between $f(x)$ and $g(x)$:

$$(f,g)_N \doteq h\sum_{j=0}^{N-1} f(x_j)\overline{g(x_j)}, \qquad (7.21)$$

which approximates (7.20) with exponential accuracy in N. Our next goal is to approximate the coefficients $\widehat{f}_k$ in (7.16) by means of the discrete formula (7.21). We define these approximate coefficients by

$$\widetilde{f}_k \doteq \frac{1}{2\pi}(f,\widetilde{\varphi}_k)_N \approx \frac{1}{2\pi}(f,\widetilde{\varphi}_k) = \widehat{f}_k. \qquad (7.22)$$

According to the scalar product introduced in (7.20), these coefficients are

$$\widetilde{f}_k = \frac{1}{2\pi}h\sum_{j=0}^{N-1} f(x_j)\overline{\widetilde{\varphi}_k(x_j)} = \frac{1}{N}\sum_{j=0}^{N-1} f(x_j)\overline{\widetilde{\varphi}_k(x_j)}, \qquad (7.23)$$

for $k = 0, 1, 2, \ldots, N - 1$. Since the shifted basis function $\tilde{\varphi}_k$ evaluated at $x_j = hj$ is

$$\tilde{\varphi}_k(x_j) = e^{i\left(k - \frac{N}{2}\right)hj} = e^{ikhj}e^{-i\frac{N}{2}hj} = e^{ikhj}e^{-ij\pi}, \tag{7.24}$$

expression (7.23) reads

$$\tilde{f}_k = \frac{1}{N} \sum_{j=0}^{N-1} f(x_j)e^{-ikhj}e^{ij\pi} \quad (k = 0, 1, 2, \ldots, N - 1). \tag{7.25}$$

To simplify the notation, we introduce the quantity $W_N \doteq e^{-ih} = e^{-i\frac{2\pi}{N}}$, so that (7.25) reads

$$\boxed{\tilde{f}_k = \sum_{j=0}^{N-1} f(x_j)\frac{W_N^{\left(k - \frac{N}{2}\right)j}}{N} \quad (k = 0, 1, 2, \ldots, N - 1).} \tag{7.26}$$

In numerical mathematics, the boxed expression (7.26) is known as the *discrete Fourier transform* of the function $f(x)$, henceforth referred to as DFT. Expression (7.26) provides the approximated Fourier coefficients $\tilde{f}_k$ in terms of the sampled values $f(x_j)$ of the function $f(x)$ at the abscissas $x_j = jh = 2\pi j/N$, for $j = 0, 1, 2, \ldots, N - 1$, originally used in the trapezoidal rule (7.19).

By substituting these coefficients in (7.15) we obtain the approximated truncated series, denoted by $F_N(x)$:

$$F_N(x) \doteq \sum_{k=0}^{N-1} \tilde{f}_k \, \tilde{\varphi}_k(x) = \sum_{k=0}^{N-1} \tilde{f}_k \, e^{i\left(k - \frac{N}{2}\right)x}, \tag{7.27}$$

which is known as the *discrete Fourier series* of order N of the function $f(x)$. We must emphasize that $F_N(x)$ is *not* the truncated Fourier series (7.15), but just an *approximation* of $f_N(x)$, built from the numerically approximated Fourier coefficients $\tilde{f}_k$ $(k = 0, 1, \ldots, N - 1)$, computed by means of the trapezoidal rule. However, the discrete Fourier series (7.27) satisfies

$$F_N(x_j) = f(x_j), \tag{7.28}$$

for $j = 0, 1, 2, \ldots, N - 1$, that is, its value at the abscissa x_j is exactly the same as the value of the original function. To prove this, first observe that since $h = 2\pi/N$,

$$\sum_{k=0}^{N-1} e^{ik(j-\ell)h} = \left\{ \begin{array}{ll} N, & (\ell = j) \\ \dfrac{1 - \left[e^{ik(j-\ell)\frac{2\pi}{N}}\right]^N}{1 - e^{ik(j-\ell)h}} = 0, & (\ell \neq j) \end{array} \right\} \equiv N\delta_{\ell j}, \tag{7.29}$$

for $0 \leq \ell, j \leq N - 1$. Let us now evaluate $F_N(x)$ in (7.27) at $x = x_j$:

$$F_N(x_j) = \sum_{k=0}^{N-1} \tilde{f}_k \, e^{i\left(k - \frac{N}{2}\right)x_j} = \sum_{k=0}^{N-1} \tilde{f}_k \, e^{ikx_j}e^{-i\frac{N}{2}x_j} = \sum_{k=0}^{N-1} \tilde{f}_k \, e^{ikjh}e^{-ij\pi}. \tag{7.30}$$

We can formally substitute the coefficient $\widetilde{f}_k$ in the last expression by its original value previously obtained in (7.25), that is,

$$F_N(x_j) = \sum_{k=0}^{N-1} \frac{1}{N} \sum_{\ell=0}^{N-1} f(x_\ell) e^{-ikh\ell} e^{i\ell\pi} e^{ikjh} e^{-ij\pi}. \tag{7.31}$$

Finally, if we permute the order of the two sums above, we obtain

$$F_N(x_j) = \sum_{\ell=0}^{N-1} f(x_\ell) \frac{1}{N} e^{i(\ell-j)\pi} \underbrace{\sum_{k=0}^{N-1} e^{ik(j-\ell)h}}_{N\delta_{\ell j}} = f(x_j), \tag{7.32}$$

for $j = 0, 1, 2, \ldots, N-1$, as we claimed. Therefore, by virtue of (7.27) and (7.32), we may write

$$f(x_j) = \sum_{k=0}^{N-1} \widetilde{f}_k \, e^{i\left(k-\frac{N}{2}\right)jh} = \sum_{k=0}^{N-1} \widetilde{f}_k \, W_N^{-j\left(k-\frac{N}{2}\right)}, \tag{7.33}$$

for $j = 0, 1, 2, \ldots, N-1$. This last expression provides the values $f(x_j)$ of the original function, henceforth denoted as f_j, at the abscissas x_j (belonging to the *physical space*) in terms of the Fourier coefficients $\widetilde{f}_k$ (belonging to the *Fourier space*). In that sense, (7.33) is the inverse of (7.26). For these reasons, the expression

$$\boxed{f_j = \sum_{k=0}^{N-1} \widetilde{f}_k \, W_N^{-j\left(k-\frac{N}{2}\right)} \quad (j = 0, 1, 2, \ldots, N-1),} \tag{7.34}$$

is known as the *inverse discrete Fourier transform*, or IDFT. While the DFT in (7.26) provides the Fourier coefficients $\widetilde{f}_k$ in terms of the values of the function $f_j = f(x_j)$ in physical space, the IDFT in (7.34) proceeds inversely. If we define the vectors $\mathbf{f} \doteq [f_0 \; f_1 \; \cdots \; f_{N-1}]^{\mathrm{T}}$ and $\widetilde{\mathbf{f}} = [\widetilde{f}_0 \; \widetilde{f}_1 \; \cdots \; \widetilde{f}_{N-1}]^{\mathrm{T}}$, the DFT can be expressed as the matrix–vector product

$$\begin{bmatrix} \widetilde{f}_0 \\ \widetilde{f}_1 \\ \vdots \\ \widetilde{f}_{N-1} \end{bmatrix} = \begin{bmatrix} \mathbf{F}_{00} & \mathbf{F}_{01} & \cdots & \mathbf{F}_{0,N-1} \\ \mathbf{F}_{10} & \mathbf{F}_{11} & \cdots & \mathbf{F}_{1,N-1} \\ \vdots & \vdots & \vdots & \vdots \\ \mathbf{F}_{N-1,0} & \mathbf{F}_{N-1,1} & \cdots & \mathbf{F}_{N-1,N-1} \end{bmatrix} \begin{bmatrix} f_0 \\ f_1 \\ \vdots \\ f_{N-1} \end{bmatrix}, \tag{7.35}$$

or simply $\widetilde{\mathbf{f}} = \mathbf{F}\,\mathbf{f}$, where the elements of the matrix $\mathbf{F}$ are $\mathbf{F}_{kj} = \frac{1}{N} W_N^{(k-\frac{N}{2})j}$, for $0 \leq k, j \leq N-1$. Similarly, the IDFT is

$$\begin{bmatrix} f_0 \\ f_1 \\ \vdots \\ f_{N-1} \end{bmatrix} = \begin{bmatrix} \mathbf{P}_{00} & \mathbf{P}_{01} & \cdots & \mathbf{P}_{0,N-1} \\ \mathbf{P}_{10} & \mathbf{P}_{11} & \cdots & \mathbf{P}_{1,N-1} \\ \vdots & \vdots & \vdots & \vdots \\ \mathbf{P}_{N-1,0} & \mathbf{P}_{N-1,1} & \cdots & \mathbf{P}_{N-1,N-1} \end{bmatrix} \begin{bmatrix} \widetilde{f}_0 \\ \widetilde{f}_1 \\ \vdots \\ \widetilde{f}_{N-1} \end{bmatrix}, \tag{7.36}$$

or $\mathbf{f} = \mathbf{P}\,\widetilde{\mathbf{f}}$, where the matrix elements of $\mathbf{P}$ are $\mathbf{P}_{jk} = W_N^{-(k-\frac{N}{2})j}$, for $0 \le k, j \le N - 1$. The transformations (7.35) and (7.36) require $O(N^2)$ operations. However, there are many more efficient ways of performing such transformations, such as the *fast Fourier transform* (FFT). The reader can find detailed analysis of these methodologies in the recommended references at the end of the chapter, in the Complementary Reading section.

Summary: DFT and IDFT

Let $N > 0$ be an even integer and let $\mathbf{f} \in \mathbb{R}^N$ be an N-dimensional vector whose components are the values $f_j = f(x_j)$ of a 2π-periodic function $f(x)$ evaluated at the abscissas $x_j = hj$ $(j = 0, 1, \ldots, N-1)$, with grid spacing $h = \dfrac{2\pi}{N}$, that is,

$$\mathbf{f} = [f_0 \ f_1 \ f_2 \ \cdots \ f_{N-1}]^\mathrm{T} = [f(0) \ f(h) \ f(2h) \ \cdots \ f([N-1]h)]^\mathrm{T}.$$

The discrete Fourier transform or DFT of $\mathbf{f}$ is the N-dimensional vector

$$\widetilde{\mathbf{f}} = [\widetilde{f}_0 \ \widetilde{f}_1 \ \widetilde{f}_2 \ \cdots \ \widetilde{f}_{N-1}]^\mathrm{T} \in \mathbb{R}^N,$$

whose kth component is given by the expression

$$\widetilde{f}_k = \sum_{j=0}^{N-1} f_j \, \frac{W_N^{(k-\frac{N}{2})j}}{N} \quad (k = 0, 1, 2, \ldots, N-1), \tag{7.37}$$

or, in matrix–vector product form, $\widetilde{\mathbf{f}} = \mathbf{F}\mathbf{f}$, where $\mathbf{F} \in \mathbb{M}_N(\mathbb{C})$ is

$$\mathbf{F}_{kj} = \frac{1}{N} W_N^{(k-\frac{N}{2})j} \quad (0 \le k, j \le N-1), \tag{7.38}$$

and where $W_N = \mathrm{e}^{-\mathrm{i}h} = \mathrm{e}^{-\mathrm{i}\frac{2\pi}{N}}$.

Conversely, the inverse discrete Fourier transform or IDFT of $\widetilde{\mathbf{f}}$ is

$$f_j = \sum_{k=0}^{N-1} \widetilde{f}_k \, W_N^{-j(k-\frac{N}{2})} \quad (j = 0, 1, 2, \ldots, N-1), \tag{7.39}$$

or $\mathbf{f} = \mathbf{P}\widetilde{\mathbf{f}}$, where $\mathbf{P} \in \mathbb{M}_N(\mathbb{C})$

$$\mathbf{P}_{jk} = W_N^{-(k-\frac{N}{2})j} \quad (0 \le k, j \le N-1). \tag{7.40}$$

Finally, the *discrete Fourier series* of order N of $f(x)$ is

$$F_N(x) \doteq \sum_{k=0}^{N-1} \widetilde{f}_k \, \widetilde{\varphi}_k(x) = \sum_{k=0}^{N-1} \widetilde{f}_k \, \mathrm{e}^{\mathrm{i}(k-\frac{N}{2})x}. \tag{7.41}$$

Codes 22 and 23 are simple matrix–vector implementations of DFT and IDFT that build matrices **F** and **P** of (7.38) and (7.40), respectively.

```
% Code 22: DFT (Matrix-vector product)
% Input:  fj (N-column vector: sampling values of f)
% Output: fk (N-column vector: Fourier coefficients)
function fk = dftmat(fj)
N = length(fj); WN = exp(-i*2*pi/N); jj = [0:N-1]; kk = jj';
F = (1/N)*WN.^((kk-N/2)*jj); fk = F*fj;
```

```
% Code 23: IDFT (Matrix-vector product)
% Input:  fk (N-column vector: Fourier coefficients)
% Output: fj (N-column vector: sampling values of f)
function fj = idftmat(fk)
N = length(fk); WN = exp(-i*2*pi/N); jj = [0:N-1]'; kk = jj';
P = WN.^(-jj*(kk-N/2)); fj = P*fk;
```

Similarly, Code 24 computes the discrete Fourier series (7.27) on an arbitrary grid for a given set of Fourier coefficients $\tilde{f}_k$, using the shifted basis (7.14).

```
% Code 24: DFS (Discrete Fourier Series)
% Input:  x (column vector containing evaluation grid)
%         fk (DFT Fourier coefficients)
% Output: FNx (DFS evaluated at grid x)
function FNx = dfs(fk,x)
N = length(fk); M = length(x); k = [0:N-1];
PHI = exp(i*x*(k-N/2)); FNx = PHI*fk;
```

As a trivial test, let us compute the DFT of

$$f(x) = \cos 2x = \frac{1}{2}e^{-i2x} + \frac{1}{2}e^{i2x},$$

using $N = 18$. This can be easily done by typing in Matlab

```
N=18; xj=(2*pi/N)*[0:N-1]'; fj=cos(2*xj); fk=dftmat(fj);
```

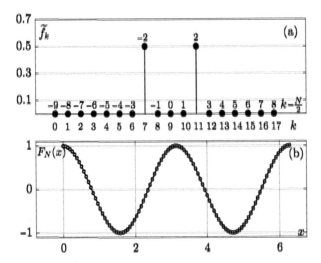

Figure 7.1 DFT of $f(x) = \cos 2x$. (a) Fourier coefficients $\widetilde{f}_k$ provided by Code 22. (b) Discrete Fourier series expansion $F_N(x)$ given in (7.27) evaluated on a finer grid using Code 24.

The resulting (in this case real) coefficients $\widetilde{f}_k$ provided by **dftmat** have been depicted in the stem plot of Figure 7.1a, indicating the two possible index-ings $k = N/2$ and k in the Fourier and shifted-Fourier bases (7.4) and (7.14), respectively. We can also use these coefficients in the discrete Fourier series expansion (7.27) to evaluate $F_N(x)$ on a finer grid containing 100 points by invoking function **dfs**:

```
x = linspace(0,2*pi,100)'; FNx = dfs(fk,x);
```

Figure 7.1b shows the result, which is indistinguishable from the original function $f(x) \equiv \cos(2x)$.

7.1.1 Time–Frequency Windows

The DFT has many applications in science and engineering. Probably, one of its most frequent usages is *discrete time signal processing*, that is, to capture discrete data from some kind of natural phenomena (such as seismic waves, tides, or pulsars) and scrutinize the internal structure in order to identify their underlying frequencies out of the noisy signal.

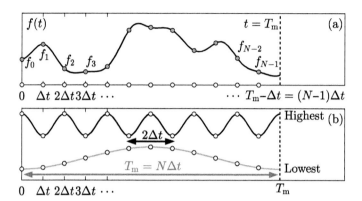

Figure 7.2 Sampling process. (a) The ordinates $\{f_0, f_1, \ldots, f_{N-1}\}$ (gray circles) are the sampled values of the signal $f(t)$ at the abscissas (white bullets) $t_j = j\Delta t$ ($j = 0, 1, 2, \ldots, N-1$). (b) Examples of simple waves with the highest (black bulleted curve) and lowest (gray bulleted curve) frequencies identifiable by the sampling.

A discrete temporal sampling is characterized by *two* quantities. The first one is the *sampling rate*, that is, the lapse of time Δt between two consecutive measurements of the signal. The second is the total number of samplings, namely N. Figure 7.2a shows an arbitrary time-dependent function $f(t)$ sampled at the time grid $t_j = j\Delta t$, for $j = 0, 1, 2, \ldots, N-1$. The last measurement of $f(t)$ is $f_{N-1} = f(t_{N-1}) = f([N-1]\Delta t)$. However, the discrete Fourier analysis of any time signal $f(t)$ implicitly assumes its *periodicity*, therefore implying that $f(t_N) = f(t_0)$ or, equivalently, that $f(N\Delta t) = f(0)$. In other words, the sampling rate Δt, along with the number of measurements N, establishes that the natural *period* or length of our physical time domain is

$$\boxed{T_{\mathrm{m}} = N\Delta t.}$$ (7.42)

To put it in another way, any time signal with a period *longer* than T_{m} cannot be captured by the sampling, so that T_{m} is the *maximum* period capable of being resolved. Figure 7.2b shows a simple signal with the longest period (gray bulleted curve) that our time discretization is capable of capturing. Conversely, the sampling rate Δt establishes a limitation in identifying highly oscillatory time signals. In particular, time signals with periods *shorter* than $2\Delta t$ cannot be resolved by our sampling either. This is illustrated in Figure 7.2b by showing a time signal (black bulleted curve) with the shortest *visible* period of length $2\Delta t$ between two consecutive valleys.

According to (7.9), the Nth-order truncated Fourier series of a T_{m}-periodic function $f(t)$ reads

$$f_N(t) = \sum_{k=-N/2}^{N/2-1} \widehat{f}_k\, \mathrm{e}^{\mathrm{i}k2\pi t/T_{\mathrm{m}}} = \sum_{k=-N/2}^{N/2-1} \widehat{f}_k\, \mathrm{e}^{\mathrm{i}\omega_k t}, \qquad (7.43)$$

where

$$\omega_k \doteq \frac{2\pi}{T_{\mathrm{m}}}\, k = \frac{2\pi}{N\Delta t}\, k, \qquad (7.44)$$

for $-N/2 \le k \le N/2 - 1$, is the *spectrum* of angular frequencies captured by $f_N(t)$. Therefore the *highest* frequency that our sampling can detect is

$$\omega_{\mathrm{max.}} = \omega_{N/2} = \frac{2\pi}{N\Delta t}\frac{N}{2} = \frac{\pi}{\Delta t}, \qquad (7.45)$$

with associated minimum period $T_{\mathrm{min.}} = 2\pi/\omega_{\mathrm{max.}} = 2\Delta t$, as expected. By the same rule, the lowest nonzero frequency of the spectrum is

$$\omega_{\mathrm{min.}} = \omega_1 = \frac{2\pi}{T_{\mathrm{m}}} = \frac{2\pi}{N\Delta t}, \qquad (7.46)$$

with obvious maximum period $2\pi/\omega_{\mathrm{min.}} = T_{\mathrm{m}}$.

Summary: Frequency Window

The sampling of a signal based on N measurements taken every Δt time units is capable of resolving frequencies ω within the range or window

$$\omega_{\mathrm{min.}} = \frac{2\pi}{N\Delta t} \le \omega \le \frac{\pi}{\Delta t} = \omega_{\mathrm{max.}}. \qquad (7.47)$$

In order to illustrate the concept of frequency windows, Figure 7.3a shows a particular time signal $f(t)$ containing three frequencies: $\alpha = 0.6$, $\beta = 2.3$, and $\gamma = 15.0$. This signal is sampled using $\Delta t = 0.4$ and $N = 240$ so that, in virtue of (7.45), the maximum frequency captured is $\omega_{\mathrm{max.}} \approx 7.8$. Figure 7.3b shows the *power spectrum* reporting the modulus of the Fourier coefficients $|\widetilde{f}_k|$ associated with their corresponding angular frequencies $\omega_k = 2k\pi/N\Delta t$, for $k = 0, 1, 2, \dots, N/2 - 1$ (for $k < 0$ the power spectrum is symmetric). As expected, two peaks appear nearby α and β ($\omega \approx 0.6$ and $\omega \approx 2.3$, respectively), but the highest frequency γ is not resolved by our discretization. According to (7.45), in order to capture ω_3, the time lapse Δt needs to be decreased to $\Delta t = \pi/\omega_3 \approx 0.21$. Figure 7.3c shows the power spectrum using $\Delta t = 0.2$, where we observe a clear peak for $\omega \approx 15.0$.

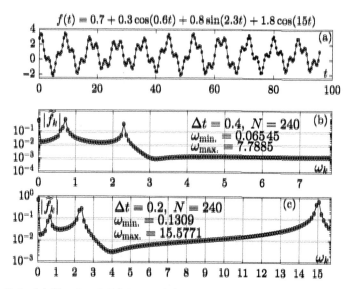

Figure 7.3 (a) The signal $f(t)$ is sampled using $N = 240$ measurements every $\Delta t = 0.4$ time units. (b) Power spectrum of the signal $f(t)$ consisting of the modulus of the Fourier coefficients $|\tilde{f}_k|$ as a function of their associated angular frequencies ω_k for $k = 0, 1, 2, \ldots, N/2 - 1$, using $\Delta t = 0.4$. (c) same as (b) using $\Delta t = 0.2$.

We have illustrated the limitations of a discrete sampling by using a known deterministic function $f(t)$. In practice, however, the problem is very different, where the explicit form $f(t)$ of the signal is in general unknown and only the sampling values f_j are available.

7.1.2 Aliasing

Consider the functions $f_1(x) = e^{ik_1 x}$ and $f_2(x) = e^{ik_2 x}$ for two arbitrary different nonzero constants k_1 and k_2. Evaluation of $f_1(x)$ at the abscissas $x_j = jh$ ($j = 0, 1, \ldots, N - 1$) leads to the values

$$f_1(x_j) = e^{ik_1 x_j} = e^{ik_1 jh}. \tag{7.48}$$

Similarly, the evaluation of $f_2(x)$ is

$$f_2(x_j) = e^{ik_2 x_j} = e^{ik_2 jh}. \tag{7.49}$$

In principle, the two functions are different when they are evaluated in $\mathbb{R}$. However, when they are sampled on a discrete grid, they may take the same values, making them *indistinguishable* or *aliased*. If, for example, k_1 and k_2 satisfy

$$k_2 = k_1 + \frac{2\pi}{h}\ell, \tag{7.50}$$

for any integer $\ell \in \mathbf{Z}$, then (7.49) reads

$$f_2(x_j) \equiv e^{i(k_1 + 2\pi\ell/h)jh} = e^{ik_1jh} e^{i2\pi\ell j} = e^{ik_1 x_j} \equiv f_1(x_j). \qquad (7.51)$$

As a result, $f_1(x)$ and $f_2(x)$ are identical when evaluated on the discrete grid $x_j \equiv jh$. Although this coincidence may initially seem purely accidental and irrelevant, it has a very important impact on the accuracy of the truncated Fourier series. To see why, let us first consider again the Fourier expansion of a 2π-periodic function $f(x)$,

$$f(x) \equiv \sum_{k=-\infty}^{+\infty} \widehat{f}_k \, e^{ikx}, \qquad (7.52)$$

along with its corresponding truncation of order N given in (7.9),[3]

$$f_N(x) \equiv \sum_{k=-\frac{N}{2}}^{\frac{N}{2}-1} \widehat{f}_k \, e^{ikx}, \qquad (7.53)$$

with Fourier coefficients given by

$$\widehat{f}_k = \frac{1}{2\pi} \int_0^{2\pi} f(x) \, e^{-ikx} \, dx. \qquad (7.54)$$

In practice, these coefficients are approximated numerically using the discrete scalar product introduced in (7.21) between two arbitrary periodic functions $f(x)$ and $g(x)$,

$$(f, g)_N \equiv h \sum_{j=0}^{N-1} f(x_j)\overline{g(x_j)} = \frac{2\pi}{N} \sum_{j=0}^{N-1} f(x_j)\overline{g(x_j)}, \qquad (7.55)$$

where $x_j \equiv 2\pi j/N$ $(j = 0, 1, \dots, N-1)$. The approximated coefficients are

$$\widetilde{f}_k \equiv \frac{1}{2\pi}(f(x), e^{ikx})_N, \qquad (7.56)$$

for $k \equiv -N/2, \dots, N/2 - 1$. The question is whether the coefficients (7.56) are good approximations of the exact quantities (7.54). By virtue of the orthogonality relation (7.29), it is trivial to verify that

$$(e^{ikx}, e^{i\ell x})_N \equiv \frac{2\pi}{N} \sum_{j=0}^{N-1} e^{ikx_j} e^{-i\ell x_j} = 2\pi\delta_{k\ell}, \qquad (7.57)$$

for $0 \le \ell, k \le N - 1$. By introducing the Fourier expansion (7.52) in the discrete products appearing in (7.56) we obtain

$$(f(x), e^{i\ell x})_N \equiv \left(\sum_{k=-\infty}^{+\infty} \widehat{f}_k \, e^{ikx}, e^{i\ell x} \right)_N \equiv \sum_{k=-\infty}^{+\infty} \widehat{f}_k \, (e^{ikx}, e^{i\ell x})_N. \qquad (7.58)$$

---

3 In this subsection we use the classical Fourier basis $\varphi_k(x) \equiv e^{ikx}$ $(k \equiv -N/2, \dots, N/2 - 1)$.

At this point, it is very tempting to exploit the orthogonality relation (7.57) in (7.58) to cancel all terms but the ℓth. However, since k in (7.58) is now *unbounded*, relation (7.57) requires a careful analysis. In particular, adding any multiple of N to ℓ reveals that

$$(\mathrm{e}^{\mathrm{i}kx}, \mathrm{e}^{\mathrm{i}(\ell+mN)x})_N = \frac{2\pi}{N} \sum_{j=0}^{N-1} \mathrm{e}^{\mathrm{i}kx_j} \mathrm{e}^{-\mathrm{i}\ell x_j} \mathrm{e}^{-\mathrm{i}mNx_j} = \frac{2\pi}{N} \sum_{j=0}^{N-1} \mathrm{e}^{\mathrm{i}kx_j} \mathrm{e}^{-\mathrm{i}\ell x_j},$$

(7.59)

so the orthogonality relation should instead read

$$(\mathrm{e}^{\mathrm{i}kx}, \mathrm{e}^{\mathrm{i}\ell x})_N = 2\pi \delta_{k,\ell \pm mN},$$

(7.60)

for $m \in \mathbf{N}$. Therefore, (7.58) becomes

$$
\begin{aligned}
(f(x), \mathrm{e}^{\mathrm{i}\ell x})_N &= \sum_{k=-\infty}^{+\infty} \widehat{f}_k \, 2\pi \delta_{k,\ell \pm mN} \\
&= 2\pi \{ \widehat{f}_\ell + \widehat{f}_{\ell-N} + \widehat{f}_{\ell+N} + \widehat{f}_{\ell-2N} + \widehat{f}_{\ell+2N} + \cdots \} \\
&= 2\pi \left\{ \widehat{f}_\ell + \sum_{m=1}^{+\infty} [\widehat{f}_{\ell-mN} + \widehat{f}_{\ell+mN}] \right\}.
\end{aligned}
$$

(7.61)

From (7.56) and (7.61) we conclude that

$$\widetilde{f}_\ell = \widehat{f}_\ell + \sum_{m=1}^{+\infty} [\widehat{f}_{\ell-mN} + \widehat{f}_{\ell+mN}].$$

(7.62)

The last expression is known as the *aliasing formula* and can also be written as

$$\widetilde{f}_\ell - \widehat{f}_\ell = \sum_{m=1}^{+\infty} [\widehat{f}_{\ell-mN} + \widehat{f}_{\ell+mN}],$$

(7.63)

where the sum on the right-hand side is known as the *aliasing error*, that is, the discrepancy between the exact and approximated Fourier coefficients.

As an example, consider the function

$$f(x) = \frac{3}{5 - 4\cos x},$$

(7.64)

with Fourier expansion

$$f(x) = \sum_{k=-\infty}^{+\infty} \widehat{f}_k \, \mathrm{e}^{\mathrm{i}kx},$$

(7.65)

whose exact coefficients are (see Problem 3)

$$\widehat{f}_k = \frac{1}{2\pi} \int_0^{2\pi} f(x)e^{-ikx}dx = 2^{-|k|},\tag{7.66}$$

for $k \in \mathbb{Z}$. The coefficients $\widehat{f}_k$ of the approximated DFT truncated series

$$f(x) = \sum_{k=-N/2}^{N/2-1} \widetilde{f}_k \, e^{ikx}\tag{7.67}$$

can be easily computed using function **dftmat** in Code 22. For $N = 8$, the reader can check that

$$\widetilde{f}_0 = 1.007\ 843\ 137\ 254\ 9 \qquad \widetilde{f}_{\pm 1} = 0.509\ 803\ 921\ 568\ 627,\tag{7.68}$$

where $\widetilde{f}_0$ and $\widetilde{f}_{\pm 1}$ are stored in **fk(5)** and **fk(4)** or **fk(6)**, respectively,[4] which clearly differ from the exact values

$$\widehat{f}_0 = 1 \qquad \widehat{f}_{\pm 1} = \frac{1}{2},\tag{7.69}$$

given by relation (7.66). The numerical discrepancies between the approximate coefficients (7.68) and the exact ones (7.69) could have been anticipated by using aliasing error formula (7.62). According to that formula, the coefficients $\widetilde{f}_0$ and $\widehat{f}_0$ satisfy

$$\widetilde{f}_0 - \widehat{f}_0 = \widehat{f}_8 + \widehat{f}_{-8} + \widehat{f}_{16} + \widehat{f}_{-16} + \cdots = \overbrace{\frac{1}{2^8} + \frac{1}{2^8}}^{0.007\ 812\ 5} + \underbrace{\frac{1}{2^{16}} + \frac{1}{2^{16}}}_{} + \cdots$$

$$\underbrace{\qquad\qquad\qquad\qquad}_{0.007\ 843\ 017\ 578\ 125}$$

$$= \frac{2}{2^8 - 1}.\tag{7.70}$$

The sum of the first overbraced and underbraced terms of the power series above reveals the quick convergence to the exact limit $2(2^8 - 1)^{-1} = 0.007\ 843\ 137\ 254\ 901\ 96$, which is, to machine precision, exactly the difference between $\widetilde{f}_0$ and $\widehat{f}_0$. As an exercise (see Problem 3), the reader can also check that formula (7.62) accurately predicts the difference between $\widetilde{f}_{\pm 1}$ and $\widehat{f}_{\pm 1}$.

---

4 Remember that **dftmat** sorts the coefficients according to the shifted Fourier basis $\widetilde{\varphi}_k(x)$.

Practical 7.1: Discrete Fourier Transform (DFT)

The figure below shows a system of masses connected via linear springs:

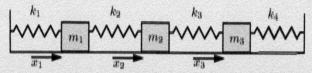

Let $x_i(t)$ and $\dot{x}_i \equiv v_i(t)$ be the displacement of the i-th block from its equilibrium position and its instantaneous speed at time t, respectively. The displacements and speeds satisfy the linear system of differential equations:

$$
\begin{bmatrix} \dot{x}_1 \\ \dot{x}_2 \\ \dot{x}_3 \\ \dot{v}_1 \\ \dot{v}_2 \\ \dot{v}_3 \end{bmatrix} = \begin{bmatrix} 0 & 0 & 0 & 1 & 0 & 0 \\ 0 & 0 & 0 & 0 & 1 & 0 \\ 0 & 0 & 0 & 0 & 0 & 1 \\ \beta_{11} & \beta_{12} & 0 & 0 & 0 & 0 \\ \beta_{21} & \beta_{22} & \beta_{23} & 0 & 0 & 0 \\ 0 & \beta_{32} & \beta_{33} & 0 & 0 & 0 \end{bmatrix} \begin{bmatrix} x_1 \\ x_2 \\ x_3 \\ v_1 \\ v_2 \\ v_3 \end{bmatrix}, \quad \begin{cases} \beta_{11} = -(k_1 + k_2)/m_1 \\ \beta_{12} = k_2/m_1, \ \beta_{21} = k_2/m_2 \\ \beta_{22} = -(k_2 + k_3)/m_2 \\ \beta_{23} = k_3/m_2, \ \beta_{32} = k_3/m_3 \\ \beta_{33} = -(k_3 + k_4)/m_3. \end{cases}
$$

Let $\mathbf{z}(t) = [x_1 \ x_2 \ x_3 \ v_1 \ v_2 \ v_3]^{\mathrm{T}} \in \mathbb{R}^6$ be the state vector at time t, and $\mathbf{B}$ the 6×6 matrix of coefficients of the ODE system above. For a given arbitrary initial condition $\mathbf{z}_0 \equiv \mathbf{z}(0)$, the solution is $\mathbf{z}(t) = e^{\mathbf{B}t}\mathbf{z}_0$. Random initial positions $x_i(0)$ and velocities $v_i(0)$ will in general lead to coupled oscillations with some characteristic frequencies or *normal modes* of oscillation. The purpose of this practical is to identify these natural frequencies using two different methods.

Take for example the numerical values $m_1 = m_2 = m_3 = 1$, $k_1 = 1$, $k_2 = \sqrt{2}$, $k_3 = \sqrt{3}$, and $k_4 = 4$, and consider the initial condition

$$\mathbf{z}_0 = [0.281 \ 0.033 \ -1.33 \ 1.12 \ 0.35 \ -0.299]^{\mathrm{T}}.$$

(a) METHOD I: Using MATLAB's command **expm**, sample the exact solution $\mathbf{z}_j \equiv \mathbf{z}(t_j) = e^{\mathbf{B}t_j}\mathbf{z}_0$ at $t_j = j\Delta t$, $(j = 0, 1, 2, \ldots, N-1)$. You just need to compute $e^{\mathbf{B}\Delta t}$ *once*:

$$\mathbf{z}_j = e^{j\Delta t\mathbf{B}}\mathbf{z}_0 = e^{\Delta t\mathbf{B}} e^{(j-1)\Delta t\mathbf{B}}\mathbf{z}_0 = e^{\Delta t\mathbf{B}}\mathbf{z}_{j-1}.$$

Integrate the system above with $\Delta t = 0.25$ and $N = 400$ and store $x_2(t_i)$. Plot $x_2(t)$ to visualize the oscillation. Using the DFT, compute and plot the power spectrum $|\tilde{f}_k|$ as a function of ω_k. Can you identify any characteristic peaks?

(b) METHOD II: The normal modes or frequencies are the imaginary parts of the spectrum of eigenvalues of $\mathbf{B}$. Using MATLAB's command **eig**, compute the eigenvalues of $\mathbf{B}$ and confirm your results obtained in (a).

7.2 Fourier Differentiation

Using the values $f_j \equiv f(2\pi j/N)$, for $j = 0, 1, 2, \ldots, N - 1$, DFT provides the Fourier coefficients $\widetilde{f}_k$ that constitute the discrete Fourier series (henceforth referred to as DFS)

$$F_N(x) \equiv \sum_{k=0}^{N-1} \widetilde{f}_k \, \widetilde{\varphi}_k(x) \equiv \sum_{k=0}^{N-1} \widetilde{f}_k \, e^{i\left(k-\frac{N}{2}\right)x}. \tag{7.71}$$

This result is essentially the same as what we already did in *interpolation* in Part I. In Part I, we saw how to obtain the coefficients $\{a_0, a_1, a_2, \ldots, a_n\}$ of the polynomial

$$\Pi_n f(x) \equiv a_0 + a_1 x + a_2 x^2 + \cdots + a_n x^n,$$

from the ordinates $f_j \equiv f(x_j)$ at a given set of nodes $\{x_0, x_1, x_2, \ldots, x_n\}$. In fact, the series (7.71) is a polynomial (a trigonometric polynomial), sometimes referred to as the *Fourier interpolant* or *trigonometric interpolant*, where the functional basis is not the set of monomials x^k, but the shifted trigonometric basis $\widetilde{\varphi}_k(x) \equiv e^{i\left(k-\frac{N}{2}\right)x}$.

Following the same reasoning as in Part I, if $F_N(x)$ is a good approximation of $f(x)$, then the derivative $F_N'(x)$ should also be a good approximation of $f'(x)$. Differentiation of (7.71) with respect to x leads to the expression

$$\frac{\mathrm{d} F_N(x)}{\mathrm{d}x} = \sum_{k=0}^{N-1} i\left(k - \frac{N}{2}\right) \widetilde{f}_k \, e^{i\left(k-\frac{N}{2}\right)x} \equiv \sum_{k=0}^{N-1} \widetilde{f}_k^{(1)} \, e^{i\left(k-\frac{N}{2}\right)x}, \tag{7.72}$$

where we have introduced the Fourier coefficients

$$\widetilde{f}_k^{(1)} \equiv i\left(k - \frac{N}{2}\right) \widetilde{f}_k, \tag{7.73}$$

for $k = 0, 1, \ldots, N - 1$. Differentiation of the DFS leads to a new function whose coefficients, in the shifted Fourier basis (7.13), are simply the original coefficients $\widetilde{f}_k$ multiplied by their corresponding factor $i(k - \frac{N}{2})$. By virtue of (7.37) and (7.38), we may write $\widetilde{f}_k^{(1)}$ as

$$\widetilde{f}_k^{(1)} \equiv i\left(k - \frac{N}{2}\right) \widetilde{f}_k = i\left(k - \frac{N}{2}\right) \sum_{\ell=0}^{N-1} \mathbf{F}_{k\ell} f_\ell, \tag{7.74}$$

so that (7.72) now reads

$$
\begin{aligned}
F_N'(x) \equiv \frac{\mathrm{d} F_N(x)}{\mathrm{d}x} &\equiv \sum_{k=0}^{N-1} i\left(k - \frac{N}{2}\right) \sum_{\ell=0}^{N-1} \mathbf{F}_{k\ell} \, f_\ell \, e^{i\left(k-\frac{N}{2}\right)x} \\
&= \sum_{\ell=0}^{N-1} \left[\sum_{k=0}^{N-1} i\left(k - \frac{N}{2}\right) \mathbf{F}_{k\ell} \, e^{i\left(k-\frac{N}{2}\right)x} \right] f_\ell. \tag{7.75}
\end{aligned}
$$

Evaluation of the last expression at any of the grid points $x_j = 2\pi j/N$ leads to the approximation

$$f'(x_j) \approx F'_N(x_j) = \sum_{\ell=0}^{N-1} \left[\sum_{k=0}^{N-1} i\left(k - \frac{N}{2}\right) \mathbf{F}_{k\ell}\, e^{i(k-\frac{N}{2})x_j} \right] f_\ell. \tag{7.76}$$

If we define the matrix

$$\mathbf{D}_{j\ell} \doteq \sum_{k=0}^{N-1} i\left(k - \frac{N}{2}\right) \mathbf{F}_{k\ell}\, e^{i(k-\frac{N}{2})x_j}, \tag{7.77}$$

for $0 \le j, \ell \le N-1$, expression (7.76) becomes the matrix–vector product

$$F'_N(x_j) = \sum_{\ell=0}^{N-1} \mathbf{D}_{j\ell} f_\ell, \tag{7.78}$$

for $j = 0, 1, \ldots, N-1$. The matrix (7.77) is actually a *differentiation matrix*, similar to the ones seen in Part I, but exclusively for periodic functions. In fact, (7.78) provides the approximation of the exact derivative $f'(x)$ at the grid points by precisely using the values of the original function $f(x)$ at the same set of points. Matrix (7.77) is known as the *Fourier differentiation matrix* and, by virtue of (7.38) and (7.77), its explicit elements are

$$\mathbf{D}_{j\ell}^{(1)} = \frac{i}{N} \sum_{k=0}^{N-1} \left(k - \frac{N}{2}\right) W_N^{-(k-\frac{N}{2})(j-\ell)} \quad (j, \ell = 0, 1, 2, \ldots, N-1), \tag{7.79}$$

where the superscript (1) indicates the order of differentiation (first in this case). Differentiating again DFS in (7.71) and evaluating at each one of the grid points leads to the *second order* Fourier matrix differentiation matrix:

$$\mathbf{D}_{j\ell}^{(2)} = -\frac{1}{N} \sum_{k=0}^{N-1} \left(k - \frac{N}{2}\right)^2 W_N^{-(k-\frac{N}{2})(j-\ell)} \quad (j, \ell = 0, 1, 2, \ldots, N-1). \tag{7.80}$$

```
% Code 25: Fourier 1st Ord Differentiation Matrix
% Input: N (even)
% Output: NxN Diff. Matrix D1
function D1 = dftdiffmat1(N)
WN = exp(-i*2*pi/N); D1 = zeros(N); k = [0:N-1];
for j = 0:N-1
  for l = 0:N-1
    D1(j+1,l+1) = (i/N)*sum((k-N/2).*WN.^(-(k-N/2)*(j-l)));
  end
end
```

```
% Code 26: Fourier 2nd Ord Differentiation Matrix
% Input: N (even)
% Output: NxN Diff. Matrix D2
function D2 = dftdiffmat2(N)
WN = exp(-i*2*pi/N); D2 = zeros(N); k = [0:N-1];
for j = 0:N-1
  for l = 0:N-1
    D2(j+1,l+1) = -sum(((k-N/2).^2).*WN.^(-(k-N/2)*(j-1)))/N;
  end
end
```

Codes 25 and 26 are simple implementations of matrix differentiation matrices (7.79) and (7.80), respectively. More compact implementations of these matrices can be found in the references at the end of the book.

We illustrate the use of Fourier differentiation matrices with a very simple example. Figure 7.4a shows the function $f(x) = e^{\sin x}$ (black solid curve) sampled on the set of points $x_j = 2\pi j/N$ ($j = 0, 1, \ldots, N-1$), shown as gray bullets, for $N = 32$. The grid evaluation of the first derivative of the DFS, $F_N'(x)$, can be performed by multiplying the Fourier matrix (7.79) by the vector of ordinates of the original function by simply typing:

```
N = 32; xj = (2*pi/N)*[0:N-1]'; fj = exp(sin(xj));
D = dftdiffmat1(N); Dfj = D*fj; plot(xj,Dfj,'ok');
```

Figure 7.4b shows the exact derivative $f'(x) = e^{\sin x} \cos x$ (black solid curve) along with the resulting approximation (white squares) $F_N'(x)$ at the nodes x_j. Figure 7.4b also shows the second derivative $f''(x)$ (dashed curve) and its corresponding approximation $F_N''(x)$ evaluated at the nodes (white triangles). Finally, Figure 7.4c depicts the pointwise absolute errors of the first and second derivative approximations, $|f'(x_j) - F_N'(x_j)|$ and $|f''(x_j) - F_N''(x_j)|$, respectively, for $j = 0, 1, \ldots, N-1$. We can observe that the error of the second derivative is slightly larger than the one corresponding to the first derivative. To some extent, this is the same kind of deterioration in the accuracy that we already observed in Part I when we studied second order differentiation matrices obtained from global interpolants.

Figure 7.5 shows the maximum pointwise errors corresponding to the first and second derivatives as a function of N. From that semi-logarithmic plot, we can clearly identify an *exponential* convergence, similar to the one observed in global polynomial interpolation addressed in Part I. In general, Fourier differentiation provides exponential convergence whenever the interpolated periodic

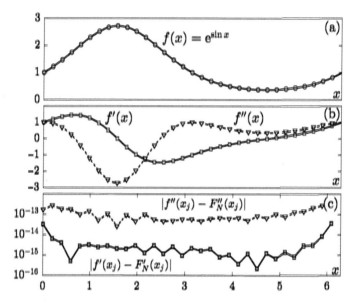

Figure 7.4 Fourier differentiation of function $f(x) \equiv e^{\sin x}$. (a) Sampled function at $N = 32$ equispaced abscissas. (b) First and second derivatives of $f(x)$ (solid black and dashed curves, respectively) and $F_N(x)$ at the nodes (white squares and triangles, respectively). (c) Pointwise errors.

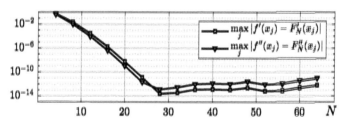

Figure 7.5 Exponential convergence of Fourier differentiation of function $f(x) \equiv e^{\sin x}$.

function has smooth derivatives. For a detailed analysis of the different types of convergence provided by the DFT, we refer the reader to the Complementary Reading section, at the end of the chapter.

Fourier differentiation matrices (7.79) and (7.80) map the values f_j of the sampled function on physical space to the approximate values $F'_N(x_j)$ and $F''_N(x_j)$ of the first and second derivatives, respectively. This is exactly what we already did in Part I with, for example, Chebyshev differentiation matrices. In that sense, we say that this differentiation procedure is carried out in *physical space*. However, sometimes it may be more convenient to proceed with the differentiation in *Fourier* space. To be more specific, recall that, by virtue of (7.72),

$$F_N'(x) \equiv \sum_{k=0}^{N-1} \widetilde{f}_k^{(1)} \, e^{i(k-\frac{N}{2})x}, \tag{7.81}$$

where

$$\widetilde{f}_k^{(1)} \equiv i\left(k - \frac{N}{2}\right) \widetilde{f}_k, \tag{7.82}$$

for $k = 0, 1, \ldots, N - 1$. In other words, if the discrete Fourier representation of $f(x)$ in S_N is given by the vector

$$\widetilde{\mathbf{f}} = [f_0 \; f_1 \; \cdots \; f_{N-2} \; f_{N-1}]^{\mathrm{T}}, \tag{7.83}$$

then, the discrete Fourier representation of $F_N'(x)$ is a vector, namely $\widetilde{\mathbf{f}}^{(1)}$, with components

$$\widetilde{\mathbf{f}}^{(1)} \equiv \left[i\left(-\frac{N}{2}\right)f_0 \; i\left(1 - \frac{N}{2}\right)f_1 \; \cdots \; i\left(\frac{N}{2} - 1\right)f_{N-1}\right]^{\mathrm{T}}. \tag{7.84}$$

Similarly, $F_N''(x)$ is represented in Fourier space by the vector

$$\widetilde{\mathbf{f}}^{(2)} \equiv \left[-\left(-\frac{N}{2}\right)^2 f_0 \; -\left(1 - \frac{N}{2}\right)^2 f_1 \; \cdots \; -\left(\frac{N}{2} - 1\right)^2 f_{N-1}\right]^{\mathrm{T}}. \tag{7.85}$$

This also allows us to write the derivatives in a matrix–vector fashion, that is,

$$\widetilde{\mathbf{f}}^{(1)} \equiv \widetilde{\mathbf{D}}^{(1)} \, \widetilde{\mathbf{f}}, \tag{7.86}$$

and

$$\widetilde{\mathbf{f}}^{(2)} \equiv \widetilde{\mathbf{D}}^{(2)} \, \widetilde{\mathbf{f}}, \tag{7.87}$$

where the differentiation matrices in Fourier space are *diagonal*, with entries

$$\widetilde{\mathbf{D}}_{j\ell}^{(1)} \equiv i\left(\ell - \frac{N}{2}\right)\delta_{\ell j} \tag{7.88}$$

and

$$\widetilde{\mathbf{D}}_{j\ell}^{(2)} \equiv -\left(\ell - \frac{N}{2}\right)^2 \delta_{\ell j}, \tag{7.89}$$

for $0 \le \ell, j \le N - 1$. Of course, in order to obtain the derivatives in physical space, we still have to pull back the vectors (7.84) and (7.85) using the IDFT (7.40). As an illustration, we compute the approximations of the first and second derivatives of $f(x) \equiv e^{\sin x}$ for $N \equiv 32$, but now performing the differentiation in Fourier space, as shown below:

```
N = 32; jj = [0:N-1]'; xj = (2*pi/N)*jj; fj = exp(sin(xj));
fk = dftmat(fj); k1 = 1i*(jj-N/2); k2 = -(jj-N/2).^2;
fk1 = k1.*fk; fk2 = k2.*fk;
Dfj = idftmat(fk1); D2fj = idftmat(fk2);
```

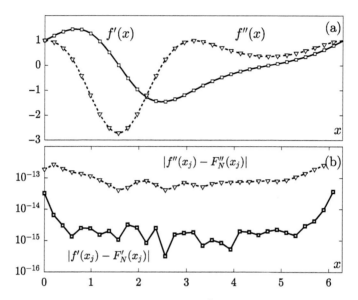

Figure 7.6 Differentiation of function $f(x) = e^{\sin x}$ in Fourier space, following Figure 7.4. (a) First and second derivatives of $f(x)$ and (b) pointwise errors.

Starting from the values of the function contained in vector `fj`, we compute its DFT using `dftmat` in order to obtain the corresponding Fourier coefficients (stored in `fk`). Then we define the vectors `k1` and `k2`, containing the factors that need to be multiplied componentwise[5] to `fk` in order to obtain the Fourier representations of the first and second derivatives, `fk1` and `fk2`, respectively. Finally, we pull back these two vectors to physical space using `idftmat`. Figure 7.6a shows the resulting derivatives `Dfj` and `D2fj`, indistinguishable to the naked eye from the ones already shown in Figure 7.4b using differentiation in physical space. Figure 7.6b depicts the pointwise errors, which are of the same order of magnitude as the ones of Figure 7.4c. Choosing between performing derivatives in physical or Fourier space strongly depends on the problem at hand. We will show how to apply both formulations in Chapter 8, when dealing with the numerical approximation of *boundary value problems*. Within that context, we will see the pros and cons of the two possible approaches.

This alternative way to approximate derivatives in Fourier space can also be extended to other sets of orthogonal functions, not necessarily periodic, such as Chebyshev polynomials. However, this approach is far beyond the scope of this introductory chapter. We refer the interested reader to the recommended bibliography in the Complementary Reading section.

---

5 In this case, we do not need to build matrices (7.88) and (7.89) explicitly.

Summary: Fourier Differentiation

Physical Space:

$$F_N'(x_j) = \sum_{\ell=0}^{N-1} \mathbf{D}_{j\ell}^{(1)} f_\ell \quad \text{and} \quad F_N''(x_j) = \sum_{\ell=0}^{N-1} \mathbf{D}_{j\ell}^{(2)} f_\ell, \tag{7.90}$$

for $j = 0, 1, \ldots, N-1$, where

$$\mathbf{D}_{j\ell}^{(1)} = \frac{\mathrm{i}}{N} \sum_{k=0}^{N-1} \left(k - \frac{N}{2} \right) W_N^{-(k-\frac{N}{2})(j-\ell)}, \tag{7.91}$$

and

$$\mathbf{D}_{j\ell}^{(2)} = -\frac{1}{N} \sum_{k=0}^{N-1} \left(k - \frac{N}{2} \right)^2 W_N^{-(k-\frac{N}{2})(j-\ell)}, \tag{7.92}$$

for $0 \leq \ell, \, j \leq N - 1$.

Fourier Space:

$$\widetilde{f}_j^{(1)} = \sum_{\ell=0}^{N-1} \widetilde{\mathbf{D}}_{j\ell}^{(1)} \widetilde{f}_\ell \quad \text{and} \quad \widetilde{f}_j^{(2)} = \sum_{\ell=0}^{N-1} \widetilde{\mathbf{D}}_{j\ell}^{(2)} \widetilde{f}_\ell, \tag{7.93}$$

where

$$\widetilde{\mathbf{D}}_{j\ell}^{(1)} = \mathrm{i} \left(\ell - \frac{N}{2} \right) \delta_{\ell j} \quad \text{and} \quad \widetilde{\mathbf{D}}_{j\ell}^{(2)} = -\left(\ell - \frac{N}{2} \right)^2 \delta_{\ell j}, \tag{7.94}$$

for $0 \leq \ell, \, j \leq N - 1$.

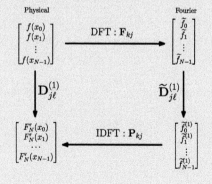

Complementary Reading

For a nice theoretical introduction to Fourier analysis, I particularly recommend Folland's *Fourier Analysis and Its Applications*. For over three decades, one of the most authoritative references of the numerical analysis of the discrete Fourier transform and its applications has been Briggs and Henson's monograph *The DFT*. This book not only addresses the underlying numerical mathematics behind the DFT, but also includes applications to a large variety of areas such diverse as seismology or tomography. However, I particularly like the monograph *Numerical Mathematics* by Quarteroni, Sacco, and Saleri, where the reader will find a beautiful and enlightening analysis of the DFT within the context of best approximation theory and trigonometric interpolation. Many parts of that analysis have inspired the approach taken in this chapter. For a very clear analysis on the aliasing phenomenon, I particularly like Trefethen's monograph *Spectral Methods in Matlab*, where the reader will also find compact implementations of the Fourier differentiation matrices, among many other aspects of the DFT. The differentiation examples shown here have been taken from that monograph. The convergence of DFT depending on the smoothness of the sampled function and on the distance to the nearest singularity is beautifully outlined in Table 2.4-1 of Fornberg's *A Practical Guide to Pseudospectral Methods*. Differentiation in Fourier space can be generalized to other types of (non-periodic) approximations, such as Chebyshev or Legendre polynomials. For a nice introduction, I particularly recommend Gottlieb and Orszag's *Numerical Analysis of Spectral Methods*.

Problems and Exercises

The symbol (A) means that the problem has to be solved analytically, whereas (N) means to use numerical Matlab codes. Problems with the symbol * are slightly more difficult.

1. (A) Show that the coefficients $\widehat{f}_k$ corresponding to the Fourier series

$$\sum_{k=-\infty}^{+\infty} \widehat{f}_k \, e^{ikx}$$

of a smooth real valued function satisfy $\widehat{f}_{-k} = \overline{\widehat{f}_k}$, for all k.

2. (A)* Show that the truncated Fourier series of any smooth 2π-periodic function $f(x)$

$$f_N = \sum_{k=-\frac{N}{2}}^{\frac{N}{2}-1} \widehat{f}_k \, e^{ikx},$$

with coefficients $\widehat{f}_k = \dfrac{1}{2\pi} \displaystyle\int_0^{2\pi} f(x) \, e^{-ikx} \, dx$, satisfies

$$\|f - f_N\| = \min_{g \in S_N} \|f - g\|,$$

that is, f_N is the optimal *least squares approximation* to the function f within the subspace S_N.

3. (AN) Consider the Fourier series $f(x) = \displaystyle\sum_{k=-\infty}^{+\infty} \widehat{f}_k \, e^{ikx}$ of the function

$$f(x) = \frac{3}{5 - 4\cos x},$$

with coefficients

$$\widehat{f}_k = (2\pi)^{-1} \int_0^{2\pi} f(x) e^{-ikx} dx.$$

Also consider the DFT of $f(x)$ for $N = 8$ and its resulting coefficients

$$\widetilde{f}_k = \frac{1}{N} \sum_{j=0}^{N-1} f(x_j) e^{ikx_j} \quad (-N/2 \le k \le N/2 - 1),$$

where $x_j = 2\pi j/N$ $(j = 0, 1, 2, \ldots, N - 1)$.
   (a) Show that $\widehat{f}_k = 2^{-|k|}$, for $k = 0, \pm 1, \pm 2, \ldots$. **Hint:** Use *Cauchy's residue theorem.*
   (b) Using aliasing error formula (7.62), compute the *exact* difference $\widetilde{f}_1 - \widehat{f}_1$. Confirm your result numerically.

8

Ordinary Differential Equations

The laws of physics, in their most abstract and general form, are generally written in the language of differential equations. These types of equations appear in every branch of the physical sciences and engineering. Ordinary differential equations (henceforth referred to as ODEs) appear in classical mechanics (Hamilton's equations), biology (Lotka–Volterra equations), or in chemical kinetic models such as the Brusselator. Partial differential equations (henceforth referred to as PDEs) appear almost in every area of physics, from electromagnetism (Maxwell's equations) or fluid dynamics (Navier–Stokes equations), to quantum mechanics (Schrödinger equation) or heat transfer (Fourier's law of thermal conduction), to mention just a few examples. Linear ODE may sometimes be solvable by means of analytical methods such as matrix exponentiation, power expansion (Frobenius method), and Laplace transform. Even when analytical techniques can provide the solution in closed form, this form may frequently be of great mathematical complexity and not always easy to use in practice. In general, if the ODE or the PDE is *nonlinear*, the only way to accurately envisage its solution (or solutions) typically requires numerical methodologies.

In this chapter is an introduction to the most fundamental techniques to approximate solutions of ODEs. We will see how to discretize *boundary value problems* (BVPs) and *initial value problems* (IVPs). The numerical solution of PDE is a far more vast and complex field and its treatment is well beyond the scope of this introductory textbook. We will only address this topic at the end of this chapter by reducing a few instances of simple PDEs to a system of ODEs. For a comprehensive analysis of the numerical analysis of PDE we refer the reader to the Complementary Reading section.

Throughout this chapter, we assume that the reader is already familiar with many concepts arising from the general theory of ODEs, such as *existence* and *uniqueness* of solutions, autonomous systems, linear stability of fixed points, Dirichlet–Neumann boundary conditions, and Sturm–Liouville problems.

Fundamentals of Numerical Mathematics for Physicists and Engineers, First Edition. Alvaro Meseguer.
© 2020 John Wiley & Sons, Inc. Published 2020 by John Wiley & Sons, Inc.

At the end of the chapter, in the Complementary Reading section, we recommend some bibliography that can be useful to review these and other concepts.

8.1 Boundary Value Problems

We start by considering a second order ODE of the form

$$\frac{\mathrm{d}^2 f}{\mathrm{d}x^2} = F\left(x, f, \frac{\mathrm{d}f}{\mathrm{d}x}\right), \tag{8.1}$$

or simply

$$f''(x) = F(x, f, f'), \tag{8.2}$$

where $f'(x)$ and $f''(x)$ will henceforth denote $\dfrac{\mathrm{d}f}{\mathrm{d}x}$ and $\dfrac{\mathrm{d}^2 f}{\mathrm{d}x^2}$, respectively. We will address how to discretize BVP in bounded, periodic, and unbounded domains. In Part I, we stressed the importance of differentiation matrices. It is precisely in this chapter where the advantages of using these types of matrices will be made more clear. For example, for bounded domains we will require Chebyshev differentiation matrices, seen in Chapter 3, whereas for periodic domains we will make use of the Fourier differentiation matrices (either in physical or frequency space) covered in Chapter 7. For unbounded domains we will make use of domain transformation techniques that reduce the problem to an open bounded interval. For this reason, we recommend the reader to review the aforementioned chapters.

8.1.1 Bounded Domains

In this section, we look for solutions of (8.2) satisfying the so-called *Robin boundary conditions*:

$$c_{11} f(a) + c_{12} f'(a) = c_{13} \tag{8.3}$$

$$c_{21} f(b) + c_{22} f'(b) = c_{23}. \tag{8.4}$$

These are generalized linear combinations of *Dirichlet–Neumann* boundary conditions. The conditions for existence and uniqueness of solutions of (8.2) subjected to (8.4) is not a simple matter. In general, if $\partial_f F > 0$, F has bounded absolute partial derivatives $|\partial_f F|$ and $|\partial_{f'} F|$ in (a, b), and the coefficients c_{ij} in (8.4) satisfy the inequalities

$$|c_{11}| + |c_{21}| > 0, \quad |c_{11}| + |c_{12}| > 0, \quad |c_{21}| + |c_{22}| > 0, \quad c_{11} c_{12} \leq 0, \quad c_{21} c_{22} \geq 0, \tag{8.5}$$

then the BVP (8.2–8.4) has a unique solution. We refer the reader to the selected bibliography at the end of the chapter, in the Complementary Reading section, for more details. For example, consider the BVP

$$f'' + \pi^2 f = 0; \quad f(0) = 1, \quad f(1) = 0, \tag{8.6}$$

where $F \equiv -\pi^2 f$ and $\partial_f F = -\pi^2 < 0$. The characteristic polynomial of the ODE above is $p(\lambda) = \lambda^2 + \pi^2$, with roots $\lambda = \pm i\pi$, so that the general solution reads

$$f(x) = C_1 e^{-i\pi x} + C_2 e^{i\pi x}. \tag{8.7}$$

The constants C_1 and C_2 must be determined by imposing the boundary conditions, that is,

$$f(0) = C_1 + C_2 = 1 \tag{8.8}$$
$$f(1) = -C_1 - C_2 = 0, \tag{8.9}$$

which is an inconsistent system. Henceforth in this book we will always assume that the BVP admits at least one solution. Multiple solutions will typically appear when solving linear eigenvalue BVP or when the equation is nonlinear, for example.

We first study how to discretize a second order linear BVP of the form

$$p(x)f'' + q(x)f' + r(x)f = g(x), \quad x \in (-1,1),$$

$$\text{with} \quad \begin{cases} c_{11} f(-1) + c_{12} f'(-1) &= e_{13}, \\ c_{21} f(1) + c_{22} f'(1) &= e_{23}, \end{cases} \tag{8.10}$$

where the given functions $p(x)$, $q(x)$, $r(x)$, and $g(x)$ are supposed to be differentiable in $(-1,1)$.[1] In order to solve (8.10) numerically, assume that the solution $f(x)$ takes the (unknown) values $\{f_0, f_1, \ldots, f_n\}$ at a given set of nodes $\{x_0, x_1, \ldots, x_n\}$ in $[-1,1]$, respectively. As seen in Part I, the derivatives $f'(x)$ and $f''(x)$ can be approximated using the corresponding differentiation matrices associated with these nodes. These matrices can be obtained from low order local (FD, CD2, etc.) or high order global (Chebyshev) interpolants. In this section, we use a *global* approach. Local strategies will be used in Section 8.2, within the context of IVPs.

To approximate derivatives, consider the Chebyshev differentiation matrix $\mathbf{D}$ seen in (3.36) associated with the Chebyshev nodes $x_j \equiv \cos(\pi j/n)$, for $j = 0, 1, \ldots, n$, and whose explicit elements are

$$\mathbf{D}_{ij} \equiv \begin{cases} (-1)^{i+j} \dfrac{\delta_j}{\delta_i (x_i - x_j)}, & (i \neq j), \\ \dfrac{(-1)^{i+1}}{\delta_i} \displaystyle\sum_{k=0 \ (k \neq i)}^{n} (-1)^k \dfrac{\delta_k}{x_i - x_k}, & (i = j), \end{cases} \tag{8.11}$$

where $\delta_0 \equiv \delta_n \equiv 1/2$ and $\delta_1 \equiv \delta_2 \equiv \cdots \equiv \delta_{n-1} \equiv 1$. To approximate the second order derivatives, we use the matrix $\mathbf{D}^{(2)} \equiv \mathbf{D}^2$ (see Practical 3.2). Our goal is to discretize the differential operator appearing on the left-hand side of (8.10)

$$\mathcal{L} = p(x) \frac{\mathrm{d}^2}{\mathrm{d}x^2} + q(x) \frac{\mathrm{d}}{\mathrm{d}x} + r(x). \tag{8.12}$$

---

1 For a domain $z \in (a,b)$, use a simple change of variable $z = a + (b-a)(x+1)/2$.

Let $\mathbf{f} = [f_0 \ f_1 \ \cdots \ f_n]^T \in \mathbb{R}^{n+1}$ be a column-vector containing the values of the solution at the Chebyshev nodes. The approximation of the first derivative of $\mathbf{f}$ at the nodes is given by the matrix-vector product

$$
\begin{bmatrix} f'(x_0) \\ f'(x_1) \\ \vdots \\ f'(x_n) \end{bmatrix} \approx \begin{bmatrix} \mathbf{D}_{00} & \mathbf{D}_{01} & \cdots & \mathbf{D}_{0n} \\ \mathbf{D}_{10} & \mathbf{D}_{11} & \cdots & \mathbf{D}_{1n} \\ \vdots & \vdots & & \vdots \\ \mathbf{D}_{n0} & \mathbf{D}_{n1} & \cdots & \mathbf{D}_{nn} \end{bmatrix} \begin{bmatrix} f_0 \\ f_1 \\ \vdots \\ f_n \end{bmatrix} = \mathbf{D}\mathbf{f}. \tag{8.13}
$$

Similarly, we can approximate the linear differential operator $q(x)\dfrac{\mathrm{d}}{\mathrm{d}x}$ at the nodes by

$$
\begin{bmatrix} q(x_0)f'(x_0) \\ q(x_1)f'(x_1) \\ \vdots \\ q(x_n)f'(x_n) \end{bmatrix} \approx \begin{bmatrix} q(x_0) & 0 & \cdots & 0 \\ 0 & q(x_1) & \cdots & 0 \\ \vdots & \vdots & & \vdots \\ 0 & 0 & \cdots & q(x_n) \end{bmatrix} \begin{bmatrix} \mathbf{D}_{00} & \mathbf{D}_{01} & \cdots & \mathbf{D}_{0n} \\ \mathbf{D}_{10} & \mathbf{D}_{11} & \cdots & \mathbf{D}_{1n} \\ \vdots & \vdots & & \vdots \\ \mathbf{D}_{n0} & \mathbf{D}_{n1} & \cdots & \mathbf{D}_{nn} \end{bmatrix} \begin{bmatrix} f_0 \\ f_1 \\ \vdots \\ f_n \end{bmatrix}, \tag{8.14}
$$

which we will simply denote by

$$
\mathrm{diag}(\mathbf{Q})\mathbf{D}\,\mathbf{f}, \tag{8.15}
$$

where $\mathrm{diag}(\mathbf{Q})$ is a diagonal matrix with entries $\mathrm{diag}(\mathbf{Q})_{jj} = q(x_j)$, for $j = 0, 1, \ldots, n$. If we proceed similarly with the other two terms of $\mathcal{L}$ in (8.12), we obtain the discrete operator given by the matrix

$$
\mathbf{L} = \mathrm{diag}(\mathbf{P})\,\mathbf{D}^{(2)} + \mathrm{diag}(\mathbf{Q})\,\mathbf{D} + \mathrm{diag}(\mathbf{R}), \tag{8.16}
$$

where $\mathrm{diag}(\mathbf{P})$ and $\mathrm{diag}(\mathbf{R})$ are diagonal matrices with entries $\mathrm{diag}(\mathbf{P})_{jj} = p(x_j)$ and $\mathrm{diag}(\mathbf{R})_{jj} = r(x_j)$, for $j = 0, 1, \ldots, n$, respectively. As a result, the discrete version of the differential equation (8.10) is

$$
\begin{bmatrix} \mathbf{L}_{00} & \mathbf{L}_{01} & \cdots & \mathbf{L}_{0,n-1} & \mathbf{L}_{0n} \\ \mathbf{L}_{10} & \mathbf{L}_{11} & \cdots & \mathbf{L}_{1,n-1} & \mathbf{L}_{1n} \\ \vdots & \vdots & & \vdots & \vdots \\ \mathbf{L}_{n-1,0} & \mathbf{L}_{n-1,1} & \cdots & \mathbf{L}_{n-1,n-1} & \mathbf{L}_{n-1,n} \\ \mathbf{L}_{n0} & \mathbf{L}_{n1} & \cdots & \mathbf{L}_{n,n-1} & \mathbf{L}_{nn} \end{bmatrix} \begin{bmatrix} f_0 \\ f_1 \\ \vdots \\ f_{n-1} \\ f_n \end{bmatrix} = \begin{bmatrix} g_0 \\ g_1 \\ \vdots \\ g_{n-1} \\ g_n \end{bmatrix}, \tag{8.17}
$$

where $g_j = g(x_j)$ are the values of the right-hand side of (8.10) at the Chebyshev nodes. However, system (8.17) *does not* incorporate Robin's boundary conditions (8.10) yet. The implementation of the boundary conditions can be done in many different ways. For example, consider the Robin condition at $x_n = -1$

$$c_{11}f(-1) + c_{12}f'(-1) = c_{13}. \qquad (8.18)$$

In our discretization, the quantity $f(-1)$ is simply[2] f_n, whereas $f'(-1)$ must be replaced by its numerical approximation $\sum_{j=0}^{n} D_{nj}f_j$, so that the discretization of the boundary condition (8.18) is

$$c_{11}f_n + c_{12}[D_{n0}f_0 + D_{n1}f_1 + \cdots + D_{n,n-1}f_{n-1} + D_{nn}f_n] = c_{13}. \qquad (8.19)$$

Similarly, the discrete version of the Robin condition at $x_0 = 1$,

$$c_{21}f(1) + c_{22}f'(1) = c_{23}, \qquad (8.20)$$

is

$$c_{21}f_0 + c_{22}[D_{00}f_0 + D_{01}f_1 + \cdots + D_{0,n-1}f_{n-1} + D_{0n}f_n] = c_{23}. \qquad (8.21)$$

The two equations (8.19) and (8.21) implicitly relate the boundary values f_0 and f_n of the approximated solution as linear combinations of the remaining *interior* values $\{f_1, f_2, \ldots, f_{n-1}\}$. These relations can be expressed as

$$\begin{bmatrix} c_{21} + c_{22}D_{00} & c_{22}D_{0n} \\ c_{12}D_{n0} & c_{11} + c_{12}D_{nn} \end{bmatrix} \begin{bmatrix} f_0 \\ f_n \end{bmatrix}$$

$$= \begin{bmatrix} c_{23} \\ c_{13} \end{bmatrix} + \begin{bmatrix} c_{22} & 0 \\ 0 & c_{12} \end{bmatrix} \begin{bmatrix} -D_{01} & \cdots & -D_{0,n-1} \\ -D_{n1} & \cdots & -D_{n,n-1} \end{bmatrix} \begin{bmatrix} f_1 \\ \vdots \\ f_{n-1} \end{bmatrix}, \qquad (8.22)$$

or

$$\mathbf{M}_2 \begin{bmatrix} f_0 \\ f_n \end{bmatrix} = \begin{bmatrix} c_{23} \\ c_{13} \end{bmatrix} + \begin{bmatrix} c_{22} & 0 \\ 0 & c_{12} \end{bmatrix} \mathbf{M}_1 \begin{bmatrix} f_1 \\ \vdots \\ f_{n-1} \end{bmatrix}, \qquad (8.23)$$

where we have introduced the matrices

$$\mathbf{M}_1 = - \begin{bmatrix} D_{01} & \cdots & D_{0,n-1} \\ D_{n1} & \cdots & D_{n,n-1} \end{bmatrix} \quad \text{and} \quad \mathbf{M}_2 = \begin{bmatrix} c_{21} + c_{22}D_{00} & c_{22}D_{0n} \\ c_{12}D_{n0} & c_{11} + c_{12}D_{nn} \end{bmatrix}. \qquad (8.24)$$

Assuming that $\mathbf{M}_2$ is non-singular, (8.23) reads

$$\begin{bmatrix} f_0 \\ f_n \end{bmatrix} = \mathbf{M}_2^{-1} \begin{bmatrix} c_{23} \\ c_{13} \end{bmatrix} + \mathbf{M}_2^{-1} \begin{bmatrix} c_{22} & 0 \\ 0 & c_{12} \end{bmatrix} \mathbf{M}_1 \begin{bmatrix} f_1 \\ \vdots \\ f_{n-1} \end{bmatrix}. \qquad (8.25)$$

---

2 Remember that the Chebyshev nodes are in reversed order, with $x_0 = 1$ and $x_n = -1$.

In Eq. (8.25), the boundary values f_0 and f_n (corresponding to the boundary nodes x_0 and x_n, respectively) are explicitly written in terms of the interior values $\{f_1, \ldots, f_{n-1}\}$. Consequently, the first and last equations of the discrete differential operator (8.17), corresponding to nodes x_0 and x_n, are no longer part of the problem. The remaining $n = 1$ equations of (8.17) must incorporate the boundary relations (8.25) in order to remove the dependence on f_0 and f_n, expressing everything in terms of the genuine $n - 1$ degrees of freedom $\{f_1, \ldots, f_{n-1}\}$. To do that, first consider the explicit linear relations (8.17) corresponding to the interior nodes,

$$\mathbf{L}_{10}f_0 + \mathbf{L}_{11}f_1 + \cdots + \mathbf{L}_{1,n-1}f_{n-1} + \mathbf{L}_{1n}f_n = g_1$$

$$\mathbf{L}_{20}f_0 + \mathbf{L}_{21}f_2 + \cdots + \mathbf{L}_{2,n-1}f_{n-1} + \mathbf{L}_{2n}f_n = g_2$$

$$\vdots$$

$$\mathbf{L}_{n-1,0}f_0 + \mathbf{L}_{n-1,1}f_1 + \cdots + \mathbf{L}_{n-1,n-1}f_{n-1} + \mathbf{L}_{n-1,n}f_n = g_{n-1}. \tag{8.26}$$

The left-hand side of the previous equation can be rearranged in order to isolate the boundary values f_0 and f_n:

$$\begin{bmatrix} \mathbf{L}_{11} & \cdots & \mathbf{L}_{1,n-1} \\ \mathbf{L}_{21} & \cdots & \mathbf{L}_{2,n-1} \\ \vdots & & \vdots \\ \mathbf{L}_{n-1,1} & \cdots & \mathbf{L}_{n-1,n-1} \end{bmatrix} \begin{bmatrix} f_1 \\ \vdots \\ f_{n-1} \end{bmatrix} + \begin{bmatrix} \mathbf{L}_{10} & \mathbf{L}_{1n} \\ \mathbf{L}_{20} & \mathbf{L}_{2n} \\ \vdots & \vdots \\ \mathbf{L}_{n-1,0} & \mathbf{L}_{n-1,n} \end{bmatrix} \begin{bmatrix} f_0 \\ f_n \end{bmatrix} = \begin{bmatrix} g_1 \\ \vdots \\ g_{n-1} \end{bmatrix}, \tag{8.27}$$

or

$$\widehat{\mathbf{L}} \begin{bmatrix} f_1 \\ \vdots \\ f_{n-1} \end{bmatrix} = -\mathbf{M}_g \begin{bmatrix} f_0 \\ f_n \end{bmatrix} + \begin{bmatrix} g_1 \\ \vdots \\ g_{n-1} \end{bmatrix}, \tag{8.28}$$

where we have introduced the matrices

$$\widehat{\mathbf{L}} \equiv \begin{bmatrix} \mathbf{L}_{11} & \cdots & \mathbf{L}_{1,n-1} \\ \mathbf{L}_{21} & \cdots & \mathbf{L}_{2,n-1} \\ \vdots & & \vdots \\ \mathbf{L}_{n-1,1} & \cdots & \mathbf{L}_{n-1,n-1} \end{bmatrix} \in \mathbb{M}_{n-1}(\mathbb{R}) \tag{8.29}$$

and

$$\mathbf{M}_g \equiv \begin{bmatrix} \mathbf{L}_{10} & \mathbf{L}_{1n} \\ \mathbf{L}_{20} & \mathbf{L}_{2n} \\ \vdots & \vdots \\ \mathbf{L}_{n-1,0} & \mathbf{L}_{n-1,n} \end{bmatrix} \in \mathbb{M}_{n-1,2}(\mathbb{R}). \tag{8.30}$$

The matrix $\widehat{\mathbf{L}}$ is the result of trimming the original discrete operator $\mathbf{L}$ appearing in (8.17) by removing its first and last rows and columns. Substitution of the boundary terms $[f_0 \ f_n]^{\mathrm{T}}$ from Eq. (8.25) into (8.28) leads to the linear system

$$\left\{ \widehat{\mathbf{L}} + \mathbf{M}_3 \mathbf{M}_2^{-1} \begin{bmatrix} c_{22} & 0 \\ 0 & c_{12} \end{bmatrix} \mathbf{M}_1 \right\} \begin{bmatrix} f_1 \\ \vdots \\ f_{n-1} \end{bmatrix} \equiv \begin{bmatrix} g_1 \\ \vdots \\ g_{n-1} \end{bmatrix} - \mathbf{M}_3 \mathbf{M}_2^{-1} \begin{bmatrix} c_{23} \\ c_{13} \end{bmatrix}$$

(8.31)

System (8.31) encompasses the discretization of the differential operator $\mathcal{L}$, along with the discrete Robin boundary conditions. In particular, for homogeneous Dirichlet boundary conditions ($c_{12} \equiv c_{22} \equiv c_{13} \equiv c_{23} \equiv 0$), system (8.31) reduces to

$$\widehat{\mathbf{L}} \begin{bmatrix} f_1 \\ \vdots \\ f_{n-1} \end{bmatrix} \equiv \begin{bmatrix} g_1 \\ \vdots \\ g_{n-1} \end{bmatrix}.$$

(8.32)

The numerical solution of (8.10) subjected to homogeneous Neumann boundary conditions may be problematic if $r(x) \equiv 0$. In that case, any constant function is a solution of the homogeneous problem and the uniqueness is lost. In that case, the operator within curly brackets in (8.31) may be singular and an additional constraint is usually required. However, whenever the solution of (8.31) is unique, it can be obtained using any of the direct solvers seen in Chapter 5, such as LU or QR.

In principle, the discretization shown in (8.31) should be valid for any consistent linear BVP within a bounded domain. In particular, for *eigenvalue* BVPs of the form

$$p(x)f'' + q(x)f' + r(x)f \equiv \lambda f, \ x \in (-1,1),$$

$$\text{with} \begin{cases} e_{11}f(-1) + e_{12}f'(-1) & \equiv 0, \\ e_{21}f(1) + e_{22}f'(1) & \equiv 0, \end{cases}$$

(8.33)

the discrete BVP (8.31) becomes

$$\left\{ \widehat{\mathbf{L}} + \mathbf{M}_3 \mathbf{M}_2^{-1} \begin{bmatrix} c_{22} & 0 \\ 0 & c_{12} \end{bmatrix} \mathbf{M}_1 \right\} \begin{bmatrix} f_1 \\ \vdots \\ f_{n-1} \end{bmatrix} \equiv \lambda \begin{bmatrix} f_1 \\ \vdots \\ f_{n-1} \end{bmatrix}.$$

(8.34)

In this case, we only need to compute the eigenvalues λ of the matrix inside the curly brackets on the left-hand side of (8.34). This can easily be performed using Matlab's **eig** command, which can provide the eigenvalues and associated eigenvectors (or eigenfunctions).

BVP (Bounded Domain)

Consider the BVP in the canonical domain $(-1, 1)$

$$p(x)f'' + q(x)f' + r(x)f = g(x), \quad \begin{cases} c_{11}f(-1) + c_{12}f'(-1) &= c_{13}, \\ c_{21}f(1) + c_{22}f'(1) &= c_{23}. \end{cases} \quad (8.35)$$

Let $x_j = \cos(j\pi/n)$, $(j = 0, 1, \ldots, n)$ be the Chebyshev nodes with associated first and second order differentiation matrices $\mathbf{D}$ and $\mathbf{D}^{(2)}$, respectively. The discretization of (8.35) on the Chebyshev grid reduces to the $(n-1) \times (n-1)$ linear system of equations

$$\left\{ \widehat{\mathbf{L}} + \mathbf{M}_3 \mathbf{M}_2^{-1} \begin{bmatrix} c_{22} & 0 \\ 0 & c_{12} \end{bmatrix} \mathbf{M}_1 \right\} \begin{bmatrix} f_1 \\ \vdots \\ f_{n-1} \end{bmatrix} = \begin{bmatrix} g_1 \\ \vdots \\ g_{n-1} \end{bmatrix} - \mathbf{M}_3 \mathbf{M}_2^{-1} \begin{bmatrix} c_{23} \\ c_{13} \end{bmatrix}, \quad (8.36)$$

where

$$\widehat{\mathbf{L}} = \begin{bmatrix} \mathbf{L}_{11} & \cdots & \mathbf{L}_{1,n-1} \\ \vdots & & \vdots \\ \mathbf{L}_{n-1,1} & \cdots & \mathbf{L}_{n-1,n-1} \end{bmatrix}, \quad \mathbf{M}_3 = \begin{bmatrix} \mathbf{L}_{10} & \mathbf{L}_{1n} \\ \vdots & \vdots \\ \mathbf{L}_{n-1,0} & \mathbf{L}_{n-1,n} \end{bmatrix}, \quad (8.37)$$

with $\mathbf{L}_{ij} = [\text{diag}(\mathbf{P})\, \mathbf{D}^{(2)} + \text{diag}(\mathbf{Q})\, \mathbf{D} + \text{diag}(\mathbf{R})]_{ij}$, and

$$\mathbf{M}_1 = -\begin{bmatrix} \mathbf{D}_{01} & \cdots & \mathbf{D}_{0,n-1} \\ \mathbf{D}_{n1} & \cdots & \mathbf{D}_{n,n-1} \end{bmatrix}, \quad \mathbf{M}_2 = \begin{bmatrix} c_{21} + c_{22}\mathbf{D}_{00} & c_{22}\mathbf{D}_{0n} \\ c_{12}\mathbf{D}_{n0} & c_{11} + c_{12}\mathbf{D}_{nn} \end{bmatrix}. \quad (8.38)$$

Similarly, the discretization of the eigenvalue problem

$$p(x)f'' + q(x)f' + r(x)f = \lambda f, \quad x \in (-1, 1), \quad (8.39)$$

subjected to homogeneous Robin's boundary conditions in (8.35) with $c_{13} = c_{23} = 0$ reduces to the $(n-1)$-dimensional eigenvalue problem

$$\left\{ \widehat{\mathbf{L}} + \mathbf{M}_3 \mathbf{M}_2^{-1} \begin{bmatrix} c_{22} & 0 \\ 0 & c_{12} \end{bmatrix} \mathbf{M}_1 \right\} \mathbf{f} = \lambda \mathbf{f}, \quad (8.40)$$

where $\mathbf{f} = [f_1 \cdots f_{n-1}]^\mathrm{T}$.

To illustrate how to implement the previous discretizations, we first approximate the solution to the linear BVP given by

$$f'' - \pi \cos(\pi x)f' + f = e^{\sin \pi x}(1 - \pi^2 \sin(\pi x)), \quad \begin{cases} f(-1) - f'(-1) &= 1 + \pi, \\ f(1) + f'(1) &= 1 - \pi. \end{cases} \quad (8.41)$$

The reader may confirm that, in this particular case, the exact solution of (8.41) is $f(x) = e^{\sin \pi x}$. The commands below make use of function `chebdiff.m` in Code 5B to compute the Chebyshev differentiation matrix for $n = 32$. The operators M_1, M_2, M_3, and L are built to form linear system (8.36), which is finally solved using \:

```
n = 32 ; [D,x] = chebdiff(n) ; D2 = D*D ; I = eye(n+1);
Q = diag(-pi*cos(pi*x)) ; L = D2 + Q*D + I ;
c11 = 1; c12 = -1; c13 = 1+pi; c21 = 1; c22 = 1; c23 = 1-pi;
g = exp(sin(pi*x)).*(1-pi^2*sin(pi*x));
M1 = -[D(1,2:n) ; D(n+1,2:n)] ; M3 = [L(2:n,1) L(2:n,n+1)];
M2 = [c21 + c22*D(1,1), c22*D(1,n+1);
         c12*D(n+1,1),   c11+c12*D(n+1,n+1)];
N = L(2:n,2:n) + (M3*inv(M2))*[c22 0;0 c12]*M1 ;
f = N\(g(2:n)-M3*inv(M2)*[c23 ; c13]);
```

The results are outlined in Figure 8.1, where we have depicted the approximated solution contained in vector $\mathbf{f} = [f_1 \ f_2 \ \cdots \ f_{n-1}]^{\mathrm{T}}$, along with the local error resulting from our Chebyshev discretization. The reader may experiment to increase n in order to improve the accuracy of the approximation.

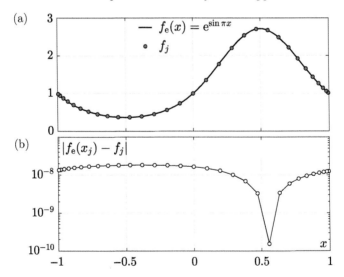

Figure 8.1 Numerical solution of (8.41) provided by the approximated linear operator (8.36). (a) Exact solution $f_e = e^{\sin \pi x}$ (solid black line) and approximated solution f_j (gray bullets) at the Chebyshev nodes $x_j = \cos(\pi j / n)$. (b) Local error $|f_e(x_j) - f_j|$.

As a second example, we show how to discretize the eigenvalue problem

$$y''(z) + \lambda y(z) = 0, \quad z \in (0,1), \quad \begin{cases} y(0) & \equiv 0, \\ y(1) + y'(1) & \equiv 0. \end{cases} \tag{8.42}$$

The second Robin condition at $z = 1$ appears in heat transfer problems, when imposing heat flux rate proportional to the temperature at the boundary. After introducing the change of variable $x \equiv 2z - 1$, with $x \in (-1,1)$, the BVP now reads

$$4f''(x) + \lambda f(x) \equiv 0, \quad z \in (0,1), \quad \begin{cases} f(-1) & \equiv 0, \\ f(1) + 2f'(1) & \equiv 0. \end{cases} \tag{8.43}$$

The short code below builds the required operators $\mathbf{M}_1$, $\mathbf{M}_2$, $\mathbf{M}_3$, and $\mathbf{L}$ that eventually lead to the discrete eigenvalue problem (8.40):

```
n = 26 ; [D,x] = chebdiff(n) ; D2 = D*D ; I = ones(n+1);
c11 = 1 ; c12 = 0 ; c13 = 0 ; c21 = 1 ; c22 = 2 ; c23 = 0 ;
L = 4*D2 ; M1 = -[D(1,2:n) ; D(n+1,2:n)] ;
M2 = [c21 + c22*D(1,1), c22*D(1,n+1);
      c12*D(n+1,1),  c11+c12*D(n+1,n+1)];
M3 = [L(2:n,1) L(2:n,n+1)] ;
N = L(2:n,2:n) + (M3*inv(M2))*[c22 0;0 c12]*M1 ;
[EVEC,EVAL] = eig(-N) ; lamb = diag(EVAL) ;
[foo,ii] = sort(lamb) ; lamb = lamb(ii) ; EVEC = EVEC(:,ii) ;
```

In previous code, the matrix N contains the discrete differential operator within curly brackets on the left-hand side of (8.40). Using Matlab's function **eig**, the code computes the eigenvalues and eigenvectors of N (negated) and stores them in matrices **EVAL** and **EVEC**, respectively. The eigenvalues are sorted and their associated eigenvectors reordered accordingly. It is left as an exercise to show that the exact eigenvalues λ_j of (8.42) are the positive solutions of the nonlinear equation

$$\sqrt{\lambda} + \tan\sqrt{\lambda} \equiv 0. \tag{8.44}$$

Table 8.1 shows the first four eigenvalues provided by the previous code, sorted and stored in **lamb(1:4)**, along with the absolute value of $\sqrt{\lambda_j} + \tan\sqrt{\lambda_j}$. This is one of those cases where, even the differential equation admits eigensolutions that can be expressed analytically in closed form; their associated eigenvalues must be approximated numerically in one way or another. Graphically, the solutions of (8.44) are the abscissas λ where the graphs $\sqrt{\lambda}$ and $-\tan\sqrt{\lambda}$ intersect,

Table 8.1 Leading approximated eigenvalues of boundary value problem (8.42).

j	λ_j	$\left|\sqrt{\lambda_j} + \tan\sqrt{\lambda_j}\right|$
1	4.115 858 365 694 1	6.6×10^{-13}
2	24.139 342 030 445	4.2×10^{-13}
3	63.659 106 550 439	8.8×10^{-13}
4	122.889 161 761 92	3.4×10^{-12}

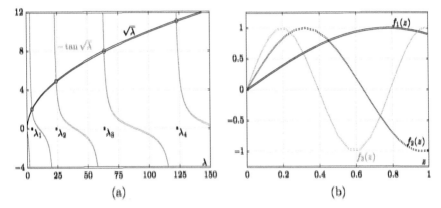

(a)   (b)

Figure 8.2 Numerical solution of (8.42). (a) Graphical representation of the location of the first four eigenvalues λ_j solution of (8.44). (b) Eigenfunctions $f_1(z)$, $f_2(z)$, and $f_3(z)$ associated with λ_1, λ_2, and λ_3, respectively.

as shown in Figure 8.2a. The reader may confirm the results shown in Table 8.1 by solving (8.44) using Newton's method. We finish our example by showing in Figure 8.2b the first three computed eigenfunctions, namely $f_1(z)$, $f_2(z)$, and $f_3(z)$, associated with the first three eigenvalues λ_1, λ_2, and λ_3, respectively. In this case, the ordinates of these eigenfunctions at the Chebyshev nodes are the first three columns of the eigenvector matrix **EVEC**. Since this BVP is linear and homogeneous, the eigenfunctions are arbitrary up to a multiplicative constant. In Figure 8.2b, we have normalized the eigenfunctions with respect to their maximum absolute values within the domain $z \in (0, 1)$.

We finish this section by illustrating how to adapt the discretization seen in (8.36) to approximate solutions of *nonlinear* BVPs of the form

$$\mathcal{L}f = \mathcal{N}(f, g), \quad \begin{cases} c_{11}f(-1) + c_{12}f'(-1) & \equiv c_{13}, \\ c_{21}f(1) + c_{22}f'(1) & \equiv c_{23}, \end{cases} \tag{8.45}$$

where $\mathcal{L}$ is the linear differential operator

$$\mathcal{L} = p(x)\frac{\mathrm{d}^2}{\mathrm{d}x^2} + q(x)\frac{\mathrm{d}}{\mathrm{d}x} + r(x), \tag{8.46}$$

and $\mathcal{N}(f, g)$ is a nonlinear operator acting on the sought solution $f(x)$, potentially also depending on a given known function $g(x)$. Consider for example the nonlinear BVP taken from the excellent classical monograph by Stoer and Bulirsch (see references at the end of the book):

$$f'' = \frac{3}{2}f^2, \quad x \in (0,1), \ f(0) = 4, \ f(1) = 1. \tag{8.47}$$

The solution to this problem has two main stages. In a first stage we need to discretize the differential equation within the domain $(-1, 1)$, reducing the problem to a finite dimensional nonlinear function. The second stage is to apply Newton's method seen in Chapter 6 on the discrete nonlinear multidimensional function. The main code below applies the multidimensional Newton solver **newtonn.m** seen in Chapter 6 on the function **fsb.m** that discretizes the nonlinear differential operator (8.47). Notice that the main code computes the Chebyshev differentiation matrices. These matrices are declared as global variables in the main and **fsm.m** function so that the last routine does not need to compute them every Newton call:

```
global n D D2 I
n = 32 ; [D,x] = chebdiff(n) ; D2 = D*D ; I = ones(n+1);
itmax = 24 ; tol = 1e-12 ;
f0 = ones(n-1,1);
xn = x(2:n);
[XK,resd,it] = newtonn(f0,tol,itmax,@fsb);
y =.5*(x+1); plot(y(2:n),XK(:,end),'-ok'); hold on

function F = fsb(f)
global n D D2 I
L = D2; c11 = 1; c12 = 0; c13 = 4; c21 = 1; c22 = 0; c23 = 1;
M1 = -[D(1,2:n) ; D(n+1,2:n)] ; M3 = [L(2:n,1) L(2:n,n+1)] ;
M2 = [c21 + c22*D(1,1), c22*D(1,n+1);
      c12*D(n+1,1) , c11+c12*D(n+1,n+1)];
N = (3/8)*f.^2; Maux = (M3*inv(M2))*[c22 0;0 c12]*M1;
F = (L(2:n,2:n) + Maux)*f + M3*inv(M2)*[c23;c13] - N;
end
```

A simple calculation shows that the nonlinear BVP (8.47) admits the solution

$$w_1(x) = \frac{4}{(1+x)^2}, \tag{8.48}$$

as it can be confirmed by formal substitution. Newton's iteration will converge to w_1 provided the starting iterate (vector `f0` in the previous code) is sufficiently close to it. In our example, we started from the initial guess `f0 = ones(n-1,1)`, that is, $f_j^{(0)} = 1$, $(j = 1, 2, \ldots, n-1)$. The resulting numerical solution f_j is shown in Figure 8.3a. In this case, Newton's iteration successfully converged to a solution clearly close to $w_1(x)$, as concluded from looking at the local error $|f_j - w_1(x_j)|$, for $j = 1, \ldots, n-1$, depicted in Figure 8.3b.

Since the differential equation in (8.47) is nonlinear, the problem may potentially admit other solutions. In general, finding solutions of a nonlinear BVP may be a formidable task unless one has some *a priori* indication of how the solution (or solutions) may look. It is left as an exercise for the reader to confirm that starting Newton's method from the initial guess function $f^{(0)}(x) = 54x^2 - 57x + 4$ leads to a new solution $\widetilde{f}_j$ (gray bulleted curve in Figure 8.3a). It can be shown that this second numerical solution is actually very close to another exact solution $w_2(x)$ that can be expressed in terms of Jacobian elliptic functions. For more details, we refer the reader to the text by Stoer and Bulirsch included in the bibliography, at the end of the book.

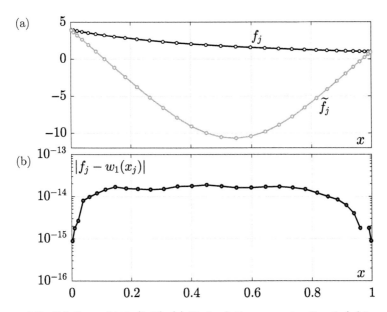

Figure 8.3 Solutions of BVP (8.47). (a) First solution approximation f_j (white bullets) converged to exact solution $w_1(x) = 4(x+1)^{-2}$ (black solid curve). A second solution $\widetilde{f}_j$ can also be found (gray bulleted curve) starting from suitable Newton's initial iterates (see text for more details). (b) Local error of the first converged solution f_j shown in (a).

Practical 8.1:   The Suspended Chain

In this practical, we study the oscillations of a
chain of length ℓ and uniform mass density ρ
suspended by one end (see Figure on the right).
We want to analyze the transversal oscillations
of the chain and to locate the nodes of its normal
modes. First, show that the tension of the chain
is $T(y) = \rho g y$ and that the wave equation for
small transversal oscillations $u(y, t)$ is

$$\frac{\partial^2 u}{\partial t^2} = g \frac{\partial}{\partial y} \left\{ y \frac{\partial u}{\partial y} \right\}.$$

(a) Using separation of variables, solve the wave
equation by assuming $u(y, t) = G(y)F(t)$.
Show that the resulting equation for $G(y)$ is

$$y \frac{\mathrm{d}^2 G}{\mathrm{d} y^2} + \frac{\mathrm{d} G}{\mathrm{d} y} + \frac{\lambda^2}{g} G = 0,$$

where λ^2 is a positive constant.

(b) In this problem, the oscillations must vanish at $y = \ell$, and also
be bounded at $y = 0$. For simplicity, take $\ell = 1$, $g = 1$ and map the
domain $y \in [0, 1]$ to the standard domain $x \in [-1, 1]$. Find the operators
M_1, M_2, M_3, and $\hat{L}$ of expressions (8.37) and (8.38) for this particular case. For $n = 26$, for example, compute the first three values
$\lambda_1 < \lambda_2 < \lambda_3$ with the smallest real part.

(c) It can be shown that the exact bounded eigenfunctions of the differential
equation above are given by the expression

$$G_k(y) = J_0(2\lambda_k \sqrt{y}),$$

where $J_0(z)$ is the 0th order *Bessel function*[3] of the first kind, and their
associated values λ_k satisfy

$$J_0(2\lambda_k) = 0.$$

Check the accuracy of your results in any mathematical table of Bessel
functions. Plot the eigenfunctions $G_1(y)$, $G_2(y)$, and $G_3(y)$ associated
with the leading eigenvalues λ_1, λ_2, and λ_3, and compute the location
of their corresponding nodes. Historically, the Bessel functions appear
for the first time in this problem.[4]

---

3 F. W. Bessel (1784–1846), German mathematician and astronomer.
4 Formulated by the Swiss mathematician and physicist D. Bernoulli (1700–1782).

8.1.2   Periodic Domains

In this section, we address how to approximate solutions to periodic BVPs within the domain $x \in [0, 2\pi]$ that have the form

$$p(x)f'' + q(x)f' + r(x)f = g(x), \quad \begin{cases} f(x) = f(x + 2\pi), \\ f'(x) = f'(x + 2\pi), \end{cases} \tag{8.49}$$

for all $x \in [0, 2\pi]$. We will henceforth assume that the functions $p(x)$, $q(x)$, $r(x)$, and $g(x)$ are also 2π-periodic and continuously differentiable. Following the formulation of Chapter 7, we approximate the solution $f(x)$ of (8.49) at the equally spaced nodes

$$x_j = \frac{2\pi}{N} j, \tag{8.50}$$

for $j = 0, 1, 2, \ldots, N - 1$, with N being an even number. In this case, the sought approximation is the N-dimensional vector

$$\mathbf{f} = [f_0 \ f_1 \ \cdots \ f_{N-1}]^{\mathrm{T}} \equiv \left[f(0) \ f\left(\frac{2\pi}{N}\right) \ \cdots \ f\left(\frac{2\pi(N-1)}{N}\right) \right]^{\mathrm{T}}. \tag{8.51}$$

As before, we first discretize the differential operator appearing in (8.49), but in this case using the Fourier differentiation matrices

$$\mathbf{D}_{j\ell}^{(1)} = \frac{1}{N} \sum_{k=0}^{N-1} \left(k - \frac{N}{2} \right) W_N^{-(k-\frac{N}{2})(j-\ell)}, \tag{8.52}$$

and

$$\mathbf{D}_{j\ell}^{(2)} = -\frac{1}{N} \sum_{k=0}^{N-1} \left(k - \frac{N}{2} \right)^2 W_N^{-(k-\frac{N}{2})(j-\ell)}, \tag{8.53}$$

for $0 \leq \ell, j \leq N - 1$, and $W_N \equiv e^{-i\frac{2\pi}{N}}$, already seen in Chapter 7. This procedure leads to the matrix

$$\mathbf{L}_{ij} = [\text{diag}(\mathbf{P}) \ \mathbf{D}^{(2)} + \text{diag}(\mathbf{Q}) \ \mathbf{D}^{(1)} + \text{diag}(\mathbf{R})]_{ij}, \tag{8.54}$$

for $0 \leq i, j \leq N - 1$, where $\text{diag}(\mathbf{P})$, $\text{diag}(\mathbf{Q})$, and $\text{diag}(\mathbf{R})$ are diagonal matrices with entries $\mathbf{P}_{jj} \equiv p(x_j)$, $\mathbf{Q}_{jj} \equiv q(x_j)$, and $\mathbf{R}_{jj} \equiv q(x_j)$, for $j = 0, 1, \ldots, N - 1$, respectively. In this case, the discrete operator $\mathbf{L}$ in (8.54) already incorporates the periodic boundary conditions, implicitly present in the structure of $\mathbf{D}^{(1)}$ and $\mathbf{D}^{(2)}$, as well as in the assumed periodicity of $p(x)$, $q(x)$, and $r(x)$. Therefore, the structure of the resulting matrix $\mathbf{L}$ does not require any type of modification or reshape, as it did in the Section 8.1.1 for non-periodic boundary problems. The resulting discrete problem is reduced to solve the linear system of equations

$$\mathbf{Lf} = \mathbf{g}, \tag{8.55}$$

where $\mathbf{g} = [g_0 \ g_1 \ \cdots \ g_{N-1}]^{\mathrm{T}} = \left[g(0) \ g\left(\dfrac{2\pi}{N} \right) \ \cdots \ g\left(\dfrac{2\pi(N-1)}{N} \right) \right]^{\mathrm{T}}$. Similarly, the discretization of a periodic eigenvalue BVP of the form

$$p(x)f'' + q(x)f' + r(x)f = \lambda f(x), \qquad \begin{cases} f(x) = f(x + 2\pi), \\ f'(x) = f'(x + 2\pi), \end{cases} \qquad (8.56)$$

reduces to the computation of the eigenvalues λ and associated eigenvectors $\mathbf{f}$ of

$$\mathbf{Lf} = \lambda \mathbf{f}. \qquad (8.57)$$

As an example, we approximate the eigensolution to what is usually known as *Mathieu's equation*[5] appearing in the quantum theory of crystals. This equation is in essence the *Schrödinger equation*[6] with a spatially periodic potential of the form

$$-\frac{\mathrm{d}^2 \Psi(x)}{\mathrm{d}x^2} + 2q \cos(2x)\Psi(x) = E \ \Psi(x), \qquad (8.58)$$

where $\Psi(x)$ is the 2π-periodic wavefunction or eigenmode associated with the energy level or eigenvalue E, and q is a positive constant. The short code below computes the spectrum of eigenvalues and eigenfunctions of Eq. (8.58) for $q = 2$ by using Fourier's second order differentiation matrix provided by function `dftdiffmat2.m` seen in Chapter 7:

```
N = 64 ; x = 2*pi*[0:N-1]'/N; D2 = dftdiffmat2(N);
q = 2.0 ; L = -D2 + 2*q*diag(cos(2*x)) ;
[EVEC,EVAL] = eig(L) ;
```

The discrete operator $\mathbf{L}$ is trivially constructed in this case by simply adding to $-\mathbf{D}^{(2)}$ the diagonal matrix $2q \operatorname{diag}(\cos(2x_j))$. Figure 8.4a depicts the five lowest energy levels E_j as a function of q, within the range $q \in [0, 5]$. Notice that for $q = 0$ we recover the spectrum $E_j = j^2$, $(j = 0, 1, 2, \ldots)$. Figure 8.4b depicts the eigenmodes associated with the first three energy levels for $q = 2$.

---

5 É. L. Mathieu (1835–1890), French mathematician, better known for his contributions to group theory and differential equations of mathematical physics.
6 E. Schrödinger (1887–1961), Austrian physicist and one of the fathers of quantum mechanics.

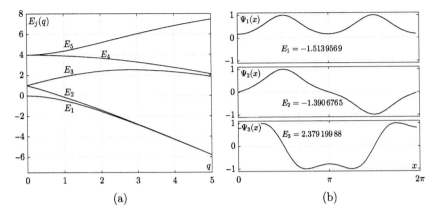

Figure 8.4 Eigenvalues and eigenfunctions of Mathieu's equation (8.58). (a) Energy spectrum as a function of q. (b) Wavefunctions associated with the three lowest energy levels for $q = 2$.

8.1.3 Unbounded Domains

In this last section we briefly address how to approximate eigenvalues λ and the corresponding eigenfunctions $f(y)$ of a homogeneous BVP of the form

$$
p(y)\frac{\mathrm{d}^2 f}{\mathrm{d}y^2} + q(y)\frac{\mathrm{d}f}{\mathrm{d}y} + r(y)f = \lambda f, \quad
\begin{cases}
\lim_{y \to +\infty} f(y) = 0, \\
\lim_{y \to -\infty} f(y) = 0.
\end{cases}
\tag{8.59}
$$

for all $y \in (-\infty, +\infty)$. We assume that the functions $p(y)$, $q(y)$, $r(y)$, and $g(y)$ are continuously differentiable on the real line. One possible way of solving (8.59) is to consider a change of variable mapping the unbounded domain $(-\infty, +\infty)$ to the canonical interval $[-1, 1]$. This can be done by introducing the change of variable

$$
y = \frac{Lx}{\sqrt{1 - x^2}} = \frac{Lx}{Q^{1/2}(x)},
\tag{8.60}
$$

where $Q(x) \doteq 1 - x^2$ and L is a positive *scaling* factor. The transformation (8.60) leads to the differential operators in the bounded variable x

$$
\frac{\mathrm{d}}{\mathrm{d}y} = \frac{Q^{3/2}}{L}\frac{\mathrm{d}}{\mathrm{d}x}
\tag{8.61}
$$

and

$$
\frac{\mathrm{d}^2}{\mathrm{d}y^2} = \frac{Q^3}{L^2}\frac{\mathrm{d}^2}{\mathrm{d}x^2} - \frac{3xQ^2}{L^2}\frac{\mathrm{d}}{\mathrm{d}x}.
\tag{8.62}
$$

Introducing these operators and defining $\overline{f}(x) \equiv f(y)$, $\overline{p}(x) \equiv p(y)$, $\overline{q}(x) \equiv q(y)$, and $\overline{r}(x) \equiv r(y)$, the transformed BVP now reads

$$\overline{p}(x)\left(\frac{Q^3}{L^2}\frac{\mathrm{d}^2}{\mathrm{d}x^2} - \frac{3xQ^2}{L^2}\frac{\mathrm{d}}{\mathrm{d}x}\right)\overline{f} + \overline{q}(x)\left(\frac{Q^{3/2}}{L}\frac{\mathrm{d}}{\mathrm{d}x}\right)\overline{f} + \overline{r}(x)\overline{f} = \lambda\overline{f}, \quad (8.63)$$

with boundary conditions

$$\lim_{x\to+1^-}\overline{f}(x) = 0 \quad \text{and} \quad \lim_{x\to-1^+}\overline{f}(x) = 0. \quad (8.64)$$

The transformed BVP (8.63) can be discretized in $[-1, 1]$ at the Chebyshev nodes $x_j \equiv \cos(j\pi/n)$, for $j = 0, 1, \ldots, n$, using the formulation described in (8.35) and (8.40) with $c_{11} \equiv c_{21} \equiv 1$ and $c_{12} \equiv c_{22} \equiv c_{13} \equiv c_{23} \equiv 0$. Although the mapping (8.60) is singular at $x_0 \equiv 1$ and $x_n \equiv -1$, the discrete eigensystem (8.40) is numerically well-posed, since it only involves the *interior* nodes. In particular, since $c_{22} \equiv c_{12} \equiv 0$, system (8.40) simply reads

$$\widehat{\mathbf{L}}\overline{\mathbf{f}} = \lambda\overline{\mathbf{f}}, \quad (8.65)$$

where $\overline{\mathbf{f}} \equiv [\overline{f}_1\ \overline{f}_2\ \cdots\ \overline{f}_{n-1}]^{\mathrm{T}}$, that is, the approximation of the eigenvector evaluated at the transformed interior nodes $y_j = Lx_jQ^{-1/2}(x_j)$, for $j = 1, 2, \ldots, n-1$.

We illustrate this methodology by computing the discrete spectrum of energies E_j and associated wavefunctions $\Psi_j(y)$ of the one-dimensional *quantum harmonic oscillator* described by the Schrödinger equation with quadratic potential $V(y) = y^2$

$$-\frac{\mathrm{d}^2\Psi}{\mathrm{d}y^2} + y^2\Psi = E\,\Psi, \quad (8.66)$$

where $\lim_{y\to\pm\infty}\Psi(y) = 0$. This a BVP that can be solved analytically and whose exact eigenvalues (energies) and associated eigenfunctions are

$$E_j = 2j + 1, \quad \Psi_j(y) = e^{-y^2/2}H_j(y), \quad (8.67)$$

for $j = 0, 1, 2, \ldots$, where $H_j(y)$ is the jth-order Hermite[7] polynomial

$$H_j(y) \equiv (-1)^j e^{y^2}\frac{\mathrm{d}^j}{\mathrm{d}y^j}e^{-y^2} \quad (j = 0, 1, 2, \ldots). \quad (8.68)$$

The code below is a simple implementation of the discretization previously described, which takes $n = 64$ and a scaling factor $L = 5$ (L0 in the lines below):

```
n = 64 ; [D,x] = chebdiff(n) ; D2 = D*D ;
L0 = 5 ; Q = 1-x.^2; V = x.^2;
L = (-diag(Q.^3)*D2+diag(3*x.*Q.^2)*D)/L0^2+diag(L0^2*V./Q);
N = L(2:n,2:n) ; [EVEC,EVAL] = eig(N) ;
```

---

7 C. Hermite (1822–1901), French mathematician who proved that e is a *transcendental* number, that is, not a root of a nonzero polynomial with integer coefficients.

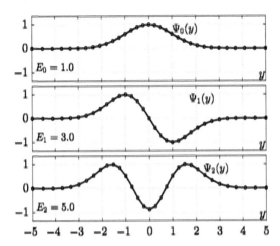

Figure 8.5 Eigenvalues and eigenfunctions of Schrödinger's equation for the harmonic oscillator (8.67). The eigenfunctions have been normalized so that $\max \Psi_j(y) \equiv 1$.

Figure 8.5 shows the resulting three lowest energies stored in **EVAL** (accurate to nearly 16 digits) and their associated eigenfunctions stored in **EVEC**. It is left as an exercise for the reader to sort eigenvalues and eigenvectors accordingly to reproduce Figure 8.5. In our computation we have set $L \equiv 5$, but we recommend the reader to explore with other values and see the accuracy of the eigenvalues. In general, if the eigenfunctions of the problem have an exponential decay, a moderate value of L will be sufficient. If, however, the eigenfunctions vanish at infinity very slowly or with strong oscillations, we may need a larger value of L (and n). The optimal value of L is, in general, problem dependent. To some extent, this is not very different from what we did in Part I when parametrizing the cotangent transformation for the quadrature formulas of improper integrals of the first kind.

8.2 The Initial Value Problem

In the Section 8.1, we have studied how to discretize second order BVPs in finite (and infinite) domains. In this section, we will address how to approximate the solution of a first order *Initial Value Problem*, henceforth referred to as IVP, of the form

$$\left.\begin{array}{l} \dfrac{\mathrm{d}u}{\mathrm{d}t} \equiv f(t, u(t)), \\[2mm] u(0) \equiv u_0, \end{array}\right\} \tag{8.69}$$

where $f(t, u) : \mathbb{R}^+ \times \mathbb{R} \to \mathbb{R}$ is a given smooth and bounded function, and u_0 is the so-called *initial condition*. The solution $u(t)$ depends on the variable t

that will typically represent time. Henceforth in this section, we will assume that the IVP (8.69) is well-posed and that the existence and uniqueness of the solution $u(t)$ is guaranteed.[8] In general, this section will deal with the numerical approximation of the n-dimensional IVP

$$
\left.\begin{aligned}
\frac{du_1}{dt} &= f_1(t, \mathbf{u}(t)), \\
\frac{du_2}{dt} &= f_2(t, \mathbf{u}(t)), \\
&\vdots \\
\frac{du_n}{dt} &= f_n(t, \mathbf{u}(t)), \\
[\, u_1(0)\ u_2(0)\ \cdots\ u_n(0)\,]^{\mathrm{T}} &= [\beta_1\ \beta_2\ \cdots\ \beta_n]^{\mathrm{T}}
\end{aligned}\right\},
\tag{8.70}
$$

or, to avoid cumbersome notation, we simply write[9]

$$
\mathbf{u}_t = \mathbf{f}(t, \mathbf{u}), \quad \mathbf{u}(0) = \mathbf{u}_0,
\tag{8.71}
$$

where $\mathbf{u}(t) \doteq [\, u_1(t)\ u_2(t)\ \cdots\ u_n(t)\,]^{\mathrm{T}}$, $\mathbf{f}(t, \mathbf{u}) \doteq [\, f_1\ f_2\ \cdots\ f_n\,]^{\mathrm{T}}$, and $\mathbf{u}_0 \doteq [\beta_1\ \beta_2\ \cdots\ \beta_n]^{\mathrm{T}}$. However, in physics and engineering, we may frequently find a second order IVP of the form

$$
\frac{d^2 v}{dt^2} = g\left(t, v, \frac{dv}{dt}\right), \quad \text{with } v(0) = v_0, \ \left(\frac{dv}{dt}\right)_{t=0} = w_0,
\tag{8.72}
$$

which, for simplicity, is written as

$$
v_{tt} = g(t, v, v_t), \quad v(0) = v_0, \ v_t(0) = w_0.
\tag{8.73}
$$

In this case, it is common practice to reduce (8.73) to a two-dimensional first order IVP of the form (8.70). This can be done by introducing the variables $u_1(t) \doteq v(t)$ and $u_2(t) \doteq v_t(t)$ so that $u_{1,t} = v_t = u_2$ and therefore (8.73) reads

$$
\left.\begin{aligned}
u_{1,t} &= u_2, \\
u_{2,t} &= g(t, u_1, u_2), \\
[\, u_1(0)\ u_2(0)\,]^{\mathrm{T}} &= [v_0\ w_0]^{\mathrm{T}}.
\end{aligned}\right\}
\tag{8.74}
$$

To simplify our analysis, we will henceforth only consider scalar initial value problems of the form

$$
u_t = f(t, u), \quad u(0) = u_0,
\tag{8.75}
$$

---

[8] For a mathematically rigorous analysis regarding existence and uniqueness of solutions see the recommended bibliography at the end of this chapter, in the Complementary Reading section.

[9] Henceforth in this chapter, we will denote $\dfrac{d\mathbf{u}}{dt}, \dfrac{d^2\mathbf{u}}{dt^2}, \ldots$, by $\mathbf{u}_t, \mathbf{u}_{tt}, \ldots$, respectively.

where it is understood that all the methods developed in this section can be generalized to the n-dimensional case.

There exists a large variety of methodologies that can be used to discretize and approximate solutions of (8.75). Choosing one method or another will depend on the context of the problem. While, for example, some methods can be excellent for integrating planetary orbits, the same methods may be inefficient to integrate ODEs arising in chemical kinetics. Conversely, a method that works perfectly in diffusive equations arising in thermal or chemical processes may not be suitable for integrating Hamiltonian systems. Therefore, the physical nature of the problem is one of the many aspects to be considered before using a given method. Other aspects, such as computational cost or accuracy, must also be considered. As a rule, there is no *ideal* method. All methods have advantages, but also drawbacks. What follows is just an introduction to the underlying concepts of numerical integration of IVP where only a few methodologies will be addressed. There are excellent monographs focused on this topic that the reader should be aware of. At the end of the chapter, we will recommend some of these authoritative treatises.

8.2.1 Runge–Kutta One-Step Formulas

One of the advantages of discretizing a second order BVP, such as the ones addressed in the Section 8.1, is that we know how the problem starts and how it finishes (even if the domain is infinite). In other words, we know in advance the boundary conditions at the *beginning* and at the *end* of the interval. In this chapter, we need to provide a numerical approximation of the solution $u(t)$ of the first order IVP

$$u_t = f(t, u), \quad u(0) = u_0, \tag{8.76}$$

within the interval $t \in [0, T]$, where T is the total integration time. While we know that the solution $u(t)$ takes the value u_0 at $t = 0$ (the so-called *initial condition*), we hardly know in advance its behavior for long times. It may occur, for example, that the solution steadily approaches a constant value U, that is, $\lim_{t \to +\infty} u(t) = U$. In other cases, however, the long-term behavior of the solution may be time-periodic, almost periodic (such as in the oscillations observed in the mechanical system of Practical 7.1), or even chaotic. Exploring the long-term behavior of the solutions of (8.76) is precisely one of the main motivations for devising numerical integrators. For this reason, the *global* approach used for solving BVP is unfeasible here. That is, approximating the solution $u(t)$ by means of a single interpolant of high degree is computationally very expensive. For example, depending on the physical problem, the solution may require *very long* transients before the asymptotic regime establishes (steady, time-periodic, etc.), thus requiring a very large total integration time T and therefore a corresponding large number of nodes for the interpolant to be accurate. In particular, the computational cost could be prohibitive when dealing with n-dimensional systems of the form (8.70), requiring a global interpolant for each component of **u**. For this reason, the numerical discretization of (8.76) needs in general to be *local*.

In order to discretize (8.76), we first need to consider a partition in the time variable t. Henceforth, we will only consider *equispaced* time grids of the form

$$t_j \doteq jh, \tag{8.77}$$

for $j = 0, 1, 2, \ldots$, where h is a small positive constant usually known as the *step size*, *grid size*, or *time-step*. In addition, we will henceforth denote

$$u_j \doteq u(t_j), \tag{8.78}$$

that is, the value of the *exact* solution $u(t)$, if available, at $t = t_j$. The methods that we are about to study start from the known initial value u_0 and subsequently provide the numerical approximations $\{v_1, v_2, v_3, \ldots\}$ of the exact values $\{u_1, u_2, u_3, \ldots\}$ of the solution of (8.76) at the grid points $\{h, 2h, 3h, \ldots\}$, respectively. This is illustrated in Figure 8.6, where we have depicted the resulting approximated values v_j (white bullets) of the exact solution $u(t_j)$ at the nodes (gray circles). In general, only $v_0 \equiv u_0$ is known exactly, and we never have access to the exact values u_j, but only to their approximations $v_j \approx u_j$ ($j = 1, 2, \ldots$) provided by our method. Therefore, it is convenient to introduce the quantities

$$f_j \doteq f(t_j, v_j), \tag{8.79}$$

for $j = 0, 1, 2, 3, \ldots$, that is, the values of the right-hand side of the differential equation (8.76) at the node t_j and the corresponding approximate value v_j at

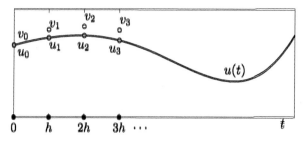

Figure 8.6 Discretization of (8.76) on an equally spaced time grid. Only $v_0 \equiv u_0$ is known exactly, the remaining values v_j provided by a discretization method being just numerical approximations of the exact values $u_j \equiv u(t_j)$, for $j \geq 1$.

that stage, and *not* at the exact solution u_j, which is in general unknown. However, let us assume by now that we have access to the exact solution $u(t)$, taking the values u_n and u_{n+1} at $t = t_n$ and $t = t_{n+1}$, respectively. These two values are related through Taylor's expansion of $u(t)$ in a neighborhood of t_n, that is,

$$u(t_n + h) \equiv u(t_n) + hu_t(t_n) + \frac{h^2}{2!}u_{tt}(t_n) + \cdots \approx u(t_n) + hu_t(t_n), \quad (8.80)$$

where we have neglected terms of order $O(h^2)$. Since $u_t \equiv f(t, u)$, the previous expression reads

$$u(t_n + h) \approx u(t_n) + hf(t_n, u_n), \quad (8.81)$$

or

$$u_{n+1} \approx u_n + hf(t_n, u_n). \quad (8.82)$$

In practice, we never have access to u_n, so we need to substitute u_n by v_n and $f(t_n, u_n)$ by $f_n \equiv f(t_n, v_n)$ on the previous right-hand side. As a result of this substitution, we obtain an approximation of u_{n+1}, namely v_{n+1}, given by

$$v_{n+1} \equiv v_n + hf_n, \quad (8.83)$$

for $n \equiv 0, 1, 2, \ldots$, and $v_0 \equiv u_0$. Iterative formula (8.83) is usually known as *Euler's Method*.[10] This recurrence is very simple to implement and provides insight on many different aspects of numerical integration, although its accuracy is rather limited. This is due to the drastic truncation performed in the Taylor series of (8.80), accumulating an error of order $O(h^2)$ at every time step. We will study the effects of these accumulated errors at the end of this section. Before that, let us try to improve the accuracy of formula (8.83) by, for example, including quadratic terms in the Taylor expansion of (8.80), that is,

$$u(t_n + h) \approx u(t_n) + hu_t(t_n) + \frac{h^2}{2}u_{tt}(t_n), \quad (8.84)$$

where we have neglected terms of order $O(h^3)$. In (8.84), $u_t(t_n) \equiv f(t_n, u_n)$ can be replaced by $f_n \equiv f(t_n, v_n)$, as in Euler's method. In what follows, we proceed to approximate the extra term $u_{tt}(t_n)$ also appearing on the right-hand side of (8.84). First, recall that if h is a small quantity, the first terms of the Taylor expansion of a function of two independent variables $f(x, y)$ reads

$$f(x + h, y + h) = f(x, y) + h(\partial_x f) + h(\partial_y f) + O(h^2). \quad (8.85)$$

Since $u_t \equiv f(t, u(t))$, we may use the chain rule of differentiation to express the second derivative u_{tt} as

$$u_{tt} \equiv \frac{\mathrm{d}}{\mathrm{d}t}u_t \equiv \frac{\mathrm{d}}{\mathrm{d}t}f(t, u(t)) \equiv \partial_t f + (\partial_u f)u_t \equiv \partial_t f + (\partial_u f)f, \quad (8.86)$$

---

10 L. Euler (1707–1783), the famous Swiss mathematician who first formulated it.

so that the right-hand side of (8.84) now reads

$$u(t_n) + hf(t_n, u_n) + \frac{h^2}{2}\{\partial_t f(t_n, u_n) + [\partial_u f(t_n, u_n)]f(t_n, u_n)\}, \qquad (8.87)$$

or, after some rearrangements,

$$u(t_n) + \frac{h}{2}f(t_n, u_n) + \frac{h}{2}\{f(t_n, u_n) + h\partial_t f(t_n, u_n) + h[\partial_u f(t_n, u_n)]f(t_n, u_n)\}.$$
$$(8.88)$$

According to (8.85), the quantity inside the curly brackets above is

$$f(t_n + h, u_n + hf(t_n, u_n)) + O(h^2), \qquad (8.89)$$

so (8.84) now reads

$$u(t_n + h) \approx u(t_n) + \frac{h}{2}f(t_n, u_n) + \frac{h}{2}\{f(t_n + h, u_n + hf(t_n, u_n)) + O(h^2)\}.$$
$$(8.90)$$

Neglecting terms of order $O(h^3)$, replacing $u(t_n)$ by v_n, $f(t_n, u_n)$ by $f_n = f(t_n, v_n)$, and $u(t_n + h)$ by v_{n+1}, leads to the recurrence

$$v_{n+1} = v_n + \frac{h}{2}f_n + \frac{h}{2}f(t_n + h, v_n + hf_n), \qquad (8.91)$$

which is usually expressed as

$$v_{n+1} = v_n + \frac{1}{2}(a + b), \quad \text{with} \quad \begin{cases} a = hf(t_n, v_n), \\ b = hf(t_n + h, v_n + a), \end{cases} \qquad (8.92)$$

and known as Heun's formula. Euler's and Heun's formulas (8.83) and (8.92) are particular instances of the so-called *Runge–Kutta methods*.[11] In particular, Euler and Heun methods are known as Runge–Kutta formulas of *first* and *second* order, henceforth referred to as RK1 and RK2, respectively. The origin of these names will be justified in later, when studying the convergence properties of these methods. To obtain RK2 Heun's formula, we neglected $O(h^3)$, whereas in Euler's RK1 formula the neglected terms were of lower order $O(h^2)$. We may therefore expect RK2 method to be more accurate than RK1. This turns out to be the case, as we will confirm numerically later in a particular case. Including more terms of the Taylor expansion in (8.80), more accurate RK methods can

---

11 Developed by the German mathematicians Carl Runge (also see *Runge's phenomenon* in Part I) and Wilhelm Kutta (also known for the *Kutta–Zhukovsky theorem* in aerodynamics) around 1900.

be obtained, although their derivation is a very tedious task and we shall not do so. The most popular of these methods is the celebrated RK4 formula:

$$v_{n+1} = v_n + \frac{1}{6}(a + 2b + 2c + d), \text{ with } \begin{cases} a = hf(t_n, v_n), \\[2mm] b = hf\left(t_n + \frac{h}{2}, v_n + \frac{a}{2}\right), \\[2mm] c = hf\left(t_n + \frac{h}{2}, v_n + \frac{b}{2}\right), \\[2mm] d = hf(t_n + h, v_n + c). \end{cases}$$

(8.93)

Before exploring the accuracy of all these formulas, it is worthwhile observing different aspects of RK methods. First, they are easily programmed once they have been derived and therefore it is left as an exercise for the reader to program simple codes implementing them. Second, they are *one-step* formulas, that is, they provide v_{n+1} in terms of v_n and t_n only. However, this step involves many evaluations of the right-hand side f at *intermediate stages* such as $t_n + h/2$ in (8.93) for example. Third, these evaluations are *sequential*, that is, they *cannot* be performed independently and simultaneously. For example, the computation of d in the fourth stage of (8.93) requires the value of c, obtained in the third stage, which also requires b, and so on. Fourth, since these formulas only require the previous step v_n for the computation of v_{n+1}, they can start the integration by just knowing v_0.[12] Fifth, RK formulas (8.83), (8.92), and (8.93) are particular instances of *explicit Runge–Kutta formulas*, that is, they provide v_{n+1} in terms of explicit evaluations of f at the previous known solution v_n. Finally, the fact that RK formulas require multiple evaluations per time-step makes them computationally expensive when compared with other approaches. Other aspects of these formulas will be addressed later.

Let us apply RK methods to approximate the solution of the IVP

$$u_t = u, \ (u_0 = 1),$$

(8.94)

within the interval $t \in [0, 1]$ and whose exact solution is $u(t) = e^t$. Table 8.2 outlines the results from the three integrations using the same time-step $h = 0.1$. In that table we clearly see that RK1 (Euler) method is by far the most inaccurate, as it can also be concluded from Figure 8.7a, where we have depicted the exact solution $u(t) = e^t$ compared with the resulting RK1 and RK4 approximations (we do not include RK2 in the plot, since it is graphically indistinguishable from RK4). As expected, decreasing h improves the numerical accuracy in the three cases. A comprehensive exploration of how the three methods converge to the exact final value $u(1) = e$ is depicted in Figure 8.7b, where we have plotted

---

12 This is an advantage with respect to other methodologies, as we will see later.

Table 8.2 Numerical solution of $u_t = u$ ($u_0 = 1$) using different Runge–Kutta methods with $h = 0.1$.

t_n	v_n^{RK1}	v_n^{RK2}	v_n^{RK4}	u_n
0.0	1.000 000 0	1.000 000 0	1.000 000 0	1.000 000 0
0.1	1.100 000 0	1.105 000 0	1.105 170 8	1.105 171 0
0.2	1.210 000 0	1.221 025 0	1.221 402 5	1.221 402 8
0.3	1.331 000 0	1.349 232 7	1.349 858 5	1.349 858 8
0.4	1.464 100 0	1.490 902 1	1.491 824 3	1.491 824 7
0.5	1.610 510 0	1.647 446 8	1.648 720 6	1.648 721 2
0.6	1.771 561 0	1.820 428 7	1.822 117 9	1.822 118 8
0.7	1.948 717 1	2.011 573 8	2.013 751 7	2.013 752 7
0.8	2.143 588 8	2.222 788 8	2.225 539 4	2.225 540 9
0.9	2.357 947 6	2.456 181 8	2.459 601 4	2.459 603 1
1.0	2.593 742 4	2.714 080 8	2.718 279 8	2.718 281 7

the error $\varepsilon \equiv |e - v_N|$ at the end of the integration as a function of h, and where $N = 1/h$ is the total number of steps used, so that the total integration time is $T = t_N = Nh = 1$. From the slopes of the three curves of Figure 8.7b, we clearly conclude that RK1 converges *linearly*, that is, its error is $O(h)$, whereas the global errors in RK2 and RK4 are $O(h^2)$ and $O(h^4)$, respectively.

From the previous example, the reader has probably noticed that while the error due to the truncation in Taylor expansion of RK1 was $O(h^2)$, its final error in Figure 8.7 is of order $O(h)$. The same occurs with RK2, with Taylor truncation error of orders $O(h^3)$ and final error of order $O(h^2)$. To explain this, first notice that the truncation error is *local*, since it is introduced locally at every stage t_j of the time integration by neglecting, say, $O(h^m)$ terms in the Taylor expansion. However, these small errors *accumulate* throughout the whole integration. When we integrate within the interval $t \in [0, T]$ using a time-step h, the total integration requires $N = T/h = O(h^{-1})$ time steps or iterations. The total error is therefore the product of the two factors $O(h^m)$ and $O(h^{-1})$, leading to a total error $O(h^{m-1})$. The error introduced due to truncation is usually known as *local truncation error*, whereas the sum of the accumulated inaccuracies at the end of the integration is known as the *global truncation error*.[13] In general, a numerical ODE time integrator is said to be of *order n* if its associated global truncation error is $O(h^n)$. We will revisit these concepts later on in this chapter when introducing other aspects of time integration of ODE such as *convergence* and *stability*.

---

13 In our analysis we neglect round-off errors due to machine precision.

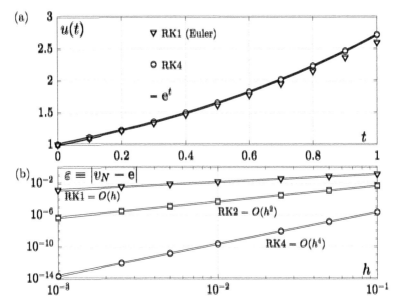

Figure 8.7 Accuracy of Runge–Kutta methods RK1, RK2 and RK4. (a) Time integration of $u_t \equiv u$ ($u_0 \equiv 1$) in $t \in [0, 1]$ with RK1 (Euler) and RK4 (RK2 is omitted here). (b) Global error $\varepsilon \equiv |e - v_N|$ at $t \equiv t_N \equiv Nh \equiv 1$.

8.2.2 Linear Multistep Formulas

In RK methods, the solution v_{n+1} was approximated by means of truncated Taylor series, suitably written in terms of sequential evaluations of $f(t, u)$ at earlier intermediate stages between t_n and t_{n+1} and just involving v_n. In this section, we formulate alternative discretization schemes that are in essence based on quadrature and numerical differentiation techniques already addressed in the last two chapters of Part I. For this reason, we recommend the reader to review the aforementioned chapters before studying this section.

We start by introducing a discretization technique, usually known as *Adams method*.[14] This method consists in integrating both sides of Eq. (8.76) between two consecutive steps t_n and t_{n+1},

$$\int_{t_n}^{t_{n+1}} u_t \, \mathrm{d}t = \int_{t_n}^{t_{n+1}} f(t, u(t)) \, \mathrm{d}t, \tag{8.95}$$

---

14 J. C. Adams (1819–1892), British mathematician and astronomer who predicted the existence and position of Neptune in 1845, independently and almost simultaneously with the French astronomer U. J. J. Le Verrier (1811–1877).

which, by virtue of the fundamental theorem of calculus and definition (8.78), reads

$$u_{n+1} = u_n + \int_{nh}^{(n+1)h} f(t, u(t)) \, dt. \tag{8.96}$$

The last expression is exact and, to some extent, it can be understood as an *iterative* formula that provides u_{n+1} using the previous iterate u_n. For example, if we take $n = 0$, the formula yields

$$u_1 = u_0 + \int_0^h f(t, u(t)) \, dt. \tag{8.97}$$

The main problem with the previous expression is that although we know the initial condition u_0, we ignore $u(t)$ within the interval $t \in [0, h]$, and therefore we cannot evaluate the integrand $f(t, u)$ and calculate the integral on the right-hand side of (8.97). However, if h is sufficiently small, we may assume that $f(t, u)$ does not experience remarkable variations within that interval. As a first (and very crude) approximation, we may assume that the integrand is approximately $f_0 = f(0, v_0)$, that is, the value of f at the lower limit of integration $t = 0$. Therefore, a first approximation of the integral appearing in (8.97) is given by

$$\int_0^h f(t, u(t)) \, dt \approx \int_0^h f_0 \, dt = h f_0. \tag{8.98}$$

As a result, we can approximate (8.97) by

$$v_1 = v_0 + h f_0, \tag{8.99}$$

where $v_0 = u_0$ is exact in this case, since it is the initial condition. For $n = 1$, however, Eq. (8.96) reads

$$u_2 = u_1 + \int_h^{2h} f(t, u(t)) \, dt, \tag{8.100}$$

and the exact value of $u_1 = u(h)$ is not available but only the nearby value v_1 recently obtained in the first iteration, as shown in Figure 8.6. If we approximate again the integrand of (8.100) by its value $f_1 = f(h, v_1)$ at the lower limit of integration $t = h$, a first approximation of (8.100) is

$$v_2 = v_1 + h f_1. \tag{8.101}$$

Finally, if we proceed similarly in (8.96), we obtain the iterative formula

$$v_{n+1} = v_n + h f_n, \tag{8.102}$$

for $n = 0, 1, 2, \ldots$, and $v_0 = u_0$. Iterative formula (8.102) is Euler's Method, already derived in the Section 8.2.1 in a much simpler way. However, the methodology used here reveals that we would obtain a more accurate discretization if we could approximate the integral of (8.96) with more precision. One way of improving the accuracy of that integral would be to have a better approximation of the integrand f. Let us assume that, starting from our numerical solution v_n at $t = t_n$, we take $s - 1$ time-steps, so overall we have the approximations $\{v_n, v_{n+1}, v_{n+2}, \ldots, v_{n+s-1}\}$ at the grid points $\{t_n, t_{n+1}, t_{n+2}, \ldots, t_{n+s-1}\}$, as shown in Figure 8.8a. The next iterate v_{n+s} at t_{n+s} is obtained as usual, integrating $u_t = f$ within the interval $[t_{n+s-1}, t_{n+s}] = [(n + s - 1)h, (n + s)h]$,

$$u_{n+s} = u_{n+s-1} + \int_{(n+s-1)h}^{(n+s)h} f(t) \, dt. \tag{8.103}$$

If we introduce the variable $\tau = t - t_n$ and define $g(\tau) \doteq f(t(\tau))$, the integral appearing on the right-hand side of (8.103) reads

$$I(g) = \int_{(s-1)h}^{sh} g(\tau) \, d\tau, \tag{8.104}$$

so that (8.103) becomes

$$u_{n+s} = u_{n+s-1} + I(g). \tag{8.105}$$

For simplicity, we introduce the quantities

$$g_j \doteq f_{n+j} = f(t_{n+j}, v_{n+j}), \tag{8.106}$$

for $j = 0, 1, 2, \ldots, s - 1$, depicted as white bullets in Figure 8.8b. Adams method consists in devising a quadrature formula to approximate (8.104) using the ordinates $\{g_0, g_1, \ldots, g_{s-1}\}$ introduced in (8.106). These ordinates are the values of $g(\tau)$, shown as a black solid curve in Figure 8.8b, at the

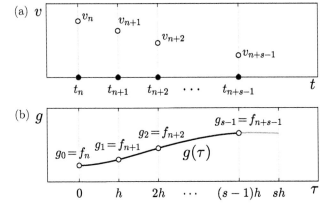

Figure 8.8 Adams method. (a) Previously computed s-steps $\{v_n, v_{n+1}, \ldots, v_{n+s-1}\}$. (b) Evaluation of f at the previously computed stages. The ordinates $g_j = f_{n+j}$ ($j = 0, 1, 2, \ldots, s - 1$) are used to extrapolate $g(\tau)$.

equispaced nodes $\tau_j \equiv jh$, for $j \equiv 0, 1, 2, \ldots, s-1$. However, to approximate (8.104) we need an approximation of $g(\tau)$ *outside* from the interval $[0, (s-1)h]$ where this function has been sampled (gray solid prolonged curve segment in Figure 8.8b). We aim to approximate (8.104) using the quadrature formula

$$I_{s-1}(g) \equiv \beta_0 g_0 + \beta_1 g_1 + \cdots + \beta_{s-1} g_{s-1}, \tag{8.107}$$

so that formula (8.105) is approximated by

$$v_{n+s} \equiv v_{n+s-1} + I_{s-1}(g). \tag{8.108}$$

The coefficients β_j in (8.107) are such that $I_{s-1}(g)$ approximates $I(g)$ with the maximum degree of exactness, that is,

$$I_{s-1}(g) \equiv I(g), \quad \forall g(\tau) \in \mathbb{R}_{s-1}[\tau]. \tag{8.109}$$

The reader must notice that rigorously, (8.107) is *not* an interpolatory quadrature formula, since we are implicitly integrating the interpolant *outside* of its domain of interpolation. In that sense, (8.107) is an *extrapolated* quadrature formula. Imposing exactness for each one of the monomials $\{1, \tau, \tau^2, \ldots, \tau^{s-1}\}$ leads to the linear system for the β_j coefficients,

$$\beta_0 + \beta_1 + \beta_2 + \cdots + \beta_{s-1} \equiv \int_{(s-1)h}^{sh} 1 \, d\tau = h$$

$$\beta_1 h + + \beta_2 2h + \cdots + \beta_{s-1}(s-1)h \equiv \int_{(s-1)h}^{sh} \tau \, d\tau \equiv \frac{h^2}{2}[s^2 - (s-1)^2]$$

$$\vdots$$

$$\beta_1 h^{s-1} + \beta_2 (2h)^{s-1} + \cdots + \beta_{s-1}[(s-1)h]^{(s-1)}$$
$$\equiv \int_{(s-1)h}^{sh} \tau^{s-1} \, d\tau \equiv \frac{h^s}{s}[s^s - (s-1)^s], \tag{8.110}$$

which, in matrix form, is of Vandermonde's type:

$$
\begin{bmatrix}
1 & 1 & 1 & \cdots & 1 \\
0 & h & 2h & \cdots & (s-1)h \\
0 & h^2 & 4h^2 & \cdots & (s-1)^2 h^2 \\
\vdots & \vdots & \vdots & & \vdots \\
0 & h^{s-1} & (2h)^{s-1} & \cdots & [(s-1)h]^{s-1}
\end{bmatrix}
\begin{bmatrix}
\beta_0 \\
\beta_1 \\
\beta_2 \\
\vdots \\
\beta_{s-1}
\end{bmatrix}
=
\begin{bmatrix}
h \\
h^2[s^2 - (s-1)^2]/2 \\
h^3[s^3 - (s-1)^3]/3 \\
\vdots \\
h^s[s^s - (s-1)^s]/s
\end{bmatrix}. \tag{8.111}
$$

For $s = 1$, the system is simply the equation $\beta_0 = h$. For $s = 2$, the system reads

$$\begin{bmatrix} 1 & 1 \\ 0 & h \end{bmatrix} \begin{bmatrix} \beta_0 \\ \beta_1 \end{bmatrix} = \begin{bmatrix} h \\ 3h^2/2 \end{bmatrix}, \tag{8.112}$$

whose solution is $\beta_0 = -h/2$ and $\beta_1 = 3h/2$. For $s \geq 3$, the two lines of Matlab code below provide the β_j coefficients in rational form (normalized with respect to the time-step h):

```
format rat; s = 3; j = [0:s-1]; k = j' + 1;
b = (j.^(j'))\(s.^k-(s-1).^k)./k); b'
5/12          -4/3          23/12
```

Once the quadrature weights β_j have been computed, we incorporate them in (8.108) to obtain

$$v_{n+s} = v_{n+s-1} + \beta_0 g_0 + \beta_1 g_1 + \cdots + \beta_{s-1} g_{s-1}, \tag{8.113}$$

or, in virtue of (8.106),

$$v_{n+s} = v_{n+s-1} + h \sum_{j=0}^{s-1} \beta_j f_{n+j}, \tag{8.114}$$

where the coefficients β_j have been normalized with respect to the time-step h. These normalized coefficients have been outlined in Table 8.3 for $s = 1, 2, 3, 4$. Expression (8.114) is known as the s-step *Adams–Bashforth*[15] formula. For

Table 8.3 Adams–Bashforth normalized coefficients

$$v_{n+s} = v_{n+s-1} + h \sum_{j=0}^{s-1} \beta_j f_{n+j}.$$

s	β_0	β_1	β_2	β_3
1	1	—	—	—
2	$-1/2$	3/2	—	—
3	5/12	$-4/3$	23/12	—
4	$-3/8$	37/24	$-59/24$	55/24

---

[15] F. Bashforth (1819–1912), British applied mathematician who first applied these formulas in the numerical solution of capillarity problems.

$s = 1, 2$, and 3, these formulas are

$$v_{n+1} = v_n + hf_n. \tag{8.115}$$

$$v_{n+2} = v_{n+1} + h\left(-\frac{1}{2}f_n + \frac{3}{2}f_{n+1}\right), \tag{8.116}$$

and

$$v_{n+3} = v_{n+2} + h\left(\frac{5}{12}f_n - \frac{4}{3}f_{n+1} + \frac{23}{12}f_{n+2}\right), \tag{8.117}$$

respectively. In particular, for $s = 1$, we recover Euler's explicit method (8.102). We will explore the numerical performance of s-step Adams–Bashforth formulas, henceforth referred to as ABs, later on in this section. The ABs formulas are *explicit* recurrence relations, since they explicitly provide the next iterate v_{n+s} as a function of the previously computed stages, that is, v_{n+s-1} and f_{n+j} $(j = 0, 1, \ldots, s-1)$.

We can reformulate the problem of approximating the solution in a very different way. Instead of approximating integrals via quadratures, as we did in ABs, let us approximate derivatives by means of differentiation matrices. Let the ordinates $\{v_n, v_{n+1}, \ldots, v_{n+s-1}, v_{n+s}\}$ be the approximations of the solution $u(t)$ at the grid points $\{t_n, t_{n+1}, \ldots, t_{n+s-1}, t_{n+s}\}$, respectively. Formal evaluation of the differential equation $u_t = f(t, u)$ at $t = t_{n+s}$ yields

$$u_t(t_{n+s}) = f(t_{n+s}, u_{n+s}) \approx f(t_{n+s}, v_{n+s}) = f_{n+s}. \tag{8.118}$$

We can approximate the exact derivative $u_t(t_{n+s})$ appearing on the left-hand side of (8.118) in terms of the ordinates v_j $(j = n, \ldots, n+s)$ as shown in Figure 8.9. Let the polynomial $\Pi_s v(t)$ be the sth degree interpolant of the given points (v_j, t_j), for $j = n, n+1, \ldots, n+s$. If $\Pi_s v(t)$ is a good approximation of the actual solution $u(t)$, we may accordingly approximate $u_t(t_{n+s})$ by the quantity

$$u_t(t_{n+s}) \approx \frac{\mathrm{d}}{\mathrm{d}t}\Pi_s v(t)\big|_{t=t_{n+s}}, \tag{8.119}$$

which is the slope of the interpolant at the rightmost node (gray solid line in Figure 8.9). To obtain that quantity, we only need to consider the differentiation

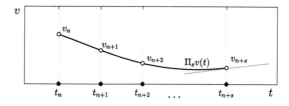

Figure 8.9 A BDF method is based on differentiating the interpolant $\Pi_s v(t)$ (black solid curve) obtained from the ordinates v_j (white bullets) and abscissas t_j, for $j = n, n+1, \ldots, n+s$.

matrix (3.15) for the $s + 1$ equispaced nodes $t_j = t_n + hj$ $(j = 0, 1, 2, \ldots, s)$:

$$
\mathbf{D}_{ij} = \begin{cases} \dfrac{1}{h} \displaystyle\sum_{\substack{k=0 \\ (k \neq j)}}^{s} \dfrac{1}{j-k} & (0 \leq i = j \leq s), \\[2em] \dfrac{1}{h} \dfrac{(-1)^{i-j}}{i-j} \dfrac{i! \, (s-i)!}{j! \, (s-j)!} & (0 \leq i, j \leq s, \quad i \neq j). \end{cases}
\tag{8.120}
$$

Assuming the vector $\mathbf{v} = [v_n \; v_{n+1} \; \cdots \; v_{n+s-1} \; v_{n+s}]^{\mathrm{T}}$ is a good approximation of the vector $\mathbf{u} = [u_n \; u_{n+1} \; \cdots \; u_{n+s-1} \; u_{n+s}]^{\mathrm{T}}$, the product $\mathbf{Dv}$ should accordingly provide a good approximation of the vector of derivatives $\mathbf{u}_t = [u_t(t_n) \; u_t(t_{n+1}) \; \cdots \; u_t(t_{n+s-1}) \; u_t(t_{n+s})]^{\mathrm{T}}$, that is,

$$
\begin{bmatrix} u_t(t_n) \\ u_t(t_{n+1}) \\ \vdots \\ u_t(t_{n+s-1}) \\ u_t(t_{n+s}) \end{bmatrix} \approx \begin{bmatrix} \mathbf{D}_{00} & \mathbf{D}_{01} & \cdots & \mathbf{D}_{0,s-1} & \mathbf{D}_{0s} \\ \mathbf{D}_{10} & \mathbf{D}_{10} & \cdots & \mathbf{D}_{1,s-1} & \mathbf{D}_{1s} \\ \vdots & \vdots & \vdots & \vdots & \vdots \\ \mathbf{D}_{s-1,0} & \mathbf{D}_{s-1,1} & \cdots & \mathbf{D}_{s-1,s-1} & \mathbf{D}_{s-1,s} \\ \mathbf{D}_{s0} & \mathbf{D}_{s1} & \cdots & \mathbf{D}_{s,s-1} & \mathbf{D}_{ss} \end{bmatrix} \begin{bmatrix} v_n \\ v_{n+1} \\ \vdots \\ v_{n+s-1} \\ v_{n+s} \end{bmatrix}.
\tag{8.121}
$$

In particular, the last row of the previous product is

$$
u_t(t_{n+s}) \approx \sum_{j=0}^{s} \mathbf{D}_{sj} v_{n+j},
\tag{8.122}
$$

which is precisely the approximated quantity that we were looking for on the left-hand side of (8.118). Using the explicit expressions of the coefficients $\mathbf{D}_{sj}$ of the last row of matrix (8.120), Eq. (8.118) reads

$$
\frac{1}{h} \left[\sum_{j=0}^{s-1} (-1)^{s-j} \frac{1}{s-j} \binom{s}{j} v_{n+j} + \left(\sum_{k=1}^{s} \frac{1}{k} \right) v_{n+s} \right] = f(t_{n+s}, v_{n+s}).
\tag{8.123}
$$

Expression (8.123) is usually written in a slightly different way. As a rule, it is customary to normalize the formula with respect to the coefficient

$$
H_s = \sum_{k=1}^{s} \frac{1}{k},
\tag{8.124}
$$

usually known as the sth *harmonic number*, multiplying v_{n+s}. Performing such normalization in (8.123) leads to the recurrence

$$
\sum_{j=0}^{s-1} \frac{(-1)^{s-j}}{(s-j)H_s} \binom{s}{j} v_{n+j} + v_{n+s} = h \frac{1}{H_s} f(t_{n+s}, v_{n+s}),
\tag{8.125}
$$

which is usually known as the *s*-step *Backwards Difference Formula*,[16] or simply BDF*s*. The reader can simply check by direct substitution in (8.125) that the first three BFD formulas are

$$v_{n+1} = v_n + hf_{n+1}, \tag{8.126}$$

$$v_{n+2} = -\frac{1}{3}v_n + \frac{4}{3}v_{n+1} + h\frac{2}{3}f_{n+2}, \tag{8.127}$$

and

$$v_{n+3} = \frac{2}{11}v_n - \frac{9}{11}v_{n+1} + \frac{18}{11}v_{n+2} + h\frac{6}{11}f_{n+3}, \tag{8.128}$$

for $s = 1$, $s = 2$, and $s = 3$, respectively. Table 8.4 outlines the formula and the corresponding coefficients of the BDF methods.

Equation (8.126) is also known as *implicit* Euler method, to be distinguished from its explicit homolog (8.102). The term implicit comes from the fact that (8.126) is actually

$$v_{n+1} = v_n + hf(t_{n+1}, v_{n+1}), \tag{8.129}$$

where v_n is assumed to be known, and v_{n+1} is the next iterate to be determined. In this case, recurrence (8.129) *does not* provide v_{n+1} explicitly in terms of the previous stage v_n. Instead, relation (8.129) defines $z = v_{n+1}$ as an implicit function of v_n through the equation

$$hf(t_{n+1}, z) - z + v_n = 0, \tag{8.130}$$

which in general will be transcendental, typically requiring Newton's method to be solved. This fact also occurs in the remaining BDF*s* formulas. For these reasons, these formulas are called *implicit*. Explicit AB and implicit BDF methods are particular instances of a wide family of formulas known as *linear multistep formulas*, henceforth referred to as LMSF. The reason for this name is that these

Table 8.4 Backward difference coefficients

$$\sum_{j=0}^{s} \alpha_j v_{n+j} = h\beta f_{n+s}, \quad \alpha_j \equiv \begin{cases} \dfrac{(-1)^{s-j}}{(s-j)H_s} \dbinom{s}{j} & (j = 0, 1, \dots, s-1) \\ 1 & (j = s) \end{cases}, \quad H_s = \sum_{k=1}^{s} \frac{1}{k}.$$

s	α_0	α_1	α_2	α_3	α_4	$\beta = 1/H_s$
1	−1	1	—	—	—	1
2	1/3	−4/3	1	—	—	2/3
3	−2/11	9/11	−18/11	1	—	6/11
4	3/25	−16/25	36/25	−48/25	1	12/25

16 First introduced in 1952 by the American physicists C. F. Curtiss and J. O. Hirschfelder within the context of chemical kinetics.

formulas are recurrences involving *linear* combinations of the approximate values v_j and f_j evaluated at *many different* steps of the time evolution. In general, LMSF formulas *cannot* self-start. For example, in order to start marching with (8.128), we need to provide not only v_0, but v_1 and v_2 as well. These two extra stages can be computed using another method, such as Runge–Kutta, for example. Logically, the reader may wonder why use LMSF formulas if they require other methods such as Runge–Kutta to be initiated. One of the reasons is that some of these formulas (BDF) have very interesting stability properties, whereas the others (AB) are computationally cheaper than RK and, at the same time, accurate.

Linear Multistep Formulas (LMSF)

The most general s-step LMSF reads

$$\sum_{j=0}^{s} \alpha_j v_{n+j} = h \sum_{j=0}^{s} \beta_j f_{n+j}. \tag{8.131}$$

For the Adams–Bashforth (ABs) formulas, the α_j coefficients are

$$\alpha_0 = \alpha_1 = \cdots = \alpha_{s-2} = 0, \; \alpha_{s-1} = -1, \; \alpha_s = 1, \tag{8.132}$$

whereas $\beta_s = 0$, so that the formula reads

$$v_{n+s} - v_{n+s-1} = h[\beta_0 f_n + \beta_1 f_{n+1} + \cdots + \beta_{s-1} f_{n+s-1}], \tag{8.133}$$

where the coefficients $\{\beta_0, \beta_1, \ldots, \beta_{s-1}\}$ are the solution of the linear system

$$\sum_{k=0}^{s-1} k^j \beta_k = \frac{1}{j+1}[s^{j+1} - (s-1)^{j+1}], \; (j = 0, 1, \ldots, s-1). \tag{8.134}$$

For the backwards difference formulas (BDF-s), the β_j coefficients are

$$\beta_0 = \beta_1 = \cdots = \beta_{s-1} = 0, \quad \beta_s = \frac{1}{H_s}, \; \left(H_s = \sum_{k=1}^{s} \frac{1}{k} \right). \tag{8.135}$$

The α_j coefficients are the elements $\mathbf{D}_{sj}$ corresponding to the last row of equispaced differentiation matrix (8.120), normalized by H_s, so that the formula reads

$$\sum_{j=0}^{s-1} \frac{(-1)^{s-j}}{(s-j)H_s} \binom{s}{j} v_{n+j} + v_{n+s} = h \frac{1}{H_s} f(t_{n+s}, v_{n+s}). \tag{8.136}$$

Let us explore the accuracy of AB and BDF formulas, as we did with Runge–Kutta one-step methods. For this purpose, we revisit the numerical approximation of the test IVP

$$u_t = u, \; (u_0 = 1), \tag{8.137}$$

also within the interval $t \in [0, 1]$, with exact solution $u(t) = e^t$. Recall that for $s > 1$, ABs and BDFs formulas cannot be started unless the previous stages $v_0, v_1, \ldots, v_{s-2}, v_{s-1}$ are initially supplied. For this reason we will make use of the *exact* solution values $v_j = u_j = e^{jh}$, $(j = 0, 1, 2, \ldots, s-1)$ to start the methods. For example, in the convergence analysis of AB3 or BDF3, the time marching schemes are initiated using $v_0 = 1$, $v_1 = e^h$, and $v_2 = e^{2h}$, so that the local errors are introduced from the third iterate onwards. If the number of total iterations N is large enough, we expect the error to be essentially the same as if the local errors were introduced from the first iteration onwards. The slopes of the resulting errors shown in Figure 8.10 clearly indicate that the error ε of these s-step formulas is $O(h^s)$. In general, it can be formally shown that s-step AB and BDF formulas have order of convergence s. This result makes AB formulas especially attractive from a computational cost point of view. As an example, take $h = 10^{-3}$ in RK2 Heun's formula (8.92), whose error shown in Figure 8.7b is approximately $\varepsilon \approx 10^{-6}$. For the same value of h, AB2 formula (8.116) provides nearly the same accuracy, as shown in Figure 8.10a. However, the computational cost has been halved in this second case, since AB2 only requires one evaluation of the right-hand side $f(t, u)$ per time-step, whereas RK2 requires two intermediate evaluations. In general, the computational cost of a RKs formula is s times larger[17] than ABs.

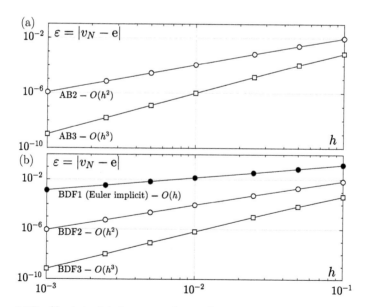

Figure 8.10 Absolute global error $\varepsilon = |e - v_N|$ at $t = t_N = Nh = 1$ of linear multistep formulas (LMSFs) applied to $u_t = u$ ($u_0 = 1$) in $t \in [0, 1]$. (a) Adams–Bashforth explicit methods AB2 and AB3. (b) Backwards difference formulas BDF1 (implicit Euler), BDF2 and BDF3.

---

17 Neglecting the cost of the $s - 1$ first steps that need to be supplied to the ABs formula.

Practical 8.2: Gravitational Motion

The figure on the right depicts a system of three particles with masses $m_1 = 1$, $m_2 = 2$, and $m_3 = 1$, located on the $\mathbb{R}^2$-plane at the coordinates $(x_1, y_1) = (0, 1)$, $(x_2, y_2) = (0, -1)$, and $(x_3, y_3) = (1, 0)$, respectively. The three particles are fixed and cannot move. Assume that the gravitational force exerted by a mass m_i at $\mathbf{r}_i$ on another mass m_j at $\mathbf{r}_j$ is

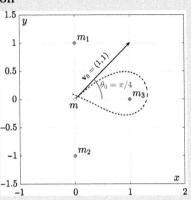

$$\mathbf{F}_{ij} = \frac{m_i m_j}{|\mathbf{r}_i - \mathbf{r}_j|^3}(\mathbf{r}_i - \mathbf{r}_j).$$

At $t = 0$, we launch from the origin a fourth particle of mass $m = 1$ with initial velocity $\mathbf{v}_0 = (1, 1)$ as depicted in the figure. Right after the launch, the particle initiates a trajectory that surrounds m_3 (dashed curve in the figure). Code your own RK4 and AB4 integrators based on (8.93) and Table 8.3, respectively, to compute the coordinates $\mathbf{r} = (x, y)$ of the moving particle at $t = 2$ (initialize AB4 using the first three steps provided by RK4).

(a) Let $\mathbf{r}_{\text{ex.}}(2)$ be the computed "*exact*" position of the particle at $t = 2$ using a very small time-step, say $h = 10^{-4}$, with any of the two methods. For a reasonably small time-step h, define the global error

$$\varepsilon(h) = |\mathbf{r}_h(2) - \mathbf{r}_{\text{ex.}}(2)|, \text{ with } h \in [10^{-3}, 10^{-1}].$$

On a log–log graph, plot $\varepsilon(h)$ as a function of h for both methods. Approximately, what time-step h is required to have $\varepsilon(h) \leq 10^{-4}$? Which method requires more evaluations of the vector field?

(b) Launching the particle from the same initial point and the same initial speed $|\mathbf{v}_0| = \sqrt{2}$, find the initial angle θ_0 such that the particle surrounds m_3 and returns to the origin before $t = 2$. Use your own RK4 method to provide the initial angle and the arrival time of the particle T_a with at least six exact figures. Suggestion: Newton's method.

(c) With the same initial speed as in (a) and (b), find initial angles θ_0 and arrival times $T_a < 2$ for trajectories surrounding the other two particles m_1 and m_2. Plot all the trajectories you have found. Do they share any common property? Explain.

8.2.3 Convergence of Time-Steppers

So far we have explained how to obtain a few particular instances of one-step RK and multistep LMSF, henceforth referred to as *time-steppers*. We also illustrated the accuracy of these methods at the end of the integration of the test IVP (8.137) by monitoring the rate at which their absolute global error $\varepsilon(h) \equiv |v_N - u_N|$ decreases when reducing the time-step h. The reader could also explore the rates of convergence of AB4 and RK4 when solving a more involved problem in Practical 8.2.

Let us consider a standard IVP

$$u_t = f(t, u), \quad u(0) = u_0, \tag{8.138}$$

within the interval $t \in [0, T]$. As always, we assume that (8.138) has a smooth exact solution $u(t)$. Consider the approximations $v_j \approx u(t_j)$ of (8.138) using any of the methods seen so far with a time-step $h = T/N$, involving N steps. For a fixed total integration time T, decreasing h obviously increases the number of steps. An obvious question is whether the limit $h \to 0$ ($N \to \infty$) leads to a sequence of approximations that *converge* to the exact solution, that is, whether the limits

$$\lim_{\substack{N \to \infty \\ Nh = T}} v_j = u(t_j), \tag{8.139}$$

for $j = 1, 2, \ldots, N$ are satisfied or not. In other words, we wonder whether the approximations v_j depicted in Figure 8.6 asymptotically approach the exact quantities u_j when h gets extremely small. This was certainly the case for RK, AB, and BDF methods, at least for the test problem (8.94), as it can be concluded from Figures 8.7 and 8.10, or from the results obtained in Practical 8.2, for example. However, we need to establish a general set of conditions required for our time-stepper to provide the right answer when decreasing h. At this point, we introduce the definition of order of convergence of a time-stepper

---

Order of Convergence of One-Step Methods

Consider a one-step method that discretizes the IVP

$$u_t = f(t, u), \quad u(0) = u_0, \tag{8.140}$$

within the time interval $t \in [0, T]$ at $t_j = jh$, ($j = 1, 2, \ldots, N$), with time-step $h = T/N$. Set $v_0 = u_0$ and let $u_j = u(t_j)$ and v_j be the exact solution and its corresponding approximation, respectively, at $t = t_j$, where the absolute global error is

$$e_j(h) = |v_j - u_j|, \tag{8.141}$$

for $j = 1, 2, \ldots, N$. The one-step method is said to be *convergent of order* p, with p being a positive integer, if

$$e_j(h) = O(h^p), \tag{8.142}$$

for $j = 1, 2, \ldots, N$.

This definition only applies to one-step methods such as Runge–Kutta. For a LMSF, things are slightly different. Linear multistep formulas must be started by supplying not only the initial condition $v_0 \equiv u_0$ but also $s - 1$ additional steps $v_1, v_2, \ldots, v_{s-1}$. These additional steps must typically be supplied by a one-step method (such as the RK4 method used in Practical 8.2 to start AB4). Therefore, any error introduced in these preliminary data will in general be propagated iteratively afterwards through the LMSF. In other words, the precision of the remaining points of the time integration $\{v_s, v_{s+1}, \ldots, v_N\}$ will in general be conditioned not only by the accuracy of the LMSF but also by the precision of the initially supplied data. For this reason, a proper definition of convergence of a LMSF requires to consider the magnitude of these initial inaccuracies:

Order of Convergence of an s-step LMSF

Consider an LMSF discretizing the IVP (8.140) within the time interval $t \in [0,T]$. Set $v_0 \equiv u_0$, $h = T/N$, and let $u_j \equiv u(t_j)$ be the exact solution at $t_j = jh$, for $j = 1, 2, \ldots, N$. Assume that the first $s - 1$ supplied steps $v_1, v_2, \ldots, v_{s-1}$ satisfy

$$|v_j - u_j| = O(h^p) \tag{8.143}$$

for $j = 1, 2, \ldots, s - 1$ and some positive integer p. The LMSF is said to be *convergent of order p* if

$$e_j(h) \equiv |v_j - u_j| = O(h^p), \tag{8.144}$$

for $j = s, s + 1, s + 2, \ldots, N$.

For a one-step method or a LMSF to be convergent, two essential conditions are required. The first one is known as *consistency*, which essentially is a requirement for the local truncation error of the method to be of order $O(h^q)$, with $q \geq 2$. The second is the commonly known 0-*stability*, requiring the recurrence formula of the method to *avoid amplification* of small inaccuracies at finite time T when $N \Rightarrow \infty$ and $h \to 0$, with $Nh \equiv T$. These two requirements for convergence are known as the *Dahlquist's equivalence theorem*.[18]

In this chapter, we have only addressed two types of LMSF, namely, explicit AB and implicit BDF methods. However, there are *many other* explicit and implicit linear multistep formulas in the literature. It can be shown that, in general,

---

[18] Germund Dahlquist (1925–2005), Swedish mathematician known for his fundamental contributions within the numerical analysis of ODEs.

any convergent s-step LMSF has order $O(h^p)$, with $p \leq s + 2$. This is also known as the *Dahlquist's first barrier theorem*. For more details regarding Dahlquist's equivalence and first barrier theorems, we refer the reader to the recommended bibliography at the end of the chapter, in the Complementary Reading section. In the rest of this section we will limit ourselves to providing the essential results concerning the convergence of one-step and multistep methods seen so far.

We start by characterizing the order of convergence of Runge–Kutta methods as a function of the number of intermediate stages, s, required per time-step:

Order of Convergence of Explicit Runge–Kutta Methods

The order of accuracy p of a s-stage explicit Runge–Kutta method is as follows:

s	1	2	3	4	5	6	7	8	9	10
p	1	2	3	4	4	5	6	6	7	7

In other words, to have order of accuracy $p = 1, 2, 3, 4$, RK methods require $s = 1, 2, 3, 4$ evaluations of the right-hand side $f(t, u)$, per time-step, respectively. However, to get $p = 5$, we need six evaluations or stages per time-step. This is one of the reasons why RK4 has become so popular among one-step methods. For the two familiar families of LMSF seen before, the convergence properties are summarized below:

Order of Convergence of Explicit AB and Implicit BDF Formulas

- The order of accuracy p of an ABs formula is $p = s$, $\forall s \geq 1$.
- The order of accuracy p of a BDFs formula is $p = s$, for $1 \leq s \leq 6$.

In other words, whereas Adams–Bashforth formulas are always convergent, BDF formulas have problems for $s = 7$ and beyond.[19] It may therefore seem that AB formulas have less limitations than BDF. However, as we will see later, convergence is *not* the only requirement for a time-stepper to be useful.

---

19 The underlying reason for this lack of convergence is that BDF formulas are *not* 0-stable for $s \geq 7$. We refer the reader to classical monographs on this topic at the end of the chapter.

8.2.4 A-Stability

All the convergence criteria seen so far were focused on the behavior of time-steppers in the limit of *very small* h and *very large* N for a *finite* time horizon $T = Nh$. In practice, however, one of the main goals of the numerical approximation of an IVP is to study the *asymptotic* behavior of its solution $u(t)$ for $t \to \infty$, that is, *very long* time integrations. Depending on the physical nature of the problem, the solution $u(t)$ may require very long *transients* before reaching a steady, time-periodic, almost periodic, or even chaotic regime, for example. Therefore, reducing the time-step h arbitrarily is out of the question, since the number of required steps N could potentially be huge, and the computational cost would be unfeasible. In practical computations, we look for time-steppers capable of providing, to some prescribed tolerance, the approximate solution $v_N \approx u(T)$, in a minimum number of time-steps N. If T is large, and we want N to be moderate, h *cannot* be arbitrarily small, as it was in the Section 8.2.3. Under these conditions, if we are forced to take a moderately small h, will the time-stepper still work? We partially answer this question in the next example.

Consider the IVP

$$u_t = -10u, \quad u(0) = 1, \tag{8.145}$$

in $t \in [0, +\infty)$ and whose solution is $u(t) = e^{-10t}$. Let us try to approximate this solution by means of the explicit Euler method (that is, AB1 or RK1) using a fixed time-step h. In this case, expression (8.83) reads

$$v_{n+1} = v_n + hf_n = v_n - 10hv_n = (1 - 10h)v_n, \tag{8.146}$$

with $v_0 = 1$. In this case, the nth term of the Euler approximation can be written explicitly:

$$v_{n+1} = (1 - 10h)v_n = (1 - 10h)^2 v_{n-1} = \cdots = (1 - 10h)^{n+1} v_0, \tag{8.147}$$

or simply

$$v_n = (1 - 10h)^n, \tag{8.148}$$

for $n = 0, 1, 2, \ldots$ Since the solution $u(t) = e^{-10t}$ decays exponentially to zero for $t \to +\infty$, we accordingly expect $v_n \to 0$ for large n, that is,

$$\lim_{n \to \infty} (1 - 10h)^n = 0. \tag{8.149}$$

This is only true if

$$|1 - 10h| < 1, \tag{8.150}$$

therefore implying that the time-step h must satisfy the inequality

$$h < \frac{1}{5}. \tag{8.151}$$

Figure 8.11 shows explicit Euler approximations of (8.145) for values of h slightly larger and smaller than the threshold value $h = 0.2$ obtained in (8.151). Taking for example $h = 0.205$, the numerical solution diverges, as shown in Figure 8.11a. For $h = 0.185$, however, the iterates v_n become damped, approaching the right asymptotic regime for large n, as shown in Figure 8.11b. Of course, the accuracy of the approximated values v_1, v_3, ..., v_{10} in this second case is very poor, but right now we are interested in the stability of the iteration rather than in its precision.

In this particular example, things would have been very different if, instead of using explicit Euler AB1, we had used the implicit Euler method BDF1 (8.126), for example. In this case, the BDF1 discretization of (8.145) reads

$$v_{n+1} = v_n + h f_{n+1} = v_n - 10 h v_{n+1}, \tag{8.152}$$

with $v_0 = 1$. Fortunately, the implicit recurrence (8.152) is so simple that it can be explicited as

$$v_{n+1} = (1 + 10h)^{-1} v_n, \tag{8.153}$$

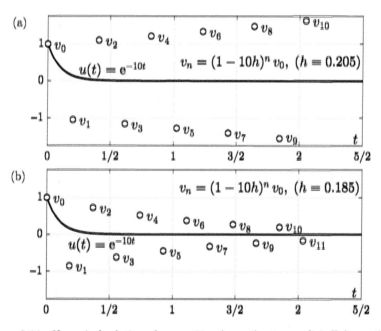

Figure 8.11 Numerical solution of $u_t = -10u$, $(u_0 = 1)$ using explicit Euler method (8.83). (a) $h = 0.205 > 1/5$ (unstable) and (b) $h = 0.185 < 1/5$ (stable).

or

$$v_n \equiv \frac{1}{(1 + 10h)^n},\tag{8.154}$$

for $n \equiv 0, 1, 2, \ldots$ In this second case, it is obvious that

$$\lim_{n \to \infty} v_n \equiv \lim_{n \to \infty} \frac{1}{(1 + 10h)^n} \equiv 0,\tag{8.155}$$

for *any* positive h. Figure 8.12 depicts this situation for $h = 0.25$. Once again, we may complain about the lack of precision of the first terms of the transient sequence $v_1, v_2, v_3, \ldots$, but the implicit Euler method does finally converge to the permanent regime $u \equiv 0$, and this regime can be reached in fewer iterations using even larger time-steps.

For a time-stepper to be reliable and useful, it must be accurate, but it must also be robust and capable of genuinely reproducing the underlying nature of the approximated solution using a moderate time-step. The numerical integration shown in Figure 8.11a exhibits an *instability* that is misleading and that has nothing to do with the *contractive* nature of forward-time integration of the solutions of $u_t \equiv -10u$. For this reason, it is common practice to study the behavior of different time-steppers when applied to the *test* problem

$$u_t \equiv \lambda u, \quad u(0) \equiv u_0,\tag{8.156}$$

for $t \in [0, +\infty)$, $\lambda \in \mathbb{C}$, $\mathrm{Re}(\lambda) < 0$, and whose solution is $u(t) \equiv u_0\, e^{\lambda t}$. Our previous test (8.145) is a particular case of (8.156), with $\lambda \equiv -10$. In general, however, it will be justified later that we need to consider the possibility of λ being complex. The discretization of (8.156) using the explicit Euler method leads to the approximate sequence

$$v_n \equiv (1 + h\lambda)^n\, u_0 \equiv (1 + z)^n\, u_0,\tag{8.157}$$

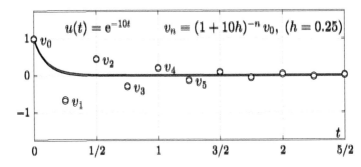

Figure 8.12 Numerical solution of $u_t \equiv -10u$, $(u_0 \equiv 1)$ using implicit Euler method BDF1 with $h \equiv 0.25$.

for $n = 0, 1, 2, \ldots$, where we have introduced the complex variable $z = h\lambda$. In this case, the explicit Euler discretization leads to bounded iterates only if

$$|1 + z| \leq 1, \tag{8.158}$$

leading to a region in the complex z-plane or $h\lambda$-plane depicted in Figure 8.13a. Only if the time-step h used in the time integration is such that the product $z = h\lambda$ falls within the gray region of Figure 8.13a, usually known as *stability region*, the iteration will lead to bounded results for large n. This has been illustrated in Figure 8.13a for the particular example (8.145) with $\lambda = -10$ and $h = 0.205$, resulting in point $z_1 = -0.205$, outside of the gray region, and leading to numerical instability. However, for $h = 0.185$, the point z_2 is clearly inside the stability region, therefore leading to numerically stable results. If (8.156) is discretized using the implicit Euler method instead, we obtain the sequence

$$v_n = \frac{u_0}{(1 - h\lambda)^n} = \frac{u_0}{(1 - z)^n}, \tag{8.159}$$

for $n = 0, 1, 2, \ldots$, which is bounded for large n only if

$$|1 - z| \geq 1. \tag{8.160}$$

The gray region of Figure 8.13b corresponds to the stability region of the implicit Euler method determined by (8.160). In particular, point z_3 is the resulting value of z used in example (8.145) for $h = 0.25$ that led to numerically stable iterations. In this case, taking larger values of h would only shift z_3 to the left, but never out of the stability region, as illustrated in Figure 8.13b.

Absolute Stability of a Time-Stepper

Consider the scalar *test* IVP

$$u_t = \lambda u, \quad u(0) = u_0, \tag{8.161}$$

where $\lambda \in \mathbb{C}$, with $\mathrm{Re}(\lambda) < 0$. A time-stepper is said to be *absolutely stable* for a certain time-step $h > 0$ if the resulting sequence v_n approximating the solution $u(nh)$ of (8.161) satisfies

$$\lim_{n \to +\infty} v_n = 0. \tag{8.162}$$

The *region of absolute stability* of a time-stepper is the set of points $z = h\lambda$ of the complex z-plane such that applying the method to (8.161) with time-step h leads to an approximate sequence v_n satisfying (8.162).

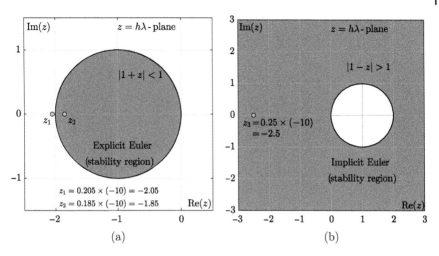

Figure 8.13 (a) Stability region of the explicit Euler method (8.158), in gray. Points z_1 and z_2 refer to the numerically unstable and stable integrations of (8.145), using $h = 0.205$ and $h = 0.185$, respectively. (b) Stability region of the implicit Euler method (8.160). Point z_3 refers to the numerically stable integration of (8.145) using $h = 0.25$.

In particular, the shaded regions depicted in Figure 8.13a,b are known as the *regions of absolute stability* for explicit Euler (RK1 or AB1) and implicit Euler (BDF1) methods, respectively.

In what follows, we proceed to obtain the regions of absolute stability of the higher order methods seen before. For example, Heun's RK2 method (8.92) applied to the test problem (8.161) leads to the recurrence

$$v_{n+1} = v_n + \frac{h}{2}[f(t_n, v_n) + f(t_n + h, v_n + hf(t_n, v_n))]$$

$$= v_n + \frac{h}{2}[\lambda v_n + \lambda(v_n + h\lambda v_n)] = \left[1 + h\lambda + \frac{1}{2}(h\lambda)^2\right]v_n, \quad (8.163)$$

or simply

$$v_{n+1} = \left(1 + z + \frac{1}{2}z^2\right)v_n = R_2(z)\,v_n, \quad (8.164)$$

where $z = h\lambda$ and $R_2(z)$ is the order-2 truncated Taylor series of the exponential function $f(z) = e^z$. The sequence (8.164) satisfies the absolute stability condition (8.162) only if $|R_2(z)| < 1$. One way to determine the region of absolute stability is to find its boundary, given by the relation

$$\left|1 + z + \frac{1}{2}z^2\right| = 1. \quad (8.165)$$

A simple way of determining geometrically the loci corresponding to the boundary (8.165) is to look for the roots of the equation

$$1 + z + \frac{1}{2}z^2 = e^{i\theta},$$

(8.166)

where $e^{i\theta}$ represents any complex number of modulus 1, parametrized in the complex plane by its phase $\theta \in (-\pi, \pi)$. The solutions of (8.166) are

$$z_{\pm}(\theta) = -1 \pm \sqrt{2e^{i\theta} - 1}.$$

(8.167)

The Matlab commands below compute the two boundary branches by evaluating the expressions $z_{\pm}(\theta)$ of (8.167) at a discrete set of phase angles $\theta_j = \pi j/N, \ (j = -N + 1, \ldots, N - 1)$.

```
N = 100; w = exp(i*pi*[-N+1:N-1]/N);
z1 =-1+(2*w-1).^.5; z2 =-1-(2*w-1).^.5;
plot(real(z1),imag(z1),'-k',real(z2),imag(z2),'-r');
axis equal
```

The resulting curves corresponding to $z_{\pm}(\theta)$ have been depicted in Figure 8.14a. It can be shown that the region of absolute stability of an explicit pth order Runge–Kutta method RKs with $s = p \leq 4$ is given by

$$|R_s(z)| = \left| 1 + z + \cdots + \frac{z^s}{s!} \right| \leq 1.$$

(8.168)

For $s = 3$ and $s = 4$, the computation of the stability boundaries $|R_s(z)| = 1$ can be carried out in many different ways. For example, Figure 8.14b shows the stability region boundaries $|R_s(z)| = 1$ for $s = 3, 4$ computed using pseudo-arclength continuation (Code 21: contstep.m) along the level curve function $f(x, y) = R_s(x + iy)\overline{R_s(x + iy)} - 1 = 0$. This is left as an exercise for the reader, who may check that Heun's boundary can also be obtained in this way.

The analysis of the absolute stability regions of AB and BDF LMSFs requires the introduction of some definitions:

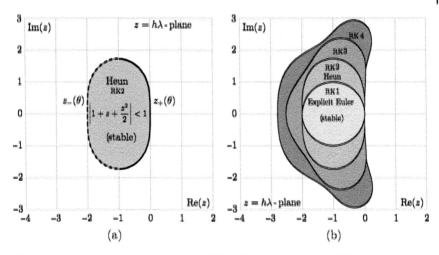

Figure 8.14 (a) Region of absolute stability (light gray region) of Heun's RK2 method. The stability boundary is the union of the solid and dashed black curves $z_\pm(\theta) = -1 \pm \sqrt{2e^{i\theta} - 1}$; see (8.167). (b) Absolute stability regions of Runge–Kutta methods RKs, for $s \leq 4$.

Characteristic Polynomials Associated with an LMSF

The *characteristic* (sometimes known as *generating*) polynomials associated with the s-step LMSF

$$\sum_{j=0}^{s} \alpha_j v_{n+j} = h \sum_{j=0}^{s} \beta_j f_{n+j} \qquad (8.169)$$

are

$$\rho(\mu) = \sum_{j=0}^{s} \alpha_j \mu^j \quad \text{and} \quad \sigma(\mu) = \sum_{j=0}^{s} \beta_j \mu^j, \qquad (8.170)$$

satisfying

$$\deg(\rho) = s \quad \text{and} \quad \deg(\sigma) = \begin{cases} s & \text{(implicit BDF)}, \\ s - 1 & \text{(explicit AB)}, \end{cases} \qquad (8.171)$$

so that in general $\rho(\mu)$, $\sigma(\mu) \in \mathbb{R}_s[\mu]$ and in particular $\sigma(\mu) \in \mathbb{R}_{s-1}[\mu]$ for AB methods.

For example, AB2 formula (8.116) written in the original format (8.169) reads

$$v_{n+2} - v_{n+1} = h\left(-\frac{1}{2}f_n + \frac{3}{2}f_{n+1}\right),$$
(8.172)

and therefore its associated characteristic polynomials are

$$\rho(\mu) = \mu^2 - \mu = \mu(\mu - 1), \quad \sigma(\mu) = \frac{3}{2}\mu - \frac{1}{2}.$$
(8.173)

Similarly, the polynomials associated with the implicit BDF2 formula (8.127)

$$v_{n+2} - \frac{4}{3}v_{n+1} + \frac{1}{3}v_n = h\frac{2}{3}f_{n+2},$$
(8.174)

are

$$\rho(\mu) = \mu^2 - \frac{4}{3}\mu + \frac{1}{3}, \quad \sigma(\mu) = \frac{2}{3}\mu^2.$$
(8.175)

To obtain the stability regions for AB and BDF methods, we must analyze the boundedness of the sequence $\{v_n\}$ rendered by the general recurrence (8.169) when applied to the test differential equation $u_t = \lambda u$, that is, for $f(t, u) = \lambda u$. In that case, (8.169) reads

$$\sum_{j=0}^{s} \alpha_j v_{n+j} = h\lambda \sum_{j=0}^{s} \beta_j v_{n+j}.$$
(8.176)

We already performed this type of analysis for the explicit Euler method, with $s = 1$, $\alpha_0 = -1$, $\alpha_1 = 1$, $\beta_0 = 1$, and $\beta_1 = 0$, leading to the recurrence

$$v_{n+1} - v_n = h\lambda\, v_n,$$
(8.177)

starting from $v_0 = u_0$, and whose nth iterate was

$$v_n = (1 + h\lambda)^n\, u_0,$$
(8.178)

formerly obtained in (8.157). From (8.178), it was almost trivial to conclude that the stability condition was $|1 + h\lambda| < 1$. However, applying AB2 or BDF2 formulas (8.172) or (8.174) to the test differential equation $u_t = \lambda u$ leads to the recurrences

$$v_{n+2} - \left(1 + \frac{3}{2}h\lambda\right)v_{n+1} - \frac{1}{2}h\,\lambda v_n = 0,$$
(8.179)

or

$$\left(1 - \frac{2}{3}h\lambda\right)v_{n+2} - \frac{4}{3}v_{n+1} + \frac{1}{3}v_n = 0,$$
(8.180)

respectively, from which it is not straightforward to foretell what the stability conditions will be.

Recurrence relations (8.179) or (8.180) are particular instances of *homogeneous linear difference equations* with constant coefficients. In what follows we provide the main results concerning the general solution of this particular type of recurrence relations. For a formal approach to these results in the most general case, including nonhomogeneous terms and non-constant coefficients, we refer the reader to the bibliography in the Complementary Reading section.

Homogeneous Linear Difference Equations

Consider the constant-coefficient s-step recurrence formula

$$v_{n+s} + a_1\, v_{n+s-1} + a_2\, v_{n+s-2} + \cdots + a_{s-1}\, v_{n+1} + a_s\, v_n = 0, \quad (8.181)$$

for $n = 0, 1, 2, \ldots$, starting from known initial values $\{v_0, v_1, \ldots, v_{s-1}\}$, along with its associated *characteristic polynomial*

$$p(\mu) = \mu^s + a_1\mu^{s-1} + a_2\mu^{s-2} + \cdots + a_{s-1}\mu + a_s \in \mathbb{R}_s[\mu]. \quad (8.182)$$

If $p(\mu)$ has s simple roots, that is,

$$p(\mu) = (\mu - \mu_1)(\mu - \mu_2)\cdots(\mu - \mu_s), \quad (8.183)$$

with $\mu_i \neq \mu_j$ for $i \neq j$, then the general solution of (8.181) is

$$v_n = c_1\, \mu_1^n + c_2\, \mu_2^n + \cdots + c_s\mu_s^n, \quad (8.184)$$

where $\{c_1, c_2, \ldots, c_s\}$ are arbitrary constants. However, if the characteristic polynomial has *multiple* roots, that is,

$$p(\mu) = (\mu - \mu_1)^{m_1}(\mu - \mu_2)^{m_2}\cdots(\mu - \mu_\ell)^{m_\ell}, \quad (8.185)$$

with $\mu_i \neq \mu_j$ for $i \neq j$, $1 \leq m_j \leq s$ for $j = 1, 2, \ldots, \ell$, with

$$m_1 + m_2 + \cdots + m_\ell = s,$$

then the general solution of (8.181) reads

$$v_n = (c_{10} + c_{11}\, n + c_{12}\, n^2 + \cdots + c_{1,m_1-1}\, n^{m_1-1})\mu_1^n$$
$$+ (c_{20} + c_{21}\, n + c_{22}\, n^2 + \cdots + c_{2,m_2-1}\, n^{m_2-1})\mu_2^n$$
$$\vdots$$
$$+ (c_{\ell 0} + c_{\ell 1}\, n + c_{\ell 2}\, n^2 + \cdots + c_{\ell,m_\ell-1}\, n^{m_\ell-1})\mu_\ell^n. \quad (8.186)$$

In both cases, the constants $\{c_1, c_2, \ldots, c_s\}$ or $\{c_{10}, c_{11}, \ldots\}$ are obtained after imposing the initial conditions.

For example, consider the recurrence

$$v_{n+2} - 5v_{n+1} + 6v_n = 0, \quad v_0 = 0, \quad v_1 = 1. \quad (8.187)$$

Using (8.187), we can recursively obtain a few first terms of the recurrence: $v_2 \equiv 5v_1 - 6v_0 = 5$, $v_3 \equiv 5v_2 - 6v_1 \equiv 19$, $v_4 \equiv 5v_3 - 6v_2 \equiv 65$, etc. Since the characteristic polynomial is $p(\mu) \equiv \mu^2 - 5\mu + 6 \equiv (\mu - 3)(\mu - 2)$, the general solution reads

$$v_n \equiv c_1\, 3^n + c_2\, 2^n. \tag{8.188}$$

To find c_1 and c_2, we impose (8.188) to satisfy the two initial iterates, namely, $v_0 = 0$ and $v_1 = 1$. This leads to the linear system

$$c_1 + c_2 = 0$$
$$3c_1 + 2c_2 = 1,$$

whose solution is $c_1 = 1$ and $c_2 = -1$, so that (8.188) now reads

$$v_n \equiv 3^n - 2^n. \tag{8.189}$$

The reader may easily check that (8.189) leads to the expected values $v_2 \equiv 5$, $v_3 \equiv 19$, $v_4 \equiv 65$, etc. As a second example, consider the three-step formula

$$2v_{n+3} + 3v_{n+2} - v_n = 0. \tag{8.190}$$

In this case, $p(\mu) \equiv 2\mu^3 + 3\mu^2 - 1 \equiv (\mu + 1)^2(2\mu - 1)$, so that $\mu_1 \equiv -1$, $m_1 \equiv 2$, $\mu_2 \equiv 1/2$, and $m_2 \equiv 1$. The general solution is therefore

$$v_n \equiv (c_{10} + c_{11}\, n)\, (-1)^n + \frac{c_{20}}{2^n}. \tag{8.191}$$

From expressions (8.184) and (8.186), it is clear that the modulus and algebraic multiplicities of the roots of $p(\mu)$ dictate the long term behavior of the general solution v_n. In particular, if $p(\mu)$ has *any* root μ_j with $|\mu_j| > 1$, we expect a *geometric growth* of the form μ_j^n. If $|\mu_j| = 1$, but its algebraic multiplicity is $m_j > 1$, then we may expect an *algebraic growth* with leading term n^{m_j-1}, such as the linearly growing factor $c_{11}n$ appearing in (8.191). In any of the two aforementioned cases, the iteration will in general lead to an unbounded sequence $\{v_n\}$, that is, to a numerical instability.

We are now ready to study the boundedness of the sequence $\{v_n\}$ rendered by the recurrence (8.176), which we write as

$$\sum_{j=0}^{s} \alpha_j v_{n+j} - h\lambda \sum_{j=0}^{s} \beta_j v_{n+j} = 0. \tag{8.192}$$

In this case, any of the roots μ must be a solution of the characteristic equation

$$\rho(\mu) - h\lambda\, \sigma(\mu) = 0, \tag{8.193}$$

where $\rho(\mu)$ and $\sigma(\mu)$ are the generating polynomials associated with the LMSF, previously introduced in (8.170). Since $\deg(\rho) = s$ and $\deg(\sigma) \leq s$, Eq. (8.193) has at most s roots (counting multiplicities). For a given time-step h, the boundedness of the sequence $\{v_n\}$ generated by (8.192) is dictated by the modulus $|\mu(h\lambda)|$ of the roots of (8.193). By solving the polynomial equation (8.193) analytically, we could find the explicit expressions of its roots $\mu_k(h\lambda)$, and obtain the stability regions from imposing the conditions $|\mu_k(h\lambda)| < 1$. Although this can be done for $2 \leq s \leq 4$, the resulting expressions are mathematically quite involved and difficult to handle. Furthermore, for $s \geq 5$, the roots of (8.193) cannot be expressed in terms of radicals, as stated by Abel–Ruffinni's theorem.

There is a simple way to determine the stability region of an LMSF, known as the *boundary locus method*, which exploits the implicit dependence between the root μ and the factor $h\lambda$ through the characteristic Eq. (8.193). Suppose that, for a given time-step h, the factor $h\lambda$ is such that the modulus $|\mu_k|$ of the kth root of (8.193) is precisely 1, that is, the marginal stability case. This root must necessarily be located at some point of the marginal curve $|\mu| \equiv 1$ in the complex μ-plane. Conversely, we can compute the value of $z \equiv h\lambda$ corresponding to *every point* of that marginal curve first by rewriting (8.193) as

$$z(\mu) \equiv h\lambda \equiv \frac{\rho(\mu)}{\sigma(\mu)}, \tag{8.194}$$

and then by evaluating that ratio along the $|\mu| \equiv 1$ loci $\mu \equiv e^{i\theta}$, that is,

$$z(\theta) \equiv \frac{\rho(e^{i\theta})}{\sigma(e^{i\theta})}, \tag{8.195}$$

for $\theta \in [0, 2\pi]$. We can sketch the stability boundary of any LMSF by simply evaluating the ratio above on a dense set of points

$$\mu_j \equiv e^{i\theta_j}, \quad \text{with } \theta_j \equiv \frac{2\pi}{N} j, \tag{8.196}$$

for $j = 0, 1, 2, \ldots, N - 1$, that is, with θ_k being N equally distributed phases along the interval $[0, 2\pi)$. For example, Figure 8.15a depicts the boundary of the Adams–Bashforth method AB2 by evaluating the ratio (8.195) of its associated generating polynomials (8.173), namely,

$$z(\mu) \equiv \frac{2\mu(\mu - 1)}{3\mu - 1}, \tag{8.197}$$

along the equally spaced phase distribution (8.196). The two Matlab command lines below generate that boundary:

```
mu = exp(i*2*pi*[0:99]/100); z = 2*mu.*(mu-1)./(3*mu-1);
plot(real(z),imag(z),'-k','linewidth',1.25);axis equal
```

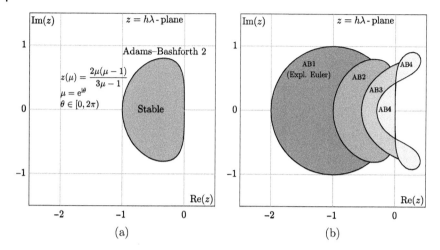

Figure 8.15 (a) Region of absolute stability (in gray) of 2-step Adams–Bashforth method AB2. The stability boundary is the curve $z(\mu) = 2\mu(\mu - 1)/(3\mu - 1)$, with $\mu = e^{i\theta}$ and $\theta \in [0, 2\pi)$; see (8.197). (b) Absolute stability regions of s-step Adams–Bashforth methods ABs, for $s \leq 4$.

Figure 8.15b shows the resulting boundaries of the s-step Adams–Bashforth formulas using the boundary locus method for $1 \leq s \leq 4$. The stable regions are shaded. Notice that the explicit Euler method AB1 has the largest stability region and that increasing s reduces the area of such domain. In particular, AB4 method has three stability sub-domains. It is left as an exercise for the reader (see Problem 8.7) to obtain the boundary curves shown in Figure 8.15b and to check that the gray areas correspond to the regions of absolute stability.[20]

The boundary locus method can also be applied to study the regions of absolute stability of the s-step BDF methods. In particular, Figure 8.16a depicts the boundary curve of BDF2 by evaluating the ratio of its generating polynomials (8.175)

$$z_2(\mu) = \frac{\mu^2 - \dfrac{4}{3}\mu + \dfrac{1}{3}}{\dfrac{2}{3}\mu^2} = \frac{3}{2} - \frac{2}{\mu} + \frac{1}{2\mu^2}, \tag{8.198}$$

along the equispaced set of angles (8.196). The reader may easily check that in this case the light gray filled area shown in Figure 8.16a corresponds to the *unstable* region of BDF2, and that the method is absolutely stable *outside* that region. Figure 8.16a also depicts the boundary for the BDF3 method obtained

---

20 **Hint**: evaluate ρ/σ along $\mu_j = 0.99\, e^{i\theta_j}$, for example.

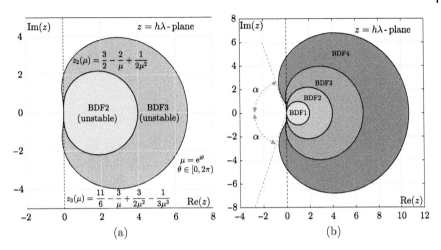

Figure 8.16 (a) Instability regions of 2-step BDF2 (light gray) and 3-step BDF3 (gray) methods, with boundary curves $z_2(\mu) = 3/2 - 2\mu^{-1} + 1/2\mu^2$, and $z_3(\mu) = 11/6 - 3/\mu + 3/2\mu^2 - 1/3\mu^3$, respectively, where $\mu = e^{i\theta}$ and $\theta \in [0, 2\pi)$; see (8.198) and (8.199). (b) Instability regions (in gray scales) of s-step BDF formulas BDFs for $s \le 4$. Although BDF4 is not A-stable, it is absolutely stable within the complex sector $S_\alpha = \{z;\ |\arg(-z)| < \alpha,\ z \ne 0\}$ for $\alpha \approx 73°$ bounded by the two gray dashed lines; see text for more details.

from evaluating the ratio

$$z_3(\mu) = \frac{11}{6} - \frac{3}{\mu} + \frac{3}{2\mu^2} - \frac{1}{3\mu^3}. \tag{8.199}$$

Throughout the white region of Figure 8.16a, the 3-step BDF method BDF3 is absolutely stable. Notice that, as opposed to the unstable region of BDF2, the unstable region of BDF3 slightly protrudes into the half-plane of negative real part, to the left of the vertical dashed line $\mathrm{Re}(z) = 0$. Finally, Figure 8.16b shows the instability regions of s-step BDF formulas BDFs for $s \le 4$. Notice that the unstable region of the s-step method contains the unstable region of the $(s-1)$-step method for $s = 2, 3, 4$. Also notice that the protrusion of the instability region of BDF4 into the half-plane $\mathrm{Re}(z) < 0$ is even more pronounced than for BDF3.

The innermost light gray region BDF1 shown Figure 8.16b is actually the unstable region of the implicit Euler method already depicted in Figure 8.13b (in white). BDF1 and BDF2 methods share a very important feature: their associated absolute stability regions *include the whole half-plane* $\mathrm{Re}(z) < 0$. As a result, the size of h required to obtain bounded solutions in the test problem (8.156) is not conditioned by the magnitude of $|\lambda|$. This leads to a very important definition and to an even more important result:

A-Stability

A LMSF is said to be *A-stable* if its region of absolute stability contains the entire left half-plane $\text{Re}(z) < 0$.

Second Dahlquist Barrier

1. An explicit time-stepper *cannot* be A-stable.
2. An A-stable (implicit) LMSF cannot have order $p > 2$.

Point 1 of the last two statements is clearly illustrated in Figures 8.14b and 8.15b, where the stability regions of the explicit Runge–Kutta and Adams–Bashforth methods, respectively, cover just a small portion of the left half-plane $\text{Re}(z) < 0$. Figure 8.16b on the other hand shows the A-stability of BDF1 and BDF2 methods, whose unstable regions are entirely within the right half-plane $\text{Re}(z) > 0$. The third and fourth order methods BDF3 and BDF4 do not qualify as A-stable, since they are unstable slightly to the left of the imaginary axis. However, these two methods are absolutely stable within the complex sector

$$S_\alpha \equiv \{z \in \mathbb{C}; \ |\arg(-z)| < \alpha, \ z \neq 0\}, \tag{8.200}$$

as shown in Figure 8.16b for BDF4, where α is approximately $73°$. For BDF3, $\alpha \approx 86°$ (not shown in Figure 8.16). In these cases, it is said that BDF3 and BDF4 are A(α)-stable.

The second Dahlquist barrier has very important consequences in the practical numerical integration of differential equations (ordinary and partial); we cannot stress this enough. First, it tells us that if we want to truly guarantee boundedness of the approximated sequences v_n for arbitrary time-step sizes, we cannot pursue orders of convergence higher than two (which sometimes may not be enough for certain computations). Second, if we want to guarantee stability, even for a modest quadratic order of convergence, the LMSF has necessarily to be *implicit*, thus requiring to solve a nonlinear problem at every iteration. Finally, assuming that we pursue quadratic convergence, say, by using BDF2, we need to initialize the time-stepping providing not only the initial condition $v_0 \equiv u_0$ but v_1 as well. This first step v_1 can only be obtained by means of a first order (also implicit) method, often with a considerably smaller time-step so that the quadratic accuracy is preserved.

Breaking the second Dahlquist barrier is possible. For example, certain *implicit* one-step Runge–Kutta methods are A-stable, and many of them have orders of convergence higher than 2. The main drawback of these methods is that every step is based on s-stages that require to solve systems of s nonlinear equations ($n \times s$ when dealing with systems of n differential equations). Addressing implicit Runge–Kutta methods is far beyond the purposes of this basic chapter, but the reader should be aware of their existence as well as of other types of strategies that can break the second stability barrier. In the recommended bibliography section we direct the reader to suitable references.

8.2.5 A-Stability in Nonlinear Systems: Stiffness

In the Section 8.2.4, we have studied the absolute stability of different time-steppers using the *scalar linear* model differential equation $u_t = \lambda u$. That analysis therefore overlooked two main important aspects. First, how can we extend the results of the Section 8.2.4 to n-dimensional systems of the form $\mathbf{u}_t = \mathbf{A}\,\mathbf{u}$, where $\mathbf{u}(t) \in \mathbb{R}^n$ and $\mathbf{A} \in \mathrm{M}_n(\mathbb{R})$, that is, a *multidimensional linear* model? Second, consider an arbitrary multidimensional IVP of the form[21]

$$\mathbf{u}_t = \mathbf{f}(t, \mathbf{u}), \ \mathbf{u}(0) = \mathbf{u}^0, \tag{8.201}$$

where $\mathbf{u}(t) \in \mathbb{R}^n$, $\mathbf{u}^0 \in \mathbb{R}^n$ is the initial condition and $\mathbf{f}(t, \mathbf{u})$ is in general a *nonlinear* vector field. What is the actual validity and relevance of the results concluded from a linear model equation (scalar or multidimensional) on the stability of the time-stepper when applied to (8.201)?

To address the first question, we consider a *system* of n linear differential equations of the form

$$\mathbf{u}_t = \mathbf{A}\,\mathbf{u}, \quad \mathbf{u}(0) = \mathbf{u}^0, \tag{8.202}$$

where $\mathbf{u}(t) \in \mathbb{R}^n$ and $\mathbf{A} \in \mathrm{M}_n(\mathbb{R})$ is a $n \times n$ real matrix with constant coefficients. Assuming that the eigenvalues $\{\lambda_1, \lambda_2, \ldots, \lambda_n\}$ of $\mathbf{A}$ are different, we can find a basis of vectors

$$\mathbf{w} = \mathbf{M}\,\mathbf{u}, \tag{8.203}$$

in which the endomorphism $\mathbf{A}$ diagonalizes, that is,

$$\mathbf{J} = \mathbf{M}\,\mathbf{A}\,\mathbf{M}^{-1}, \tag{8.204}$$

where $\mathbf{J} = \mathrm{diag}(\lambda_1, \lambda_2, \ldots, \lambda_n)$ is a diagonal matrix containing the eigenvalues of $\mathbf{A}$ so that (8.202) reads

$$\mathbf{w}_t = \mathbf{J}\,\mathbf{w}, \ \mathbf{w}^0 = \mathbf{M}\mathbf{u}^0, \tag{8.205}$$

and can be split into each invariant eigensubspace:

$$\mathbf{w}_{j,t} = \lambda_j \mathbf{w}_j, \ \mathbf{w}_j(0) = \mathbf{w}_j^0, \tag{8.206}$$

for $j = 1, 2, \ldots, n$, so that the solution is

$$\mathbf{w}(t) = \mathrm{e}^{\mathbf{J}t}\mathbf{w}^0, \tag{8.207}$$

---

21 To avoid confusion with the indexing of vector components, throughout this section we will use superscripts to denote the kth time iteration, that is, $\mathbf{v}^k \approx \mathbf{u}(t_k) = \mathbf{u}^k$.

or, since $e^{\mathbf{J}t}$ is also a diagonal matrix,

$$\mathbf{w}_j(t) = e^{\lambda_j t} \mathbf{w}_j^0, \ (j = 1, 2, \ldots, n), \tag{8.208}$$

so that the asymptotic growth or decay of the jth component of $\mathbf{w}(t)$ is exclusively dictated by the sign of the real part of λ_j.

In what follows, let us assume that all the eigenvalues of $\mathbf{A}$ lie on the left-hand side of the complex plane, that is,

$$\text{Re}(\lambda_j) < 0, \ (j = 1, 2, \ldots, n). \tag{8.209}$$

In this case,

$$\lim_{t \to +\infty} \mathbf{w}_j(t) = \lim_{t \to +\infty} e^{\lambda_j t} \mathbf{w}_j^0 = 0, \tag{8.210}$$

for $j = 1, 2, \ldots, n$, and therefore the asymptotic behavior of any solution $\mathbf{u}(t)$ of (8.202) satisfies

$$\lim_{t \to +\infty} \|\mathbf{u}(t)\| = \lim_{t \to +\infty} \|\mathbf{M}^{-1}\mathbf{w}(t)\| \le \|\mathbf{M}^{-1}\| \lim_{t \to +\infty} \|\mathbf{w}(t)\| = 0, \tag{8.211}$$

where $\|\mathbf{M}^{-1}\|$ stands for the standard induced spectral norm of $\mathbf{M}^{-1}$ satisfying (5.182). A reliable numerical approximation of $\mathbf{u}(t)$ should therefore reproduce limit (8.211). Let $\boldsymbol{\omega}^k \approx \mathbf{w}(t_k)$ be the numerical approximation of the solution of (8.205) at $t_k = kh$ provided by the explicit Euler iteration

$$\boldsymbol{\omega}^{k+1} = (\mathbf{I}_n + h\mathbf{J})\boldsymbol{\omega}^k, \tag{8.212}$$

with $\boldsymbol{\omega}^0 = \mathbf{w}^0$, or simply

$$\boldsymbol{\omega}^k = (\mathbf{I}_n + h\mathbf{J})^k \mathbf{w}^0, \tag{8.213}$$

which componentwise[22] reads

$$\omega_j^k = (1 + h\lambda_j)^k \mathbf{w}_j^0 = (1 + z_j)^k \mathbf{w}_j^0, \tag{8.214}$$

for $j = 1, 2, \ldots, n$, where we have introduced the complex numbers $z_j = h\lambda_j$. By the same arguments used in (8.157) and (8.158) for the one-dimensional case, the explicit Euler discretization (8.214) leads to bounded iterates only if

$$|1 + z_j| \le 1, \tag{8.215}$$

for $j = 1, 2, \ldots, n$, that is, only if *all* the complex numbers $\{z_1, z_2, \ldots, z_n\}$ are contained *within* the stability region of the Euler explicit method shown in Figure 8.13a. In that case,

$$\lim_{k \to +\infty} \omega_j^k = 0. \tag{8.216}$$

Of course, the explicit Euler discretization is carried out on the original system (8.202) and not on the diagonalized system (8.205). Let $\mathbf{v}^k \approx \mathbf{u}(t_k)$ be the

---

22 We denote ω_j^k the jth component of the approximated solution $\boldsymbol{\omega}$ at $t_k = kh$.

numerical approximation of the solution of (8.202) at $t_k = kh$, provided by the explicit Euler method iteration

$$\mathbf{v}^{k+1} = \mathbf{v}^k + h\mathbf{A}\,\mathbf{v}^k = (\mathbf{I}_n + h\mathbf{A})\mathbf{v}^k. \tag{8.217}$$

By virtue of the inverse change of basis, $\mathbf{v}^k = \mathbf{M}^{-1}\boldsymbol{\omega}^k$, so that premultiplication of (8.217) by $\mathbf{M}$ leads to

$$\mathbf{M}\mathbf{v}^{k+1} = \mathbf{M}(\mathbf{I}_n + h\mathbf{A})\mathbf{v}^k = \mathbf{M}(\mathbf{I}_n + h\mathbf{A})\mathbf{M}^{-1}\boldsymbol{\omega}^k, \tag{8.218}$$

or

$$\boldsymbol{\omega}^{k+1} = (\mathbf{I}_n + h\mathbf{M}\mathbf{A}\mathbf{M}^{-1})\boldsymbol{\omega}^k, \tag{8.219}$$

which, since $\mathbf{J} = \mathbf{M}\mathbf{A}\mathbf{M}^{-1}$, is precisely the explicit Euler iteration for the diagonalized system (8.212), as expected. Since $\boldsymbol{\omega}^k = \mathbf{M}\mathbf{v}^k$, assuming all the eigenvalues of $\mathbf{A}$ have negative real part and that conditions (8.215) are satisfied, (8.216) implies that

$$\lim_{k \to +\infty} v_j^k = 0, \tag{8.220}$$

for $j = 1, 2, \ldots, n$, being consistent with (8.211). For the implicit Euler method, it can be easily seen that there is no restriction for the time-step h in order to satisfy the stability condition in every eigenspace.

We have seen the stability analysis of the implicit and explicit Euler methods when applied to a linear system of ODE. The same analysis could be extrapolated to any of the other time-steppers previously addressed. We summarize this as follows:

Absolute Stability of a Time-Stepper: Linear Systems

Consider the IVP

$$\mathbf{u}_t = \mathbf{A}\,\mathbf{u}, \; \mathbf{u}(0) = \mathbf{u}^0, \tag{8.221}$$

where $\mathbf{u}(t) \in \mathbb{R}^n$ and $\mathbf{A} \in M_n(\mathbb{R})$ is a $n \times n$ constant real matrix with distinct eigenvalues $\{\lambda_1, \lambda_2, \ldots, \lambda_n\}$ satisfying $\operatorname{Re}(\lambda_j) < 0$, for $j = 1, 2, \ldots, n$. Let $\mathbf{v}^0 = \mathbf{u}^0$ and let $\mathbf{v}^k \approx \mathbf{u}(t_k)$ be the kth approximation of the exact solution of (8.221) at $t_k = kh$ provided by a given time-stepper whose absolute stability region is $S \subset \mathbb{C}$. If

$$z_j = h\lambda_j \in S \tag{8.222}$$

for $j = 1, 2, \ldots, n$, then,

$$\lim_{k \to +\infty} \|\mathbf{v}^k\| = 0. \tag{8.223}$$

In what follows, we address the second question regarding the applicability of the statement above for nonlinear systems.

Let $\tilde{\mathbf{u}}(t)$ be the solution to the IVP

$$\mathbf{u}_t = \mathbf{f}(t, \mathbf{u}), \ \mathbf{u}(0) = \mathbf{u}^0, \tag{8.224}$$

depicted in Figure 8.17 as a solid black curve. Let us assume that the solution $\tilde{\mathbf{u}}(t)$ is *asymptotically stable*, that is, any neighboring solution

$$\mathbf{u}(t) = \tilde{\mathbf{u}}(t) + \boldsymbol{\delta}(t), \tag{8.225}$$

with $\boldsymbol{\delta}(t)$ representing a tiny deviation from $\tilde{\mathbf{u}}(t)$, converges to $\tilde{\mathbf{u}}$ for large times (gray solid curve in Figure 8.17). When a time-stepper approximates $\tilde{\mathbf{u}}(t)$, it does it by providing a sequence $\mathbf{v}^k$ that is (or should be) locally close to $\tilde{\mathbf{u}}(t_k)$. In that sense, the difference $\mathbf{v}^k - \tilde{\mathbf{u}}(t_k)$ can be understood as a small disturbance $\boldsymbol{\delta}(t_k)$ that the time-stepper should damp out after a sufficient number of iterations. This is indeed what the explicit and implicit Euler methods achieved in Figures 8.11b and 8.12, respectively, when approaching the asymptotically stable scalar solution $\tilde{u} = 0$. In other words, the time-stepper has to resolve not only the time scales associated with the dynamics of the solution $\tilde{\mathbf{u}}(t)$ itself but also the *local* dynamics related to the evolution of small disturbances $\boldsymbol{\delta}(t)$ added to it. We can obtain an equation for this local dynamics by writing (8.225) as

$$\boldsymbol{\delta}(t) = \mathbf{u}(t) - \tilde{\mathbf{u}}(t), \tag{8.226}$$

which, after differentiating both sides with respect to time, reads

$$\boldsymbol{\delta}_t = \mathbf{u}_t - \tilde{\mathbf{u}}_t = \mathbf{f}(t, \mathbf{u}) - \mathbf{f}(t, \tilde{\mathbf{u}}). \tag{8.227}$$

Since $\mathbf{u} = \tilde{\mathbf{u}} + \boldsymbol{\delta}$ and $\|\boldsymbol{\delta}\|$ is assumed to be small, we can expand the term $\mathbf{f}(t, \mathbf{u})$ appearing in (8.227) in its Taylor series in $\mathbf{u}$, in a neighborhood of the solution $\tilde{\mathbf{u}}$:

$$\mathbf{f}(t, \mathbf{u}) = \mathbf{f}(t, \tilde{\mathbf{u}} + \boldsymbol{\delta}) = \mathbf{f}(t, \tilde{\mathbf{u}}) + [\tilde{\mathbf{D}}_{\mathbf{u}}\mathbf{f}] \, \boldsymbol{\delta} + O(\|\boldsymbol{\delta}\|^2), \tag{8.228}$$

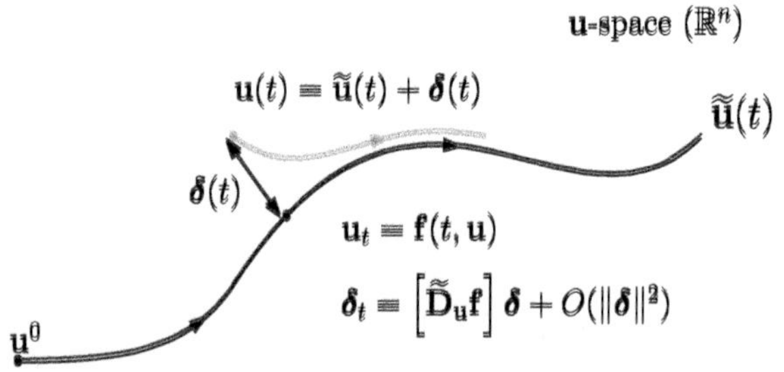

Figure 8.17 Solution $\tilde{\mathbf{u}}(t)$ to the IVP (8.224) starting from $\tilde{\mathbf{u}}(0) = \mathbf{u}^0$ (solid black). If $\tilde{\mathbf{u}}(t)$ is stable, any small disturbed solution $\mathbf{u}(t) = \tilde{\mathbf{u}}(t) + \boldsymbol{\delta}(t)$ (gray solid) asymptotically approaches $\tilde{\mathbf{u}}(t)$, that is, $\|\boldsymbol{\delta}(t)\| \to 0$ ($t \to +\infty$), where $\|\boldsymbol{\delta}(t)\|$ measures the size of the disturbance at any time.

where $[\widetilde{\mathbf{D}}_{\mathbf{u}}\mathbf{f}]$ is the Jacobian matrix of the vector field $\mathbf{f} \equiv [f_1 \ f_2 \ \cdots \ f_n]^{\mathrm{T}}$ with respect to the spatial coordinates $\mathbf{u} \equiv [u_1 \ u_2 \ \cdots \ u_n]$ evaluated at $\widetilde{\mathbf{u}}(t)$, and whose matrix elements are

$$[\widetilde{\mathbf{D}}_{\mathbf{u}}\mathbf{f}]_{ij} \equiv [\partial_{u_j} f_i]_{\widetilde{\mathbf{u}}}. \tag{8.229}$$

Introducing the expansion (8.228) in (8.227) leads to

$$\boldsymbol{\delta}_t = \mathbf{f}(t, \widetilde{\mathbf{u}}) + [\widetilde{\mathbf{D}}_{\mathbf{u}}\mathbf{f}] \, \boldsymbol{\delta} + O(\|\boldsymbol{\delta}\|^2) - \mathbf{f}(t, \widetilde{\mathbf{u}}), \tag{8.230}$$

which, at first order in $\|\boldsymbol{\delta}\|$, reads

$$\boldsymbol{\delta}_t = [\widetilde{\mathbf{D}}_{\mathbf{u}}\mathbf{f}] \, \boldsymbol{\delta}. \tag{8.231}$$

The equation above controls the asymptotic behavior of small deviations from $\widetilde{\mathbf{u}}$ and, to some extent, is similar to the model equation $\mathbf{u}_t = \mathbf{A} \, \mathbf{u}$ appearing in (8.202), where the role of the matrix $\mathbf{A}$ is now played by the Jacobian $[\widetilde{\mathbf{D}}_{\mathbf{u}}\mathbf{f}]$. In this case, however, this matrix is no longer constant, but *time dependent*, since it is being evaluated along $\widetilde{\mathbf{u}}(t)$. An additional problem is that in general we never have access to the actual values of $\widetilde{\mathbf{u}}(t)$ so the actual computation of $[\widetilde{\mathbf{D}}_{\mathbf{u}}\mathbf{f}]$ is never done in practice. To simplify the analysis, it is common to replace $[\widetilde{\mathbf{D}}_{\mathbf{u}}\mathbf{f}]$ by $[\mathbf{D}_{\mathbf{u}}\mathbf{f}]_{\mathbf{v}^k}$ (i.e. the Jacobian matrix evaluated at $t = t_k$, the kth stage of the time integration) and replace the non-autonomous linearized system (8.231) by

$$\boldsymbol{\delta}_t = [\mathbf{D}_{\mathbf{u}}\mathbf{f}]_{\mathbf{v}^k} \, \boldsymbol{\delta}, \tag{8.232}$$

that is, *freeze* the Jacobian from $t = t_k$ onwards.

Linearized Criterion for Stability

Consider the nonlinear IVP in $\mathbb{R}^n$,

$$\mathbf{u}_t = \mathbf{f}(t, \mathbf{u}), \ \mathbf{u}(0) = \mathbf{u}^0. \tag{8.233}$$

Let $\mathbf{v}^k \approx \mathbf{u}(t_k)$ be the approximation of the exact solution of (8.233) at $t_k \equiv kh$, provided by a given time-stepper with time-step h, and whose region of absolute stability is $S \subset \mathbb{C}$. Let $\mathbf{A} \in \mathrm{M}_n(\mathbb{R})$ be the Jacobian matrix

$$\mathbf{A} = [\mathbf{D}_{\mathbf{u}}\mathbf{f}]_{\mathbf{v}^k}, \tag{8.234}$$

with eigenvalues $\{\lambda_1, \lambda_2, \ldots, \lambda_n\}$. To avoid numerical instabilities, it is necessary to use a time-step h such that

$$z_j = h\lambda_j \in S, \tag{8.235}$$

for $j = 1, 2, \ldots, n$. That is, the complex quantities z_j must lie within the region of absolute stability of the time-stepper.

Nonlinear ODE arising in science or engineering are models intended to capture the essential dynamic features of a given phenomena that have been previously observed in experiments or in nature. If the ODE model is faithful to the real system and the time integration is free from numerical instabilities, we should observe the same kind of dynamics as the experimental ones. Frequently, these regimes (steady, time-periodic, or chaotic) are very robust and can be easily reproduced numerically by the model, often regardless of the initial conditions. The underlying reason is that many of these regimes are *stable*, and may have attracting properties in the phase **u**-space. In other words, if these regimes were unstable, they would hardly be observed in the experiments, or visited in the phase space by our time-stepper.

Sometimes, the topological structure of the trajectories arising from a nonlinear ODE may be really complex. Nonlinear physics provides a wide variety of examples showing this complexity, such as the *Lorenz attractor*. However, it is far beyond the scope of this course to study the so-called *invariant manifolds*, such as *equilibria*, *limit cycles*, or *homoclinic orbits*. The interested reader may have a look at any monograph on dynamical systems included in the bibliography. What is relevant within the context of this section is that at any point $\mathbf{v}^k$ of the numerical trajectory approaching these attracting objects, the matrix $[\mathbf{D_u f}]_{\mathbf{v}^k}$ will typically have eigenvalues λ_j with *negative* real part. This is quite common in ODE models arising from *dissipative phenomena* where, in particular, these eigenvalues may have *very large* absolute magnitude, that is, $|\mathrm{Re}(\lambda_j)| \gg 1$.

We illustrate the previous ideas by analyzing the following nonlinear system of differential equations in $\mathbb{R}^2$:

$$
\begin{aligned}
\dot{x} &= -y + x[a - b(x^2 + y^2)],\\
\dot{y} &= x + y[a - b(x^2 + y^2)],
\end{aligned}
\tag{8.236}
$$

where a and b are two positive constants. The dynamics of this system in $\mathbb{R}^2$ is very simple if we use polar coordinates

$$
\begin{aligned}
x &= \rho \, \cos\theta,\\
y &= \rho \, \sin\theta.
\end{aligned}
\tag{8.237}
$$

A simple calculation shows that the differential equations for the radial $\rho(t)$ and angular $\theta(t)$ coordinates are

$$
\begin{aligned}
\dot{\rho} &= \rho(a - b\rho^2),\\
\dot{\theta} &= 1.
\end{aligned}
\tag{8.238}
$$

The system above clearly shows that the radial and angular dynamics are decoupled. In particular, the equation for θ tells us that all points in the $\mathbb{R}^2$-plane rotate with the same angular speed $\omega = 1$. The equation for $\dot{\rho}$ tells us that any initial condition starting from the origin $\rho = 0$ will remain there, and that the radial speed of any point in the plane is positive if $0 < \rho < \sqrt{a/b}$ and negative if $\rho > \sqrt{a/b}$. Along the radius $\rho = \sqrt{a/b}$ we have therefore a radial equilibrium, which, according to the sign of $\dot{\rho}$, must be *attracting*. Figure 8.18a shows the phase portrait of system (8.236) for $a = 40$ and $b = 0.3$, where the attracting equilibrium radius is $r = (a/b)^{1/2} \approx 5.8$. The phase portrait depicts many trajectories starting from different initial points in the plane (white bullets). Initial conditions with large initial radii, such as point A in Figure 8.18a, show a very rapid initial decay in the radial coordinate toward the equilibrium circle (almost at a constant angle), eventually landing on the attracting set. By contrast, any initial condition along this equilibrium (such as point B) evolves smoothly, following a periodic motion. The difference between the time evolution of a trajectory starting from points A or B is better illustrated in Figure 8.18b, showing $x(t)$ for both cases. Starting from A, we can clearly identify a very rapid exponential decay before eventually relaxing along the equilibrium circle (zoomed initial region in Figure 8.18b). We can therefore identify *two* very different time scales, namely, the one related to the rapid exponential decay until reaching the equilibrium radius ($\tau_1 \approx 0.2$) and another one related to the periodic motion ($\tau_2 \approx 6.3$).

Let us integrate (8.236) using explicit and implicit Euler methods. Figure 8.19a shows the integration of (8.236) using the implicit Euler method

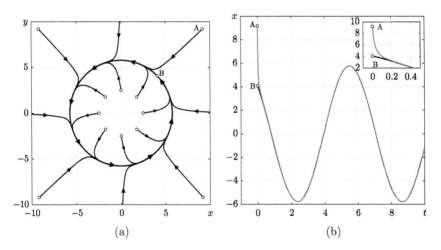

Figure 8.18 Nonlinear system (8.236) for $a = 40$ and $b = 0.3$. (a) Phase portrait showing trajectories starting from different initial conditions (white bullets). The arrows indicate the orientation of the flow. (b) Time evolution $x(t)$, starting from points A (solid gray) and B (solid black). The zoomed region depicts the initial stage of the evolution to show the rapid exponential decay starting from A.

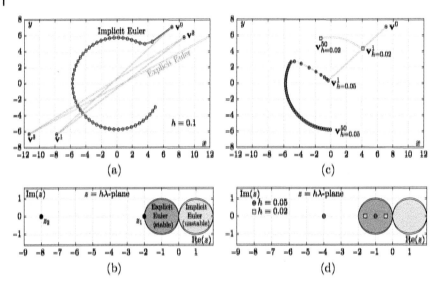

Figure 8.19 Numerical integration of nonlinear system (8.236) for $a = 40$ and $b = 0.3$. (a) For $h = 0.1$, the implicit Euler method works fine, but the explicit is unstable. (b) For $h = 0.1$, $z_{1,2} = h\lambda_{1,2}$ are outside of the stability region of explicit Euler method. (c) Explicit Euler integrations for smaller time-steps $h = 0.05$ (gray bullets) and $h = 0.02$ (white squares). (d) $z_{1,2} = h\lambda_{1,2}$ for $h = 0.05$ (dark gray circles) and $h = 0.02$ (white squares).

BDF1 (white bullets) with time-step $h = 0.1$, starting from $\mathbf{v}^0 = [5\sqrt{2}\ 5\sqrt{2}]^\mathrm{T}$. We can see how the resulting iterates $\mathbf{v}^k$ smoothly approach the equilibrium circle, eventually tracing a trajectory at a constant radius, as expected. Figure 8.19a also shows the same integration carried out with the explicit Euler method (same time-step and initial condition), revealing a clear numerical instability. Figure 8.19b provides the underlying explanation for that instability by depicting the complex quantities $z_{1,2} = h\lambda_{1,2}$ (black bullets) corresponding to the two eigenvalues of the Jacobian[23] matrix (8.234) of the right-hand side of (8.236) at the initial condition, that is, $[\mathbf{D_u f}]_{\mathbf{v}^0}$. For $h = 0.1$, $z_2 \approx -7.998$ is clearly outside of the stability region of the explicit Euler method ($z_1 \approx -2.002$ is just slightly outside as well) but obviously within the stable region of its implicit homolog. In order to stabilize the explicit Euler iteration, we need to diminish the time-step until both quantities $z_{1,2}$ fall within its region of absolute stability. Figure 8.19c shows two more integrations for $h = 0.05$ (dark gray bullets) and $h = 0.02$ (white squares). For $h = 0.05$, the quantity z_1 is already inside the stability region (Figure 8.19d) but z_2 is still outside,

---

23 Computed using, for example, Code 19 `jac.m` and Matlab's **eig** function

therefore leading to a first numerically unstable step $\mathbf{v}^1$ that, by chance, lies near the origin (Figure 8.19e). From $\mathbf{v}^1$ onwards, the explicit Euler method works fine (although the numerical integration is ultimately wrong) approaching the circle from inside. The absence of instabilities from $\mathbf{v}^1$ will be explained in short. For $h = 0.02$, both quantities $z_{1,2}$ lie within the stability region of the explicit method (white squares in Figure 8.19d), resulting in a numerically stable (and correct) integration depicted with white squares in Figure 8.19c.

The previous numerical experiment reveals a very interesting phenomenon related to the time scales of the problem and their implication in the performance of the explicit Euler method. While using a time-step of $h = 0.02$ may sound reasonable in order to describe the transient initial decay to the equilibrium circle, it is offensively small afterwards, once the time-stepper needs to trace a periodic motion with a period $T = 2\pi$ requiring more than 300 time-steps per cycle. By contrast, the implicit Euler with $h = 0.1$ manages to solve both problems at once: the rapid decay at the beginning (in just three steps) and computational economy and reasonable accuracy at the end, just involving 63 steps per cycle. The problem with the explicit Euler method is that the time-step $h = 0.02$ has been chosen not in order to satisfy some accuracy requirements, but to have stability. With the implicit Euler method, the choice of h is exclusively based on accuracy grounds.

What we have just described is a symptom that is present in many IVP containing widely varying time scales. This feature is usually known as *stiffness*. It is not easy to give a precise definition of stiffness, as this concept is typically described in many different ways in the literature of numerical differential equations. In this chapter, we introduce the following definition:

Stiffness

Consider the numerical discretization of a nonlinear IVP in $\mathbb{R}^n$,

$$\mathbf{u}_t = \mathbf{f}(t, \mathbf{u}), \; \mathbf{u}(0) = \mathbf{u}^0. \tag{8.239}$$

Let $\mathbf{v}^k \approx \mathbf{u}(t_k)$ be the approximation of the exact solution of (8.239) at $t_k = kh$. Assume that the eigenvalues $\{\lambda_1, \lambda_2, \ldots, \lambda_n\}$ of the Jacobian matrix

$$[\mathbf{D_u f}]_{\mathbf{v}^k} \tag{8.240}$$

have negative real part, that is, $\mathrm{Re}(\lambda_j) < 0$. Let μ and ν be the leftmost and rightmost eigenvalues in the complex plane, respectively, that is

$$\mathrm{Re}(\mu) \leq \min_{1 \leq j \leq n} \mathrm{Re}(\lambda_j), \tag{8.241}$$

(Continued)

and

$$\mathrm{Re}(\nu) \geq \max_{1 \leq j \leq n} \mathrm{Re}(\lambda_j). \tag{8.242}$$

If

$$s = \frac{|\mathrm{Re}(\mu)|}{|\mathrm{Re}(\nu)|} \gg 1, \tag{8.243}$$

we say that (8.244) is *stiff* at $\mathbf{v}^k$. In particular, the numerical discretization of the scalar IVP

$$u_t = f(t, u), \ u(0) = u_0 \tag{8.244}$$

is said to be stiff at the kth stage $v_k \approx u(t_k)$ if $\lambda = \partial_u f(t_k, v_k)$ is such that $\mathrm{Re}(\lambda) < 0$ and $|\mathrm{Re}(\lambda)| \gg 1$.

The quantity s introduced in (8.243) is generally known as the *stiffness ratio*. A problem is therefore qualified as stiff if $s \gg 1$. However, some authors dislike this criterion since s can be very large if for example $|\mathrm{Re}(\nu)| \ll 1$. For this reason, in this chapter we will consider that a system is *genuinely* stiff when s is very large because $|\mathrm{Re}(\mu)| \gg 1$, i.e. as in the scalar case.

In general, stiffness characterizes the *local dynamics* of a certain solution of $\mathbf{u}_t = \mathbf{f}(t, \mathbf{u})$ in *a certain region* the phase space. For example, the trajectory shown in Figure 8.19c for $h = 0.05$ clearly illustrates a numerical instability at $\mathbf{v}^0$ but not at $\mathbf{v}^1$. The reason is that whereas the initial condition $\mathbf{v}^0$ lies in a stiff region characterized by a strong radial shrinking (i.e. $\mathrm{Re}(\lambda_j) < 0$ and $|\mathrm{Re}(\lambda_j)| \gg 1$), $\mathbf{v}^1$ happens to be near the origin, where the radial dynamics is expansive, *non-stiff*, and where the eigenvalues of $[\mathbf{D_u f}]_{\mathbf{v}^1}$ have *positive* real parts (and nonzero imaginary part). That is, while $\mathbf{v}^0$ is abruptly *repelled* toward $\mathbf{v}^1$ by a numerical instability, $\mathbf{v}^1$ is naturally driven outwards by a *genuine instability* of the mathematical model.

As the reader may easily check, the eigenvalues of the Jacobian $[\mathbf{D_u f}]$ of system (8.236) at the initial condition $\mathbf{v}^0 = [5\sqrt{2} \ 5\sqrt{2}]^T$ are $\lambda_1 \approx -20.016$ and $\lambda_2 \approx -79.98$. These eigenvalues are not alarmingly large. We have simply chosen them to have manageable numbers with orders of magnitude capable of being easily calibrated and to provide a qualitative idea of the potential effects in an actual computation. In practice, however, stiffness may embrace eigenvalues with *huge* real parts, typically encompassing many orders of magnitude. We end the section with a final example that perfectly illustrates one of the paradigms of stiffness, namely, *thermal diffusion*. For that, we need to step slightly beyond the framework of ODE and to consider one of the most fundamental *partial differential equations*, namely, *Fourier's heat equation*,

$$\partial_t f(x, t) = \alpha \partial_{xx}^2 f(x, t), \tag{8.245}$$

where $f(x,t)$ represents the *temperature* at a point $x \in \mathbb{R}$ of a certain thermally conducting material (a cooper wire, for example) at an arbitrary instant of time t. The constant α is the so-called *thermal diffusivity* of the material, which, for simplicity, is taken as $\alpha = 1$. In addition, assume that our physical domain is periodic in the spatial variable x (think of a cooper ring, for example) within the domain $[0, 2\pi)$. In that case, at every instant of time t, the temperature profile $f(x,t)$ can be approximated by a discrete Fourier series (DFS) (7.71) of the form

$$f_N(x,t) = \sum_{k=0}^{N-1} \widetilde{f}_k(t) \, e^{i(k - \frac{N}{2})x}, \qquad (8.246)$$

where in this case we allow the Fourier coefficients $\widetilde{f}_k$ to depend on time. That is, at each time t, the temperature profile $f_N(x,t)$ is prescribed by some Fourier coefficients whose dynamics is still to be determined. Introducing the time-dependent Fourier expansion (8.246) in (8.245) for $\alpha = 1$ reads

$$\sum_{k=0}^{N-1} \frac{d\widetilde{f}_k}{dt} \, e^{i(k - \frac{N}{2})x} = - \sum_{k=0}^{N-1} \left(k - \frac{N}{2}\right)^2 \widetilde{f}_k(t) \, e^{i(k - \frac{N}{2})x}. \qquad (8.247)$$

After identifying Fourier coefficients to the left and right of the previous expression, we obtain the following system of ODE for the time-dependent amplitudes $\widetilde{f}_k(t)$:

$$\frac{d\widetilde{f}_k}{dt} = -\left(k - \frac{N}{2}\right)^2 \widetilde{f}_k, \qquad (8.248)$$

for $k = 0, 1, 2, \ldots, N-1$. This is a linear N-dimensional ODE that, in matrix form, reads

$$
\begin{bmatrix} \dot{\widetilde{f}}_0 \\ \dot{\widetilde{f}}_1 \\ \vdots \\ \dot{\widetilde{f}}_{N-1} \end{bmatrix}
=
\begin{bmatrix}
-\dfrac{N^2}{4} & 0 & \cdots & 0 \\
0 & -\dfrac{(N-2)^2}{4} & \ddots & \vdots \\
\vdots & \ddots & \ddots & 0 \\
0 & \cdots & 0 & -\dfrac{(N-2)^2}{4}
\end{bmatrix}
\begin{bmatrix} \widetilde{f}_0 \\ \widetilde{f}_1 \\ \vdots \\ \widetilde{f}_{N-1} \end{bmatrix}. \qquad (8.249)
$$

In this case, the kth Fourier mode decays exponentially according to (8.248), that is,

$$\widetilde{f}_k(t) = C_k \, e^{-\left(k - \frac{N}{2}\right)^2 t}, \qquad (8.250)$$

for $k = 0, 1, \ldots, N-1$, where the constants C_k are uniquely determined after imposing the initial condition. Since system (8.248) is linear and time independent, the Jacobian $[\mathbf{D_u f}]$ along any trajectory is precisely *Fourier's differentiation matrix* (7.89) seen in Chapter 7

$$\mathbf{D_u f} = \text{diag}\left(-\frac{N^2}{4} \ldots, -4, -1, 0, -1, -4, \ldots, -\frac{(N-2)^2}{4}\right). \qquad (8.251)$$

The leftmost and rightmost eigenvalues are therefore $\mu = -N^2/4$ and $\nu = -1$, respectively.[24] According to (8.243), the stiffness ratio of (8.248) is $s = O(N^2)$. Consequently, the explicit Euler method requires a time-step of order $h = O(N^{-2})$ in order to ensure that $z = h\mu$ lies within its absolute stability region. For example, discretizing (8.245) with a moderate number of Fourier modes, say $N = 50$, requires a time-step $h = 3.2 \times 10^{-3}$, but when $N = 200$ the time-step must be reduced to $h = 2 \times 10^{-4}$ to avoid numerical instabilities. In general, to improve the spatial accuracy of $f_N(x, t)$, we need to increase N. This has a double effect on the computational efficiency. On the one hand, the number of ODE to be integrated grows proportionally with N. On the other hand, the time-step h required for a numerically stable integration of these ODE decreases as the inverse of N^2. This may lead to a prohibitive computational cost and demands the use of an A-stable time-stepper such as implicit Euler or BDF2. Code 27 implements the BDF1 implicit Euler method using an inner Newton's iteration with absolute tolerance 10^{-12}.

```
% Code 27: BDF1 (implicit Euler time-stepper)
% Solution for u_t = f(t,u) with u(t0) = v0 (n-dimensional)
% Input: fun (function name) ; t0 (initial time)
% h (time-step) ; v0 (initial condition) ; N (no. steps)
% Output: T (time vector: 1 x N+1)
%         Y (solution matrix: n x N+1)
function [T,Y] = bdf1(fun,t0,h,v0,N)
T = zeros(1,N+1); n = length(v0); I = eye(n);
Y = zeros(length(v0),N+1); DF = zeros(n); H = sqrt(eps)*I;
T(1) = t0; Y(:,1) = v0;
for j = 1:N
 z0 = Y(:,j) ; tplus = t0 + h*j; T(j+1) = tplus;
 dz = 1; z = z0 ;
  while norm(dz) > 1e-12
   f1 = feval(fun,tplus,z);
    for kk = 1:n
     f2 = feval(fun,tplus,z+H(:,kk));
     Df(:,kk) = (f2 - f1)/H(kk,kk);
    end
     dz = -(I-h*Df)\(z-z0-h*f1); z = z + dz;
  end
   Y(:,j+1) = z;
end
```

---

24 In this particular case we ignore the eigenvalue $\lambda = 0$ corresponding to the Fourier coefficient $\widetilde{f}_{N/2}$, which remains constant since $\widetilde{f}_{N/2}' = 0$.

Practical 8.3: Shock Waves (Method of Lines)

In this practical, we study the numerical solutions of what is usually known as *Bateman–Burgers equation*[25]:

$$\partial_t f = \alpha\, \partial_{xx}^2 f + f \partial_x f, \tag{8.252}$$

where $f(x,t)$ is a time-dependent scalar 2π-periodic function in the spatial variable x. This is a nonlinear partial differential equation that contains the two key elements of the Navier–Stokes equations for viscous fluids: first, the diffusive effects, played by the linear term $\alpha\, \partial_{xx}^2 f$; see (8.245); second, the advective effects, played by the nonlinear term $f\partial_x f$, which allows the generation of propagating waves. A simple way to integrate (8.252) is the so-called *method of lines*, where we consider the spatial discretization of $f(x,t)$ on the standard Fourier grid:

$$\mathbf{f}(t) = [f_0(t)\ f_1(t)\ \cdots f_{N-1}(t)]^{\mathrm{T}}, \quad \text{with } f_j(t) = f\left(\frac{2\pi j}{N}, t\right), \tag{8.253}$$

for $j = 0, 1, \ldots, N-1$, at any time t. We can introduce the vector $\mathbf{f}(t)$ in (8.252) to obtain an ODE for each one of its components by approximating the partial derivatives $\partial_x f$ and $\partial_{xx}^2 f$ using Fourier differentiation matrices $\mathbf{D}_{j\ell}^{(1)}$ and $\mathbf{D}_{j\ell}^{(2)}$ from (7.91) and (7.92), respectively. Therefore,

$$\frac{df_j}{dt} = \sum_{\ell=0}^{N-1} \mathbf{D}_{j\ell}^{(2)} f_\ell(t) + f_j(t) \sum_{\ell=0}^{N-1} \mathbf{D}_{j\ell}^{(1)} f_\ell(t), \tag{8.254}$$

at the jth node, for $j = 0, 1, 2, \ldots, N-1$. In this practical, take $\alpha = 10^{-1}$, $N = 128$, and initial condition $f(x,0) = e^{-10\cos^3(x/2)}$. Our goal is to compute $f(x,t)$ later on at $t = 2$ and $t = 4$.

(a) Integrate (8.254) using the explicit Euler method with time-steps $h = 0.1$ and $h = 0.01$. Are these time integrations stable? Compute the eigenvalues λ_j of the Jacobian of (8.254) evaluated at the initial condition and plot the complex quantities $z_j = h\lambda_j$. Explain the stability or instability of the two time integrations plotting the stability region of the method and the time-step used in each case.

(b) Using the implicit Euler method (Code 27), is the time integration stable for $h = 0.1$? If yes, explain why and plot the solution at the requested values of t.

---

25 Originally introduced by H. Bateman (1882–1946), English mathematician and later studied by J. M. Burgers (1895–1981), Dutch physicist.

Complementary Reading

(1) General References on ODE:

There are dozens of excellent monographs on the theory of ordinary differential equations. For a comprehensive review on the most fundamental concepts (existence and uniqueness theorems, linear stability, etc.), Boyce, DiPrima, and Meade's *Elementary Differential Equations and Boundary Value Problems* is an excellent introduction. For a more advanced approach, full of insights and very powerful analytical methodologies I recommend Bender and Orszag's *Advanced Mathematical Methods for Scientists and Engineers*, which also contains a comprehensive theory on the solutions of linear difference equations (including nonhomogeneous and non-constant coefficients cases not covered here). For those readers interested in the qualitative nonlinear theory of IVPs I strongly recommend Wiggin's *Introduction to Applied Nonlinear Dynamics and Chaos*.

(2) Boundary Value Problems:

There are not many monographs specialized in the numerical solution of BVPs for ODE. An outstanding introduction, including alternative techniques to the ones addressed here (such as *shooting methods*), is Ascher, Mathheij, and Russell's *Numerical Solution of Boundary Value Problems for Ordinary Differential Equations*. For many years, the classical reference has been Keller's *Numerical Solution of Two Point Boundary Value Problems*, where the reader will find existence and uniqueness theorems for second order BVPs with Robin's conditions. A more advanced approach is that of Boyd's *Chebyshev & Fourier Spectral Methods*, which addresses challenging topics such as semi-infinite domains, or *behavioral* boundary conditions arising in polar or spherical coordinates. Finally, Trefethen's *Spectral Methods in Matlab* is a must-read, particularly if you are interested in how to implement boundary conditions in higher order problems.

(3) Initial Value Problems:

For a nice introduction to the theory of numerical methods for initial value problem, my first options are Ascher and Petzold's *Computer Methods for Ordinary Differential Equations and Differential-Algebraic Equations* and Griffiths and Higham *Numerical Methods for Ordinary Differential Equations–Initial Value Problems*, the last one introducing many interesting special topics such as Hamiltonian integrators or stochastic equations. Specially delicate topics such as 0-stability, consistence and Dahlquist Equivalence Theorem are beautifully treated in Süli and Mayers' *An Introduction to Numerical Analysis*, or in Henrici's classical *Discrete Variable Methods in Ordinary Differential Equations*. More advanced texts are Lambert's *Numerical Methods for Ordinary Differential Systems–The Initial Value Problem* or Butcher's *Numerical Methods for Ordinary Differential Equations*, where the reader will find a comprehensive study of Runge–Kutta methods. For more than three decades, the classical reference has been the two-volume monograph by Hairer, Nørsett, and Wanner's *Solving Ordinary Differential Equations, Volumes (I) and (II)*. In that treatise, the reader will find almost everything, including many time-steppers not mentioned here (Nyström, Crank–Nicolson, Adams–Moulton, Dormand, and Prince, or implicit Runge–Kutta, for example).

Finally, a very important topic, not addressed in this textbook, is that of *symplectic* or *Verlet* integrators and their application in Hamiltonian and molecular dynamics. As mentioned before, Griffiths and Higham's monograph introduces the topic very nicely. The classical references are Sanz-Serna and Calvo's *Numerical Hamiltonian Problems* and Hairer, Lubich, and Wanner's *Geometric Numerical Integration: Structure-Preserving Algorithms for Ordinary Differential Equations*.

(4) Partial Differential Equations:

In this chapter, we have just seen two very particular examples of PDE. For an outstanding theoretical introduction to this field, my preference is John's *Partial Differential Equations*. A more applied book, mainly focused on analytical techniques is Haberman's *Applied Partial Differential Equations with Fourier Series and Boundary Value Problems*, already in its fifth edition. The numerical discretization of partial PDE can be performed in many different ways. Three more standard approaches are *finite differences*, *finite elements*, and *spectral methods*. For a general introduction I recommend Iserle's *A First Course in the Numerical Analysis of Differential Equations* and Trefethen's *Finite Difference and Spectral Methods for Ordinary and Partial Differential Equations* (this second one unpublished, but freely available online). For a more advanced monograph on finite difference techniques, LeVeque's *Finite Difference Methods for Ordinary and Partial Differential Equations* is an excellent choice. For a

(Continued)

nice introduction to *pseudo*-spectral collocation methods, Trefethen's *Spectral Methods in Matlab* is the reference to look at. For a slightly more advanced reader, Boyd's book *Chebyshev & Fourier Spectral Methods* is full of insight and useful tips for the practitioner. However, for more than three decades, the classical reference on spectral methods has been the monograph by Canuto, Hussaini, Quarteroni, and Zang, *Spectral Methods in Fluid Dynamics*, recently updated and published in two volumes: *Spectral Methods: I-Fundamentals in Single Domains* and *II-Evolution to Complex Geometries and Applications to Fluid Dynamics*.

Problems and Exercises

The symbol (A) means that the problem has to be solved analytically, whereas (N) means to use numerical Matlab codes. Problems with the symbol * are slightly more difficult.

1. (N) The nonlinear BVP

    $$u''(z) + e^{u+1} = 0,$$
    $$u(0) = u(1) = 0$$

    admits two solutions.
    (a) Using Newton's method, find at least one of these two solutions $u_1(z)$ and $u_2(z)$ and plot them. **Hint**: Use initial guess $u^{(0)}(z) = 1$.
    (b*) Homotopy: In the absence of a good initial guess, we may generalize the problem via *embedding* by considering the parameter-dependent BVP

    $$u''(z) + \lambda e^u = 0,$$
    $$u(0) = u(1) = 0,$$

    where $\lambda \in \mathbb{R}$. For $\lambda = e$, we recover the original problem, whereas for $\lambda = 0$ the problem only admits the trivial solution $u(z) = 0$. Use continuation to compute possible solutions for $\lambda \in [-4, 4]$. Define the norm of a solution as

    $$\|u(z)\| = \sqrt{\int_0^1 u^2(z) \, dz}.$$

    On a λ-$\|u(z)\|$ graph, plot the branches of solutions. Is there any value of λ above which solutions cannot be found?

2. (AN) *Mathieu's Equation in Frequency Space.* Let us consider again Mathieu's periodic BVP

$$-\frac{d^2\Psi(x)}{dx^2} + 2q\cos(2x)\Psi(x) = E\,\Psi(x), \quad \text{with} \quad x \in [0, 2\pi] \quad \text{and} \quad q \geq 0,$$

where $x \in [0, 2\pi]$, $q \geq 0$, $\Psi(x + 2\pi) = \Psi(x)$, and $\Psi'(x + 2\pi) = \Psi'(x)$. For even N, consider the discrete Fourier approximation of Ψ

$$\Psi_N(x) = \sum_{k=-N/2}^{N/2-1} \hat{\Psi}_k\, e^{ikx},$$

represented in $\mathbb{S}_N = \text{span}\left\{e^{-i\frac{N}{2}x}, \ldots, e^{-ix}, 1, e^{ix}, \ldots, e^{i(\frac{N}{2}-1)x}\right\}$ by the vector

$$\hat{\Psi}_N = [\hat{\Psi}_{-N/2}\ \hat{\Psi}_{-N/2+1}\ \cdots\ \hat{\Psi}_{N/2-1}]^{\mathrm{T}};$$

see (7.12) in Chapter 7.

(a) For arbitrary q and N, sketch the structure of the matrix version of Mathieu's equation for the unknown vector $\hat{\Psi}_N$,

$$\begin{bmatrix} * & * & * & * & \cdots \\ * & * & * & * & \cdots \\ * & * & * & * & \cdots \\ * & * & * & * & \cdots \\ \vdots & \vdots & \vdots & \vdots & \end{bmatrix} \begin{bmatrix} \hat{\Psi}_{-N/2} \\ \hat{\Psi}_{-N/2+1} \\ \vdots \\ \hat{\Psi}_{N/2-2} \\ \hat{\Psi}_{N/2-1} \end{bmatrix} = E \begin{bmatrix} \hat{\Psi}_{-N/2} \\ \hat{\Psi}_{-N/2+1} \\ \vdots \\ \hat{\Psi}_{N/2-2} \\ \hat{\Psi}_{N/2-1} \end{bmatrix}.$$

Provide the explicit expression of the elements of the first four rows and columns of the matrix above.

(b) For $N = 24$ and $q = 3/2$, generate the matrix appearing in (a) with a short Matlab code and compute the five lowest energy levels. Reproduce Figure 8.4b.

3. (N) Consider a double-well model potential $V(y)$ for the ammonia molecule (NH_3) depicted in the figure on the right. The energy levels E_n and their corresponding wavefunctions $\Psi_n(y)$ are obtained from solving the Schrödinger eigenvalue problem on the real axis $y \in (-\infty, +\infty)$:

$$-\frac{\hbar^2}{2m}\frac{d^2}{dy^2}\Psi_n + V(y)\Psi_n = E_n\Psi_n, \quad \text{with}$$

$$V(y) = \frac{1}{2}ky^2 + \alpha e^{-\beta y^2}.$$

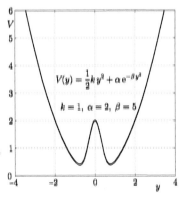

$$V(y) = \frac{1}{2}ky^2 + \alpha e^{-\beta y^2}$$

$$k = 1,\ \alpha = 2,\ \beta = 5$$

For simplicity, take $\hbar = m = k = 1$.

(a) For $\alpha = 2$ and $\beta = 5$, provide the three lowest energy levels $E_0 < E_1 < E_2$, with at least six exact digits.

(b) Define the *normalized* wavefunctions $\widehat{\Psi}_n(y) \equiv \dfrac{1}{S}\Psi_n(y)$, with

$$S = \left\{ \int_{-\infty}^{\infty} |\Psi_n(y)|^2\, \mathrm{d}y \right\}^{1/2}.$$

Plot the probability densities $|\widehat{\Psi}_n(y)|^2$ for the first three energy levels obtained in (a), indicating the locations of the abscissas y where the probabilities are local *maxima* and *minima*, also indicating their corresponding values.

4. Solve the Neumann BVP

$$u'' + u' + 100u = x, \quad u'(-1) = u'(1) = 0,$$

with $x \in [-1, 1]$.

5. (N*) The figure on the right shows three particles of masses $m_1 = 1$, $m_2 = 1$, and $m_3 = 2$, located on the $\mathbb{R}^2$-plane at the coordinates $(x_1, y_1) = (0, 1)$, $(x_2, y_2) = (0, -1)$, and $(x_3, y_3) = (1, 0)$, respectively. These particles are fixed at their positions. At $t = 0$, we launch a fourth particle of mass $m = 1$ from a certain point of the plane and with a certain initial speed. The moving particle performs a periodic orbit that surrounds m_3, as depicted in the figure.

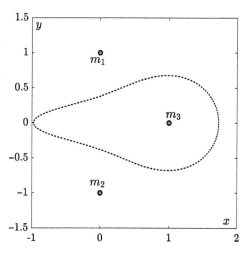

As in Practical 8.2, assume that the gravitational force exerted by a mass m_i at $\mathbf{r}_i$ on another mass m_j at $\mathbf{r}_j$ is

$$\mathbf{F}_{ij} = \frac{m_i m_j}{\|\mathbf{r}_i - \mathbf{r}_j\|^3}(\mathbf{r}_i - \mathbf{r}_j).$$

(a) Find the period T of the orbit with at least six exact digits.

(b) Plot the coordinates of the moving particle $x(t)$ and $y(t)$, as well as its corresponding horizontal and vertical velocities $u(t)$ and $v(t)$, respectively, as a function of time over two periods of the orbit.

6. (AN) Using arclength continuation, reproduce the regions of absolute stability of the Runge–Kutta time-steppers shown in Figure 8.14.

7. (AN) Using the boundary locus method, reproduce the regions of absolute stability of the Adams–Bashforth time-steppers shown in Figure 8.15.

8. (AN) Consider the two-dimensional Hamiltonian:

$$H(x, y, p_x, p_y) = \frac{1}{2}(p_x^2 + p_y^2) + (x^2 + y - 11)^2 + (x + y^2 - 7)^2,$$

and its corresponding Hamilton ODEs:

$$\dot{x} = \frac{\partial H}{\partial p_x}$$

$$\dot{p}_x = -\frac{\partial H}{\partial x}$$

$$\dot{y} = \frac{\partial H}{\partial p_y}$$

$$\dot{p}_y = -\frac{\partial H}{\partial y}$$

Use the fourth order Runge–Kutta time-stepper to integrate the system of ODE, starting from the initial condition

$$x(0) = 3.0, \ y(0) = 2.1, \ p_x(0) = 0, \ p_y(0) = 0,$$

with a time step $k = 0.01$. Integrate the system of differential equations within the time interval $t \in [0, 60]$.

(a) Plot $x(t)$ for $t \in [0, 5]$.

(b) Perform a DFT analysis of $x(t)$ over the complete time integration $t \in [0, 60]$. Represent on a semi-logarithmic plot the absolute amplitudes $|\tilde{f}_k|$ resulting from the DFT and identify the two leading frequencies (that is, the approximate values of ω_k with largest $|\tilde{f}_k|$).

(c) In this problem, the potential is Himmelblau's function (see Problem 1, Chapter 6), which has a relative minimum at $(x, y) = (3, 2)$. The frequencies that you have obtained in (b) should be very close to the ones obtained by linearizing the system at that point. Provide the Jacobian matrix resulting from that linearization and its eigenvalues.

Solutions to Problems and Exercises

Chapter 1

1. (a) $\log_2\left(\dfrac{b-a}{\varepsilon}\right) - 1$.

   (b) More than three iterations.

2. (a) $\alpha \approx 0.860\ 333\ 589$ $(K = 36)$.

   (b) $\alpha \approx 0.641\ 185\ 744\ 5$ $(K = 38)$.

   (c) $\alpha \approx 1.829\ 383\ 601\ 9$ $(K = 38)$.

3. $C \equiv \left|1 - \dfrac{f'(\alpha)}{q}\right|$ where $q \equiv \dfrac{f(b) - f(a)}{b - a}$, $qf'(\alpha) > 0$, and $f'(\alpha) < 2q$.

4. $C \equiv \left|\dfrac{f''(\alpha)}{2f'(\alpha)}\right|^{(\sqrt{5}-1)/2}$.

6. (a) The method is linear $(p \equiv 1)$.

7. Since $\dfrac{f(x_k)}{f'(x_k)} \equiv zx_k^2 - x_k$, we conclude that $\dfrac{d}{dx}\log(f(x)) \equiv \dfrac{1}{x(zx - 1)}$.

   Integrate to obtain $f(x) \equiv K\left(z - \dfrac{1}{x}\right)$, with K arbitrary integration constant. The purpose is to solve $x^{-1} - z = 0$, whose solution is z^{-1}, i.e. computing the inverse of z. The quantity z^{-1} is the limit obtained in the iteration if we assume $x_{k+1} \equiv x_k$ as $k \to \infty$.

Fundamentals of Numerical Mathematics for Physicists and Engineers, First Edition. Alvaro Meseguer.
© 2020 John Wiley & Sons, Inc. Published 2020 by John Wiley & Sons, Inc.

8. Approximately 36 iterations. The first 33 are required for x_k to be of order 1 (the approximate iteration being $x_{n+1} \approx x_n/2$). The next three iterations must be added to reduce the error (with approximate iteration $e_{n+1} \approx e_n^2/2$) from, say, 10^{-1} to 10^{-8}.

9. Since $x_{n+1} = \sqrt{p + x_n}$ and, in the limit $n \to \infty$, $x_{n+1} = x_n = x$, $x = \sqrt{p + x}$. Therefore, $x = (1 + \sqrt{1 + 4p})/2$.

10. For simplicity, we define $f_n = f(x_n)$, $f'_n = f'(x_n)$, $f''_n = f''(x_n)$, ...
  (a) $1/m$.
  (b) $s_n = \dfrac{f_n f'_n}{(f'_n)^2 - f_n f''_n}$.
  (c) $x_{n+1} = x_n - \dfrac{1}{2 \tan x_n}$, $x_{12} = 1.570\ 892\ 4...$ $(n \geq 12)$.
  (d) $x_{n+1} = x_n + \sin x_n \cos x_n$, $x_2 = 1.570\ 883\ 6...$ $(n \geq 2)$.

11. (a) $p \approx 2$.
  (b) $p \approx 1$.
  (c) $p = 2$, $C = \dfrac{1}{2} \left| f''(\alpha) \left(1 + \dfrac{1}{f'(\alpha)} \right) \right|$.

12. (a) $x_{n+1} = x_n - \dfrac{2 f_n f'_n}{2(f'_n)^2 - f''_n f_n}$.
  (b) $p \approx 3$.

13. (a) Use conservation of mechanical energy.
  (b) Two possible angles: $\theta_1 \approx 1.029$ rad ($58.97°$) and $\theta_2 \approx 0.440$ rad ($25.21°$).
  (c) $\theta_{\text{optimal}} \approx 0.715$ rad ($40.96°$) with maximum range $G_{\text{max}} \approx 23.94$ m.

Practical 1.1:

γ	x (m)
0.1	0.859
0.25	0.775
0.5	0.750
0.9	0.912

Applying chain rule:

$$\ddot{x} = 2g \left(\frac{1 - f}{1 + (f')^2} \right)^{1/2} \frac{d}{dx} \left(\frac{1 - f}{1 + (f')^2} \right)^{1/2} \frac{dx}{dt} =$$

$$= -g \frac{f'[1 + (f')^2 + 2(1 - f)f'']}{[1 + (f')^2]^2}.$$

Practical 1.2:

(b) $x_{\text{imp}} \approx 76.9537$ m.

(c)

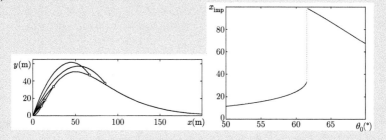

(d) $\theta_{\text{min}} \approx 61.59°$.

(e) For $\theta = \theta_{\text{min}}$, the parabola becomes tangent to the hill's profile at x_{imp}, which becomes a double root, therefore leading to an ill-conditioned problem.

Chapter 2

2. Apply remainder's formula for $n = 1$. According to the estimation, the error is smaller than 1.104×10^{-7}. The actual error is 1.101×10^{-7}.

3. For an arbitrary pair of neighboring nodes, x_i and x_{i+1}, it is easy to show that $|x - x_i||x - x_{i+1}| < h^2/4$. For x in $[x_i, x_{i+1}]$, the absolute nodal polynomial can be bounded as follows:

$$|x - x_0||x - x_1| \cdots \underbrace{|x - x_{i-2}||x - x_{i-1}|}_{<5h} \frac{h^2}{4} \underbrace{|x - x_{i+2}||x - x_{i+3}|}_{<2h} \cdots |x - x_n|$$
$$\phantom{|x - x_0||x - x_1| \cdots |x - x_{i-2}|}\underbrace{\phantom{|x - x_{i-1}|}}_{<3h}\phantom{\frac{h^2}{4}|x - x_{i+2}|}\underbrace{\phantom{|x - x_{i+3}|}}_{<4h}$$

4. The logarithm of the nodal polynomial is

$$\log \ell(x) = \log \prod_{k=0}^{n} (x - x_k) = \sum_{k=0}^{n} \log(x - x_k).$$

Differentiating on both sides, $\dfrac{\ell'(x)}{\ell(x)} = \displaystyle\sum_{k=0}^{n} \dfrac{1}{x - x_k}$, or $\ell'(x) \equiv$

$\ell(x) \displaystyle\sum_{k=0}^{n} \dfrac{1}{x - x_k}$. Therefore, the derivative of $\ell(x)$ is

$$\ell'(x) \equiv \prod_{k \neq 0}(x - x_k) + \prod_{k \neq 1}(x - x_k) + \cdots + \prod_{k \neq j}(x - x_k) + \cdots + \prod_{k \neq n}(x - x_k).$$

Evaluation of $\ell'(x)$ at x_j cancels all the products above, but not the one with $k \neq j$, so that $\ell'(x_j) = \displaystyle\prod_{k \neq j}(x_j - x_k)$. This quantity is the normalizing factor in the jth cardinal polynomial and therefore we can write

$$\ell_j(x) \equiv \frac{1}{\ell'(x_j)} \prod_{k=0(k \neq j)}^{n}(x - x_k) \equiv \frac{\ell(x)}{\ell'(x_j)(x - x_j)}.$$

5. The inverse of the j-th barycentric weight is

$$\lambda_j^{-1} = \prod_{k \neq j}^{n}(x_j - x_k) = \left(\frac{2}{n}\right)^n \prod_{k \neq j}^{n}(j - k)$$

$$= \left(\frac{2}{n}\right)^n \underbrace{j[j-1]\cdots[j-(j-1)]}_{j!}\underbrace{[j-(j+1)]}_{(-1)}\underbrace{[j-(j+2)]}_{(-2)}\cdots\underbrace{[j-(j+n-j)]}_{(-1)(n-j)}$$

$$= (-1)^{n-j}\left(\frac{2}{n}\right)^n j!(n-j)! \equiv (-1)^{n-j}\left(\frac{2}{n}\right)^n \frac{n!}{\binom{n}{j}}.$$

Therefore, $\quad \lambda_j \equiv (-1)^{n-j}\dfrac{1}{n!}\left(\dfrac{n}{2}\right)^n \binom{n}{j} \equiv (-1)^{n-j}\dfrac{1}{n!h}\binom{n}{j}, \quad$ with $h = 2/n$ being the spacing between neighboring nodes. Similarly, for an arbitrary equispaced distribution $x_j \equiv a + (b-a)j/n$, we have

$$\prod_{k \neq j}^{n}(x_j - x_k) \equiv \left(\frac{b-a}{n}\right)^n (-1)^{n-j}j!(n-j)!,$$

so that the barycentric weights are

$$\lambda_j \equiv (-1)^{n-j}\frac{1}{n!h^n}\binom{n}{j}, \quad \text{with } h \equiv \frac{b-a}{n}.$$

6. Differentiating the result of Problem 4,

$$\ell'_j(x) = \frac{1}{\ell'(x_j)} \frac{\ell'(x)(x - x_j) - \ell(x)}{(x - x_j)^2}, \text{ so that, for } i \neq j,$$

$$\ell'_j(x_i) = \frac{\lambda_j}{\lambda_i} \frac{1}{x_i - x_j}.$$

For the case $i = j$ it is better to consider the original expression of $\ell_j(x)$ and take logarithm before differentiating:

$$\log \ell_j(x) = \log \left[\lambda_j \prod_{k \neq j} (x - x_k) \right] = \log \lambda_j + \sum_{k \neq j} \log(x - x_k), \text{ so that}$$

$$\frac{\ell'_j(x)}{\ell_j(x)} = \sum_{k \neq j} \frac{1}{x - x_k} \quad \text{and therefore} \quad \ell'_j(x_j) = \sum_{k \neq j} \frac{1}{x_j - x_k}.$$

Also see Chapter 3 on numerical differentiation for more details.

7. **(a)** $\Pi_2 f(x) = 1 + 3x/2 + x^2$.
   (b) $\Pi_2 f(x) = (1/2)2x(x - 1/2) + (1)4(1/4 - x^2) + (2)2x(x + 1/2)$.
   (c) $\sqrt[3]{4} = f(1/3) \approx \Pi_2 f(1/3) = 29/18$.

8. **(a)** Imposing $H(x_j) = f(x_j)$ and $H'(x_j) = f'(x_j)$ for $j = 0, 1$ we obtain a linear system of equations for the coefficients a_j:

$$\begin{bmatrix} 1 & 0 & 0 & 0 \\ 1 & h & h^2 & h^3 \\ 0 & 1 & 0 & 0 \\ 0 & 1 & 2h & 3h^2 \end{bmatrix} \begin{bmatrix} a_0 \\ a_1 \\ a_2 \\ a_3 \end{bmatrix} = \begin{bmatrix} f_0 \\ f_1 \\ f'_0 \\ f'_1 \end{bmatrix}$$

The determinant of the 4×4 matrix appearing in this system is h^4 and therefore the polynomial exists and is unique.

   (b) $a_0 = f_0$, $a_1 = f'_0$, $a_2 = \frac{1}{h^2}[3(f_1 - f_0) - h(2f'_0 + f'_1)]$,

   $a_3 = \frac{1}{h^2}\left[f'_1 + f'_0 + \frac{2}{h}(f_0 - f_1) \right]$

   (c) $H(x) = \frac{4}{\pi^3}x^3 - \frac{6}{\pi^2}x^2 + 1$. Yes: $H(x) = \left(x - \frac{\pi}{2}\right)\left(\frac{4}{\pi^3}x^2 - \frac{4}{\pi^2}x - \frac{2}{\pi}\right)$.

9. **(a)** Since $\Pi_n 1 = 1$, use the first barycentric form to show that

$$\ell(x) \sum_{j=0}^{n} \lambda_j (x - x_j)^{-1} = 1$$

or, equivalently, that

$$\lambda_0 \prod_{k \neq 0} (x - x_k) + \lambda_1 \prod_{k \neq 1} (x - x_k) + \cdots + \lambda_n \prod_{k \neq n} (x - x_k) = 1.$$

Grouping $O(x^n)$ terms we have

$$(\lambda_0 + \lambda_1 + \cdots + \lambda_n) x^n + O(x^{n-1}) = 1 \quad \text{and hence}$$
$$\lambda_0 + \lambda_1 + \cdots + \lambda_n = 0.$$

(b) Differentiate $\displaystyle\sum_{j=0}^{n} \ell_j(x) = 1$.

10. **(a)** $g_j = f_j / (x_j^2 - 1)$.
    (b) $a_0 = -(2/3)(f_0 + f_1)$ and $a_1 = (4/3)(f_0 - f_1)$.
    $\Pi f(x) = (2\sqrt{2}/3)(1 - x^2)$.

Practical 2.3:

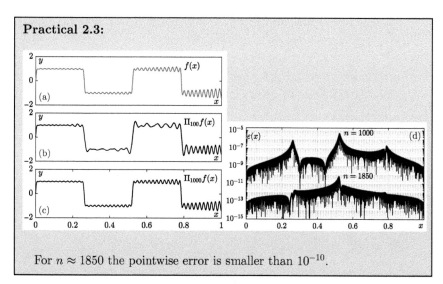

For $n \approx 1850$ the pointwise error is smaller than 10^{-10}.

Chapter 3

1. **(a)** Use Taylor expansion.
   (b) Also $O(h^4)$.

Practical 3.1:

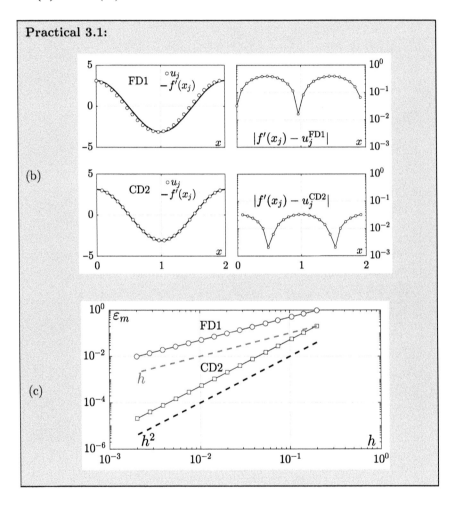

Practical 3.2:

The plot below shows the errors for parts (a), (b) and (c) as a function of n.

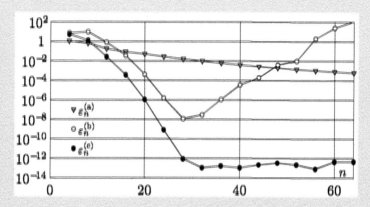

(b) Yes, ill-conditioning of the global equispaced differentiation for $n > 28$.
(a) In general, the Chebyshev global differentiation requires a smaller number of evaluations. In particular, for $\delta = 10^{-3}$, Chebyshev global differentiation requires nearly $n = 16$ function evaluations, closely followed by equispaced global differentiation ($n = 20$). Finite difference formula requires $n = 60$.
(d) The maximum error in the approximation of the second derivative is slightly above 10^{-11}:

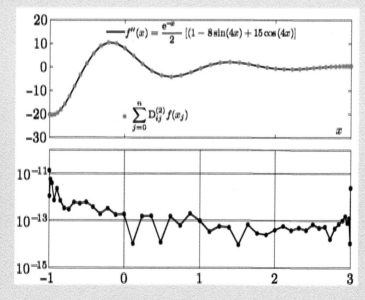

2. The second formula is more accurate.

3. $O(h)$ and $O(h^2)$, respectively.

4. The cardinal polynomials are
$$\ell_0(x) \equiv 3(x - 1/2)(x - 1/3), \quad \ell_1(x) \equiv -12(x - 1)(x - 1/3),$$
$$\ell_2(x) \equiv 9(x - 1)(x - 1/2).$$

The differentiation matrix is
$$\mathbf{D} \equiv \begin{bmatrix} 7/2 & -8 & 9/2 \\ 1/2 & 4 & -9/2 \\ -1/2 & 8 & -15/2 \end{bmatrix}.$$

Chapter 4

3. **(a)** $w_{0,3} = 1/9$, $w_{1,2} = 8/9$.

4. Yes, they are symmetric, that is, $w_{n-k} = w_k$, for $k = 0, 1, 2, \ldots$. Only half of them need to be computed.

5. The chain has a total length $\ell = 0.4 + \displaystyle\int_0^{x_0} \sqrt{1 + [f'(x)]^2}\, dx$, where $f(x) = e^{-x^2}$. Since $\ell = 1$, x_0 must be the solution of the equation
$$F(z) = \int_0^z \sqrt{1 + 4x^2 e^{-2x^2}}\, dx - 0.6 = 0$$

Devise a Matlab function for $F(z)$ using Clenshaw–Curtis quadrature and then apply Newton's method to solve $F(z) = 0$. The abscissa is at $x_0 = 0.533\ 626\ 607\ 834\ 99$.

6. **(a)** The perimeter of the curve is $\ell = \displaystyle\int_0^{2\pi} \sqrt{\dot{x}^2 + \dot{y}^2}\, dt$. Using Clenshaw–Curtis with 49 evaluations, $\ell_{CC} \approx 7.510\ 18$. Using composite trapezoidal $\ell_{CTP} \approx 7.510\ 181\ 57$. Trapezoidal is more accurate. **(b)** Clenshaw–Curtis is much more accurate than trapezoidal ($\ell_{CC}^{A-C} \approx 2.523\ 024\ 414\ 526\ 69$, $\ell_{CTP}^{A-C} \approx 2.52$; only converged figures are included in each case). The reason is that in this case the domain of integration ($t_{A-C} \in [0, \pi/2]$) *does not* coincide with a complete period of the integrand.

7. **(a)** Yes, the integrand is periodic. **(b)** Yes, the integral becomes singular (and divergent) near the ring. **(c)** Even the singularity has been removed, the exponential convergence is lost because the domain of integration is no longer a period of the integrand.

8. **(a)** For $\gamma = 1$, the two curves intersect at $x_c \approx 0.739\,085\,1$, which can be obtained by solving $x - \cos x = 0$ with Newton's method, for example. The length of the first arc is therefore
$$\ell_{\mathrm{I}} = \int_0^{x_c} \sqrt{1 + \sin^2 x}\, dx \approx 0.795\,721\,49 \quad \text{(obtained with Clenshaw–Curtis for } n = 14).$$

(b) Define $\ell = \ell_{\mathrm{I}} + \ell_{\mathrm{II}} = \int_0^{\pi/2} \sqrt{1 + \sin^2 x}\, dx \approx 1.910\,098\,894\,5$. For a positive γ, the equation $\gamma x - \cos x = 0$ defines implicitly $x_c = x_c(\gamma)$. Devise a Matlab function $F(\gamma) = \int_0^{x_c(\gamma)} \sqrt{1 + \sin^2 x}\, dx - \ell/2 = 0$, and solve $F(\gamma) = 0$ using Newton's method ($\gamma \approx 0.743\,953\,432$).

(c) Proceed similarly as in (b) to obtain $\gamma \approx 0.699\,626\,659\,9$.

9. **(a)** $\Delta T_{\mathrm{BC}} = 0.375\,169\,2$ s. **(b)** Introducing the change of variable $1 - x = z^4$, the integral becomes $I(1) = 4\int_0^1 (2 - z^4)^{-3/4}\, dz$. Using Clenshaw–Curtis with $n = 16$, $I(1) \approx 2.622\,057\,5$ and $t_{\mathrm{D}} \approx 1.183\,920\,97$ s. **(c)** $T = 2t_{\mathrm{D}}$.

11. **(a)** $w_0 = -17/36$, $w_1 = -5/9$, $w_2 = 1/36$. **(b)** The system reads
$$\begin{bmatrix} 1 & 1 & 1 & \cdots & 1 \\ 0 & 1/n & 2/n & \cdots & 1 \\ 0 & 1/n^2 & 4/n^2 & \cdots & 1 \\ \vdots & \vdots & \vdots & & \vdots \\ 0 & 1/n^n & (2/n)^n & \cdots & 1 \end{bmatrix} \begin{bmatrix} w_0 \\ w_1 \\ w_2 \\ \vdots \\ w_n \end{bmatrix} = \begin{bmatrix} -1 \\ -1/4 \\ -1/9 \\ \vdots \\ -1/(n+1)^2 \end{bmatrix}.$$

(c) $I_1 \approx -0.822\,472\,85$, $I_2 \approx -0.208\,760\,25$.

12. $I_{\mathrm{D}} \approx 0.777\,504\,63$, $I_{\mathrm{F}} \approx 1.954\,902\,7$ (hyperbolic tangent rule with $n = 44$ and $c = 5$).

Practical 4.1:

(a) For $k = 7, 4$, the exponent is -4 (same as in Figure 4.4). For $k = 3$ the error is zero (to machine precision) since Simpson's quadrature is exact in $\mathbb{R}_3[x]$.

(b) For $m = 43$, the approximated value is $3.141\,592\,653\,589\,69$, with an absolute error smaller than 10^{-13}. It required $2m + 1 = 87$ function evaluations.

(c) The most accurate approximation is $3.141\,592\,653\,589\,87$, and it takes place for $n = 23$ (i.e. 24 function evaluations), with an absolute error 7.5×10^{-14}. Increasing n beyond deteriorates the accuracy due to ill-conditioning of equidistant interpolation. In addition, the Vandermonde matrix is also ill-conditioned (as Matlab warns you) as we will see in Part II, so the numerical accuracy of the weights is not reliable. Ironically, the integrand $1/(1 + x^2)$ is Runge's function (whose equispaced interpolation may be divergent). Overall, the quadrature formula based on equidistant nodes *should be avoided* in general for large n.

Practical 4.2:

$V(1.5, 1) \approx -3.453\,849$. For $n \geq 23$. (a)

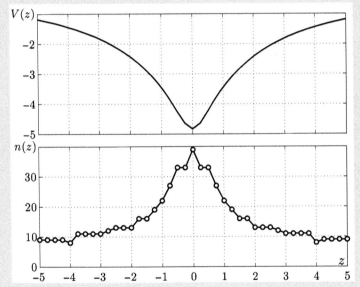

The potential function is singular at $(y, z) = (1, 0)$. This deteriorates convergence rates of the Clenshaw–Curtis quadrature when the integration domain is close to that singularity. Away from it, the required number of nodes diminishes.

(Continued)

(b)

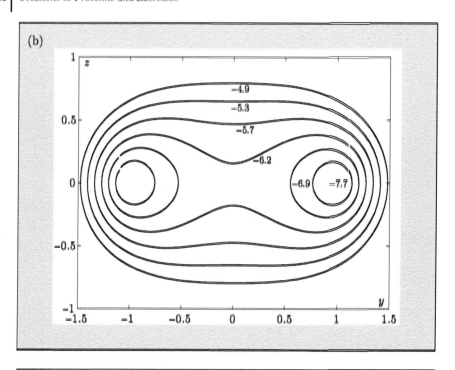

Practical 4.3:

(b) Assuming $f'(0) = f_0' < 0$ and $|f_0'|$ bounded, the numerator of the integrand $\{1 + [f'(z)]^2\}^{1/2}$ is regular at $z = 0$. The denominator $[f_0 - f(z)]^{-1/2}$ makes the integrand singular in the limit $z \to 0^+$ since $\lim_{z \to 0^+}[f_0 - f(z)]^{-1/2} = +\infty$.

Let $f(z) = f_0 + f_0' z + \frac{1}{2}f_0'' z^2 + \cdots$ be the Taylor series of $f(z)$ at $z = 0$. The integrand is therefore

$$\{1 + [f'(z)]^2\}^{1/2}\left\{-f_0' z - \frac{1}{2}f_0'' z^2 - \cdots\right\}^{-1/2},$$

or

$$\{1 + [f'(z)]^2\}^{1/2}z^{-1/2}\left\{-f_0' - \frac{1}{2}f_0'' z - \cdots\right\}^{-1/2} \approx \sqrt{-\frac{1 + (f_0')^2}{f_0'}}\,z^{-1/2},$$

to leading order. Since the exponent is $-1/2$, the integral is convergent. However, if $f_0' = 0$, then the integrand reads

$$\{1 + [f'(z)]^2\}^{1/2}z^{-1}\left\{-\frac{1}{2}f_0'' - \frac{1}{6}f_0''' z - \cdots\right\}^{-1/2} \approx \sqrt{-\frac{1}{2}f_0''}\,z^{-1},$$

and since the exponent is -1 the integral (and therefore time) is divergent, i.e. the bead will not move if initially released from a flat segment of the wire. Therefore, the condition is $f'_0 \neq 0$.

(c) $T(a) = \sqrt{2(a^2 + h^2)/gh}$.

(d) $T(2) = (2g)^{-1/2}I(2)$, where $I(2) = \int_0^2 \left(\dfrac{1 + e^{-2z}}{1 - e^{-z}} \right)^{1/2} dz$. This integral is singular. Fejér's quadrature with $n = 64$ provides $I(2) \approx 3.8$. Tanh-rule with $n = 48$ and $c = 5$ provides $I(2) \approx 3.893\,133$. By introducing the change of variable $1 - e^{-z} = x^2$, the integral becomes

$$I(2) = 2 \int_0^{\sqrt{1-e^{-2}}} \frac{\sqrt{1 + (1 - x^2)^2}}{1 - x^2} \, dx.$$

Since $\sqrt{1 - e^{-2}} \approx 0.9298 < 1$, this integral is regular. Clenshaw–Curtis with $n = 44$ (or Fejér's with $n = 48$) provides $I(2) \approx 3.893\,133\,885\,739\,8$.

Chapter 5

4. The sum of the first n squares is the sum of the areas of the gray rectangles (see the figure below):

$$\sum_{k=0}^{n} k^2 = \int_0^n x^2 \, dx + E_n,$$

where E_n is the area of the hatched region:

$$E_n = \sum_{k=1}^{n} \left(k^2 - \int_{k-1}^{k} x^2 \, dx \right) = n(n+1)/2 - n/3.$$

Therefore, $\displaystyle\sum_{k=0}^{n} k^2 = \frac{n^3}{3} + \frac{n(n+1)}{2} - \frac{n}{3} = \frac{n(n+1)(2n+1)}{6}$.

For $k = 1$, we compute $n - 1$ multipliers of the form $m_{ik} \equiv a_{ik}^{(k)}/a_{kk}^{(k)}$. For $k = 2$, we compute $n - 2$, etc. Overall, the computation of multipliers involves $1 + 2 + \cdots + (n-1) = n(n-1)/2$ divisions. Simultaneously, at $k = 1$ we update the $(n-1) \times (n-1)$ matrix with the Gaussian transformation $a_{ij}^{(2)} = a_{ij}^{(1)} - m_{i1} a_{1j}^{(1)}$, involving $(n-1)^2$ products and $(n-1)^2$ subtractions, i.e. $2(n-1)^2$ operations. For $k = 2$ we have $2(n-2)^2$, etc. Overall, neglecting the computational cost of the multipliers, the total cost of GEM is, to leading order,

$$2[(n-1)^2 + (n-2)^2 + \cdots + 2^2 + 1^2] \approx \frac{2}{3}n^3.$$

6. The case $n = 1$ is trivial. Assume that it is valid for any non-singular matrix in $\mathbb{M}_n(\mathbb{R})$ and take $\mathbf{L} \in \mathbb{M}_{n+1}(\mathbb{R})$ lower triangular with $\det \mathbf{L} \neq 0$. Split matrix $\mathbf{L}$

$$\mathbf{L} = \begin{bmatrix} \mathbf{L}_{11} & 0 \\ \mathbf{L}_{21} & \mathbf{L}_{22} \end{bmatrix}, \text{ with } \mathbf{L}_{11} \in \mathbb{M}_m(\mathbb{R}), \ \mathbf{L}_{22} \in \mathbb{M}_p(\mathbb{R}),$$

where $\mathbf{L}_{11}$ and $\mathbf{L}_{22}$ are lower triangular, m, $p \leq n$ and $m + p = n + 1$. Let $\mathbf{M} \doteq \mathbf{L}^{-1}$ be the matrix

$$\mathbf{M} = \begin{bmatrix} \mathbf{M}_{11} & \mathbf{M}_{12} \\ \mathbf{M}_{21} & \mathbf{M}_{22} \end{bmatrix}, \text{ with } \mathbf{M}_{11} \in \mathbb{M}_m(\mathbb{R}), \ \mathbf{M}_{22} \in \mathbb{M}_p(\mathbb{R}).$$

Then

$$\mathbf{LM} = \begin{bmatrix} \mathbf{L}_{11} & 0 \\ \mathbf{L}_{21} & \mathbf{L}_{22} \end{bmatrix} \begin{bmatrix} \mathbf{M}_{11} & \mathbf{M}_{12} \\ \mathbf{M}_{21} & \mathbf{M}_{22} \end{bmatrix} = \mathbf{I}_{n+1},$$

or $\mathbf{L}_{11}\mathbf{M}_{11} = \mathbf{I}_m$ and $\mathbf{L}_{11}\mathbf{M}_{12} = [0]$. Therefore, $\mathbf{M}_{11} = \mathbf{L}_{11}^{-1}$, which is lower triangular and $\mathbf{M}_{12} = [0]$ (since $\mathbf{L}_{11}$ is non-singular). Finally, $\mathbf{M}_{22} = \mathbf{L}_{22}^{-1}$, which is also lower triangular.

8.
$$\begin{aligned} -k_1 x_1 - k_2(x_1 - x_2) + F_1 &= 0 \\ k_2(x_1 - x_2) - k_3(x_2 - x_3) + F_2 &= 0 \\ k_3(x_2 - x_3) - k_4 x_3 + F_3 &= 0 \end{aligned}.$$

$x_1 = 0.625$ m, $x_2 = 1.0$ m, $x_3 = 0.875$ m.

Practical 5.1:

(a) Intensities given in Ampere

I_1	I_2	I_{15}
0.122 008 467 927 583	0.089 316 397 478 167	0.000 012 094 735 645

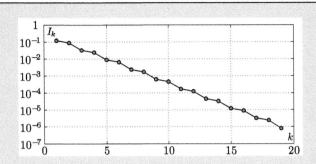

The intensities decay exponentially.

(b) In the limit $n \to \infty$, the equivalent resistance R_e between a and b is the same as between a' and b' (see figure below).

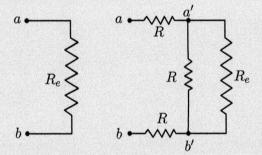

As a result, R_e satisfies the equation $R_e^2 - 2RR_e - 2R^2 = 0$, whose positive solution is $R_e = R(1 + \sqrt{3})$.

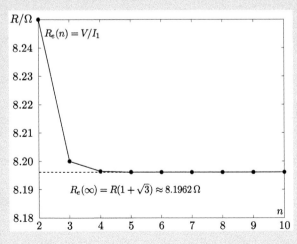

14. Use the vector $\mathbf{w} = [1\ 0\ -1]^{\mathrm{T}}$ so that $\mathbf{Q} = \mathbf{I} - \mathbf{w}\mathbf{w}^{\mathrm{T}} = \begin{bmatrix} 0 & 0 & 1 \\ 0 & 1 & 0 \\ 1 & 0 & 0 \end{bmatrix}$.

15. Since $\mathbf{F}\mathbf{w} = -\mathbf{w}$, then $\lambda_1 = -1$, with $\dim\{\ker(\mathbf{F} - \lambda_1\mathbf{I}) = 1\}$. Also, since $\mathbf{F}\mathbf{v} = \mathbf{v}$, $\forall \mathbf{v} \perp \mathbf{w}$, $\mathbf{v} \in \mathbb{R}^n$, then $\lambda_2 = 1$. The subspace orthogonal to $\mathbf{w}$ is $(n-1)$-dimensional, $\dim\{\ker(\mathbf{F} - \lambda_2\mathbf{I}) = n - 1\}$.

16.

$$\mathbf{Q} = \begin{bmatrix} 1/3 & 2/9 & -8/9 & -2/9 \\ -2/3 & 5/9 & -2/9 & 4/9 \\ -2/3 & -4/9 & -2/9 & -5/9 \\ 0 & -2/3 & -1/3 & 2/3 \end{bmatrix}, \quad \mathbf{A} = \begin{bmatrix} 1/3 & 16/9 & =43/9 & -16/3 \\ -2/3 & 13/9 & =4/9 & -1/3 \\ -2/3 & -32/9 & =58/9 & -25/3 \\ 0 & -10/3 & =20/3 & -7 \end{bmatrix}$$

Practical 5.2:

(b)

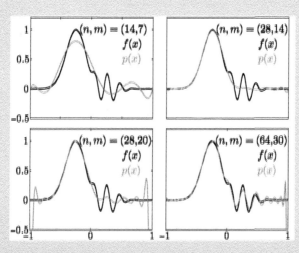

Yes, oscillations near the edges of the interval (ill-conditioning due to equally spaced nodes).

(c)

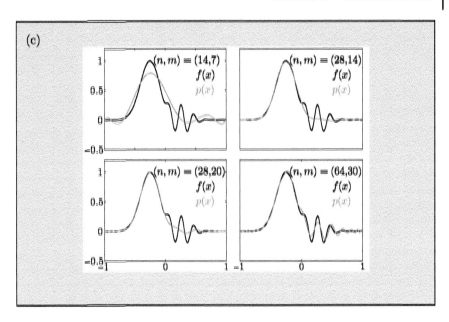

17. $p(x) \equiv 1 + x + x^2$.

18. $R \approx 61.89 \ \Omega$.

19. Longest day: 20th June, shortest day: 19th December.

21. The solution to the original system is $\mathbf{x} \equiv [1\ 1\ 1\ 1]^{\mathrm{T}}$. The solution to the modified system is $\hat{\mathbf{x}} \approx [-3.34\ 8.75\ -5.85\ -2.34]^{\mathrm{T}}$ so $\|\Delta\mathbf{x}\| \equiv \|\hat{\mathbf{x}} - \mathbf{x}\| \approx 11.284$. Assuming the matrix is exact,

$$\|\Delta\mathbf{x}\| \leqslant M \equiv \|\mathbf{x}\|\|\mathbf{b}\|^{-1}\kappa(\mathbf{A})\|\Delta\mathbf{b}\| \approx 21.64.$$

$$\|\Delta\mathbf{x}\| \approx 11.284, \|\Delta x\| \leq \|\mathbf{x}\|\kappa(\mathbf{A})\|\delta\mathbf{b}\|/\|\mathbf{b}\| = 21.639$$

Practical 5.3:

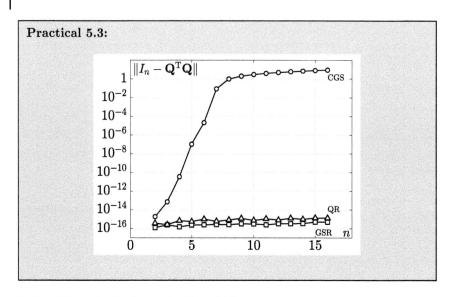

Practical 5.4:

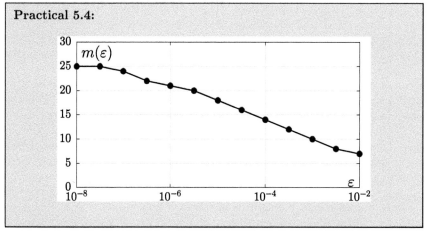

Chapter 6

1. The condition of critical point is given by the system of equations

$$\nabla V = \begin{bmatrix} \partial_x V \\ \partial_y V \end{bmatrix} = \begin{bmatrix} 4x(x^2 + y - 11) + 2(x + y^2 - 7) \\ 4y(x + y^2 - 7) + 2(x^2 + y - 11) \end{bmatrix} = \begin{bmatrix} 0 \\ 0 \end{bmatrix}$$

x	y	Type
$-0.270\ 844\ 591$	$-0.923\ 038\ 556$	Max.
3	2	Min.
$-2.805\ 118\ 086$	$3.131\ 312\ 518$	"
$-3.779\ 310\ 253$	$-3.283\ 185\ 991$	"
$3.584\ 428\ 340$	$-1.848\ 126\ 527$	"
$-0.127\ 961\ 347$	$-1.953\ 714\ 980$	Sad.
$0.086\ 677\ 505$	$2.884\ 254\ 701$	"
$-3.073\ 025\ 751$	$-0.081\ 353\ 044$	"
$3.385\ 154\ 184$	$0.073\ 851\ 880$	"

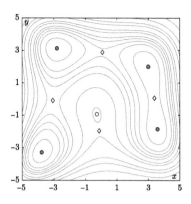

In the plot on the right, gray bullets are minima, the white bullet is the only maximum and the diamonds are saddles.

2. The abscissas of the points A and B satisfy

$$f'(x_A) - f'(x_B) = 0$$
$$f'(x_A)(x_B - x_A) - f(x_B) + f(x_A) = 0,$$

where $f'(x) = -2e^{-x^2} - (x-2)e^{-(x-2)^2}$. The first equation imposes that the tangent at both abscissas must be the same. The second equation imposes that this tangent must precisely coincide with the slope of the chord $[f(x_B) - f(x_A)]/(x_B - x_A)$. Solving the system with a Newton method provides $(x_A, y_A) \approx (0.158\ 188\ 702\ 3, 0.992\ 102\ 972\ 7)$ and $(x_B, y_B) \approx (2.226\ 676\ 592, 0.481\ 984\ 079\ 7)$.

3. (a) $C = 2/n$.

   (b) For $n = 4$, the nodes abscissas must satisfy the relations

$$x_1 + x_2 + x_3 + x_4 = 0$$
$$x_1^2 + x_2^2 + x_3^2 + x_4^2 = 4/3$$
$$x_1^3 + x_2^3 + x_3^3 + x_4^3 = 0$$
$$x_1^4 + x_2^4 + x_3^4 + x_4^4 = 4/5,$$

   whose solution (found using Newton's iteration started from an equispaced distribution) is $x_1 \approx -0.794\ 654\ 472\ 291\ 766$, $x_2 \approx -0.187\ 592\ 474\ 085\ 08$, $x_3 = -x_2$, and $x_4 = -x_1$.

   (c) $x_5 \approx 0.832\ 497\ 487\ 000\ 982$, $x_6 \approx 0.866\ 246\ 818\ 107\ 821$, $x_7 \approx 0.883\ 861\ 700\ 758\ 049$.

   (d) Yes, $n = 9$, with $x_9 \approx 0.911\ 589\ 307\ 728\ 434$. For $n \geq 10$ the nodes are complex.

4. **(a)** For arbitrary q_j particles, the equilibrium equations are

$$\sum_{i=1(i \neq j)}^{n} \frac{q_i q_j}{x_j - x_i} + \frac{q_j Q}{x_j + 1} + \frac{q_j Q}{x_j - 1} = 0,$$

for $j = 1, 2, \ldots, n$. Replace q_i and q_j by 1.

(b) $x_1 \approx -0.923\,880$, $\quad x_2 \approx -0.382\,683$, $\quad x_3 \approx 0.382\,683$, $\quad$ and $\quad x_4 \approx 0.923\,880$.

(c) Taking for example $\gamma = 0.1$ provides $x_{12} = 0.991\,445$.

(d) The equilibrium positions $\{x_1, x_2, \ldots, x_n\}$ are the roots of $T_n(x)$, that is, the Chebyshev nodes of the first kind (4.93) or Chebyshev classical abscissas $x_k = \cos((2k - 1)\pi/2n)$, for $k = 1, 2, \ldots, n$. For $Q = 1$ the equilibrium abscissas are the *Legendre* nodes. In general, *potential theory* shows that the equilibrium abscissas x_k are the roots of the *Jacobi* polynomial $P_n^{(2Q-1, 2Q-1)}(x)$.

5. **(a)** Let $\mathbf{X} = \begin{bmatrix} x_1 & x_2 \\ x_3 & x_4 \end{bmatrix}$. Since $\mathbf{X}^2 = \mathbf{A}$, then

$$x_1^2 + x_2 x_3 - 7 = 0$$
$$x_1 x_2 + x_2 x_4 - 10 = 0$$
$$x_3 x_1 + x_4 x_3 - 15 = 0$$
$$x_3 x_2 + x_4^2 - 23 = 0.$$

Starting Newton's iteration from the *close* solution $\mathbf{X}^{(0)} = \begin{bmatrix} 1 & 2 \\ 3 & 4 \end{bmatrix}$, we

obtain $\mathbf{X} \approx \begin{bmatrix} 0.761\,985 & 2.068\,712 \\ 3.103\,074 & 4.071\,931 \end{bmatrix}$. This is one possible solution.

(b) Proceed in a similar way. We suggest *not* to write the explicit equations, but use Matlab's `reshape` command instead:

```
function y = fun(x)
A = [37 54; 81 119]; X = reshape(x,[2 2])';
y = reshape((X^3-A)',[4 1]);
```

One possible solution is $\mathbf{X} \approx \begin{bmatrix} 1.708\,789 & 1.669\,025 \\ 2.503\,537 & 4.243\,232 \end{bmatrix}$.

6. **(a)** Let $f(x,y) = \left(x + \dfrac{y}{10} - 2\right)^2 + \left(y - \dfrac{x}{4} - \dfrac{3}{2}\right)^4 + \dfrac{xy^2}{20} = 1$. The geometric conditions for points A and B are

$$f(x,y) = 0$$
$$[x\ y] \cdot \nabla f = 0.$$

The first constraint means that the point (x,y) must belong to the asteroid's boundary. The second imposes the tangency condition, i.e. the gradient of $f(x,y)$ at the points must be orthogonal to the vector position.

(b) Using Newton's methods we find $x_A \approx 2.199\ 935$, $y_A \approx 1.114\ 312$, $x_B \approx 0.950\ 365$, and $y_B \approx 2.267\ 022$.

(c) Points A' and B' are the intersections of the straight lines $y = m_1 x$ and $y = m_2 x$ with Alderaan's boundary, where $m_1 = y_A/x_A$ and $m_2 = y_B/x_B$. The coordinates are $x_{A'} \approx 4.176\ 193$, $y_{A'} \approx 2.115\ 326$, $x_{B'} \approx 2.007\ 455$, and $y_{B'} \approx 4.788\ 631$.

Practical 6.1:

(a) We provide some selected values in the table below:

T	v_ℓ	v_g	p
0.975	0.755 140 600 754	1.436 931 258 000	0.902 985 170 975
0.950	0.684 122 113 656	1.727 071 192 256	0.811 879 243 364
0.925	0.637 851 638 159	2.024 621 393 007	0.726 585 053 520
0.900	0.603 401 903 178	2.348 842 376 202	0.646 998 351 872
0.875	0.576 016 046 000	2.712 408 236 896	0.573 007 253 114
0.850	0.553 360 458 440	3.127 639 292 441	0.504 491 649 787

(b) In this case, the nonlinear system reads

$$I_n(v_\ell, v_g, T) - (v_g - v_\ell)\, p(v_\ell, T) = 0$$

$$\frac{1}{2v_g - 1}\exp\left(2 - \frac{2}{v_g T}\right) - \frac{1}{2v_\ell - 1}\exp\left(2 - \frac{2}{v_\ell T}\right) = 0,$$

where I_n is a suitable quadrature formula (such as Clenshaw–Curtis) that approximates the integral arising from the Maxwell construction

(Continued)

through evaluation of the integrand on a set of nodes v_j:

$$I_n(v_\ell, v_g, T) = \sum_{j=0}^{n} w_j \, p(v_j, T) \approx \int_{v_\ell}^{v_g} p(v, T) \, dv.$$

T	v_ℓ	v_g	p
0.975	0.781 111 623 680	1.352 987 818 610	0.927 216 952 747
0.950	0.716 033 340 121	1.560 710 036 605	0.858 736 883 360
0.925	0.673 844 398 916	1.754 631 313 648	0.794 366 307 368
0.900	0.642 840 743 503	1.947 036 581 542	0.733 915 739 056
0.875	0.618 652 754 732	2.142 917 821 447	0.677 199 845 629
0.850	0.599 112 607 104	2.345 112 466 347	0.624 037 602 450

Notice that in the limit $T \to 0^+$ the integrand becomes singular at $v = v_\ell$.

Practical 6.2:

(a) The determinant of the Jacobian at $(\phi_1, \phi_2) = (0, 0)$ is the quadratic $2\alpha^2 - 4\alpha + 1$, shown in the figure below (the black dashed curve I).

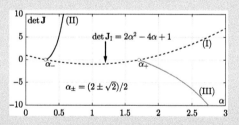

Therefore, for $\alpha = \alpha_\pm$, uniqueness of solutions is not guaranteed and new branches may potentially emerge at those values.

(c) A second branch, namely branch (II), emerges at $\alpha = \alpha_- = (2 - \sqrt{2})/2$, where ϕ_1 and ϕ_2 are both positive. The third branch (III) is born later on at $\alpha = \alpha_+ = (2 + \sqrt{2})/2$, where $\phi_2 < 0$. The determinant of the Jacobian along branches (II) and (III) are depicted in previous figure as solid black and gray curves, respectively. In both cases the determinant only vanishes at the known critical values $\alpha_\pm$, so no new equilibrium branches are expected to emerge from (II) or (III). The angles corresponding to the new branches are depicted below as a function of α. We also provide a physical representation of these new solutions for $\alpha = 2.14$.

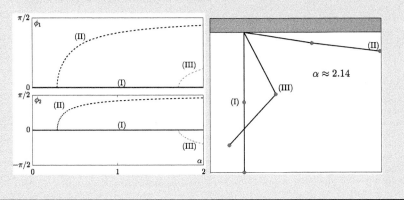

Chapter 7

1. Since $f \in \mathbb{R}$, $\overline{f} = f$, therefore $\sum_k \overline{\widehat{f}_k} e^{-ikx} = \sum_k \widehat{f}_k e^{ikx}$. Identify Fourier coefficients of the two expansions.

2. The difference $f - f_N$ is not in S_N since it only contains terms $\phi_k(x) = e^{ikx}$ with $k < -N/2$ and $k > N/2 - 1$. Therefore,

$$(f - f_N, g) = \sum_{-\infty}^{-N/2-1} f_k(\phi_k, g) + \sum_{N/2}^{+\infty} f_k(\phi_k, g) = 0$$

On the other hand,

$$\begin{aligned}
\|f - f_N\|^2 &= (f - f_N, f - f_N) = (f - f_N, f - g + g - f_N) \\
&= (f - f_N, f - g) + (f - f_N, g - f_N) \\
&= (f - f_N, f - g) \le \|f - f_N\| \, \|f - g\|.
\end{aligned}$$

Therefore, $\|f - f_N\| \le \|f - g\|$, for any $g \in S_N$.

3. (a) $\widehat{f}_k = \dfrac{3}{2\pi} \displaystyle\int_0^{2\pi} \dfrac{e^{-ik\theta}}{5 - 4\cos\theta} \, d\theta$. Introduce $z = e^{i\theta}$, so that

$$\widehat{f}_k = \frac{3}{2\pi i} \oint_{|z|=1} \frac{z^{-k}}{5z - 2z^2 - 2} \, dz.$$

Assume $k \leq 0$, the integrand has two simple poles at the roots of $-2z^2 + 5z - 2$, located at $z = 1/2$ and $z = 2$. Only $z = 1/2$ lies inside $|z| = 1$, and its residue is

$$\lim_{z \to 1/2} (z - 1/2) \frac{z^{-k}}{-2(z - 1/2)(z - 2)} = \frac{2^k}{3}.$$

Apply Cauchy's residue theorem, so that $\widehat{f}_k = \dfrac{3}{2\pi i} 2\pi i \dfrac{2^k}{3} = 2^k = 2^{-|k|}$, since $k \leq 0$. For $k > 0$, apply conjugation: since $f(x)$ is real, $\overline{\widehat{f}_k} = \widehat{f}_{-k}$, therefore $\widehat{f}_k = 2^{-|k|}$, for $k > 0$ as well.

(b) $\widetilde{f}_1 - \widehat{f}_1 = 5[2(2^8 - 1)]^{-1} \approx 0.009\ 803\ 921\ 568\ 627\ 45.$

Practical 7.1:

(a)

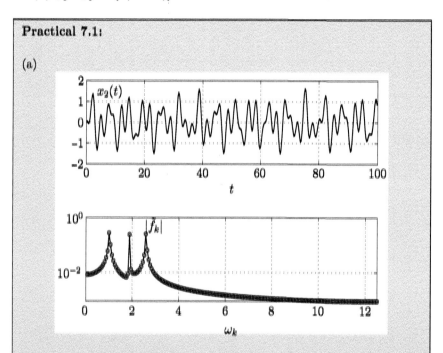

There are three peaks clearly identified at the frequencies $\omega_1 \approx 1$, $\omega_2 \approx 2$, and $\omega_3 \approx 2.6$. **(b)** Computing the eigenvalues of **B** with **eig**, the exact frequencies are $\omega_1 = 1.022\ 365\ 727\ 22$, $\omega_2 = 1.882\ 670\ 163\ 6$, $\omega_3 = 2.588\ 986\ 310\ 2$.

Chapter 8

1. There are no solutions above $\lambda \approx 3.514$. For accurate computation of $\|u\|$, use Clenshaw–Curtis quadrature.

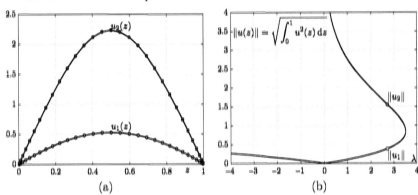

(a)   (b)

See Ascher and Petzold's (1998) for more details.

2. **(a)** Introduce Fourier's expansion in Mathieu's equation to get

$$\left\{-\frac{d^2}{dx^2} + 2q\cos(2x)\right\} \sum_k \widehat{\Psi}_k e^{ikx}$$

$$\equiv -\sum_k \widehat{\Psi}_k \frac{d^2}{dx^2} e^{ikx} + q(e^{i2x} + e^{-i2x}) \sum_k \widehat{\Psi}_k e^{ikx}$$

$$\equiv \sum_k k^2 \widehat{\Psi}_k e^{ikx} + q \sum_k \widehat{\Psi}_{k-2} e^{ikx} + q \sum_k \widehat{\Psi}_{k+2} e^{ikx}.$$

In matrix form:

$$
\begin{bmatrix}
\left(-\frac{N}{2}\right)^2 & 0 & q & 0 & \cdots \\
0 & \left(-\frac{N}{2}+1\right)^2 & 0 & q & \cdots \\
q & 0 & \left(-\frac{N}{2}+2\right)^2 & 0 & \cdots \\
0 & q & 0 & \left(-\frac{N}{2}+3\right)^2 & \cdots \\
\vdots & \vdots & \vdots & \vdots & \ddots
\end{bmatrix}
\begin{bmatrix}
\widehat{\Psi}_{-N/2} \\
\widehat{\Psi}_{-N/2+1} \\
\vdots \\
\widehat{\Psi}_{N/2-2} \\
\widehat{\Psi}_{N/2-1}
\end{bmatrix}
$$

(b) $E_1 = -0.936\ 818$, $E_2 = -0.733\ 265$, $E_3 = 2.165\ 939\ 9$,
$E_4 = 3.814\ 290\ 8$, $E_5 = 4.746\ 779$.

3. **(a)** $E_0 = 1.142\ 834\ 271\ 507\ 19$, $E_1 = 1.618\ 608\ 497\ 608\ 88$,
$E_2 = 2.867\ 122\ 888\ 684\ 2$.

(b) Compute S using Clenshaw–Curtis in the transformed domain:

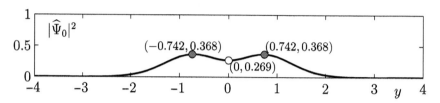

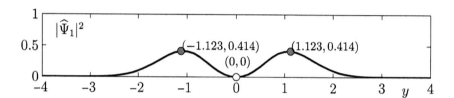

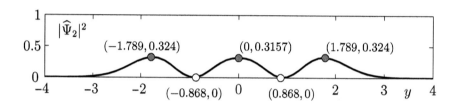

4. Using the bounded discretization formulation (8.35–8.38) with $n = 32$ and
$c_{11} = c_{21} = c_{13} = c_{23} = 0$, we obtain

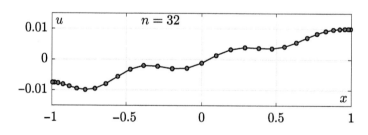

See Trefethen's (2012) for more details.

Practical 8.1:

(b) $c_{21} = 1$, $c_{22} = c_{23} = 0$, $\mathbf{M}_1 = -[\mathbf{D}_{01} \cdots \mathbf{D}_{0n}]$, $\mathbf{M}_2 = [1]$, $\mathbf{M}_3 = [\mathbf{L}_{10} \cdots \mathbf{L}_{n0}]^{\mathrm{T}}$,

$$\widehat{\mathbf{L}} = \begin{bmatrix} \mathbf{L}_{11} & \cdots & \mathbf{L}_{1n} \\ \vdots & & \vdots \\ \mathbf{L}_{n1} & \cdots & \mathbf{L}_{nn} \end{bmatrix}.$$

j	λ_j	$2\lambda_j$
1	1.202 412 778 847 9	2.404 825 557 695 74
2	2.760 039 055 143 2	5.520 078 110 286 32
3	4.326 863 956 455 5	8.653 727 912 911 01

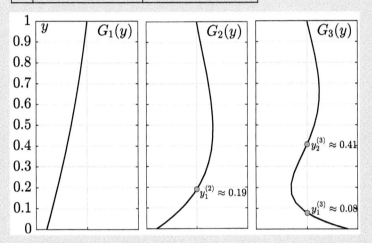

The exact location $y_j^{(k)}$ of the jth node corresponding to the kth eigenmode $G_k(y)$ satisfies $\mathrm{J}_0\left(2\lambda_k\sqrt{y_j^{(k)}}\right) = 0$, or

$$y_j^{(k)} = (\lambda_j/\lambda_k)^2, \quad (j = 1, 2, \ldots, k).$$

Ignoring the trivial node $y = 1$:

$y_1^{(2)} = (\lambda_1/\lambda_2)^2 \approx 0.189\ 791\ 479\ 516\ 748$	
$y_1^{(3)} = (\lambda_1/\lambda_3)^2 \approx 0.077\ 225\ 492\ 255\ 422\ 9$	
$y_2^{(3)} = (\lambda_2/\lambda_3)^2 \approx 0.406\ 896\ 518\ 495\ 226$	

Practical 8.2:

Introduce the velocity vector $\mathbf{v} \equiv (u, v) = \dot{\mathbf{r}} \equiv (\dot{x}, \dot{y})$, so that the IVP reads

$$\dot{x} = u,$$

$$\dot{u} = \sum_{i=1}^{3} m_i \frac{x_i - x}{\|\mathbf{r}_i - \mathbf{r}\|^3},$$

$$\dot{y} = v,$$

$$\dot{v} = \sum_{i=1}^{3} m_i \frac{y_i - y}{\|\mathbf{r}_i - \mathbf{r}\|^3}.$$

(a) $\mathbf{r}_{10^{-4}}(2) \approx (-0.063\,562\,309\,47, 0.104\,920\,886\,92)$.

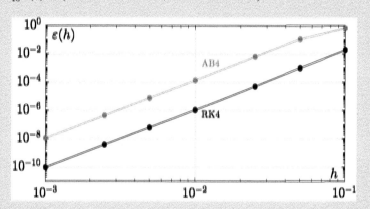

$h_{AB4} \lesssim 9 \times 10^{-3}$, $h_{RK4} \lesssim 3 \times 10^{-2}$. RK4 requires more function evaluations.

(b) Impose the coordinates (provided by your RK4) of the particle launched with an initial angle θ_0 after T time units to satisfy

$$x(\theta_0, T) = 0,$$

$$y(\theta_0, T) = 0.$$

Use Newton's method with initial guess $(\theta, T) = (\pi/4, 2)$.
Solution: $(\theta_0, T_a) = (0.777\,454\,979\,237, 1.911\,586\,213\,96)$

(e)	θ_0	T_a
(I)	0.777 454 979 237	1.911 586 213 96
(I)	−0.284 329 842 225	1.911 586 213 96
(II)	2.246 794 769 65	1.436 360 267 91
(II)	1.491 809 231 64	1.436 360 267 91
(III)	3.446 142 093 41	1.599 214 849 37
(III)	5.111 159 983 15	1.599 214 849 37

Three reversible orbits (same T_a).

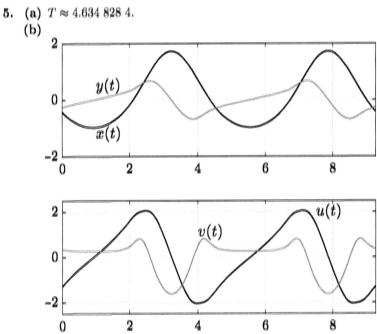

5. (a) $T \approx 4.634\ 828\ 4.$

(b)

8.

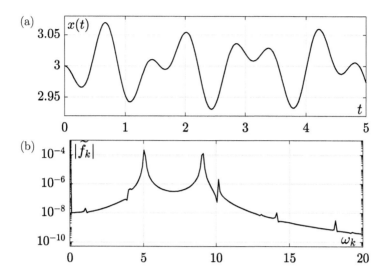

(a)

(b)

The DFT analysis shows two peaks at $\omega_1 \approx 5.0$ and $\omega_2 \approx 9.1$.

(c) The Jacobian of the system at $(3, 2)$ is

$$\begin{bmatrix} 0 & 1 & 0 & 0 \\ -74 & 0 & -20 & 0 \\ 0 & 0 & 0 & 1 \\ -20 & 0 & -34 & 0 \end{bmatrix},$$

with exact eigenvalues

$$\lambda_\pm = \pm i\sqrt{2(27 - 10\sqrt{2})} \approx \pm 5.071\,067\,811\,865\,48\,i$$

and

$$\mu_\pm = \pm i\sqrt{2(27 + 10\sqrt{2})} \approx \pm 9.071\,067\,811\,865\,48\,i,$$

whose imaginary parts agree with the measured frequencies in (b).

Practical 8.3:

(a) For $h = 0.1$, the explicit Euler integration is unstable, since the spectrum z_j (black bullets in the figure below) is partially outside the region of absolute stability of the method (gray disk). For $h = 0.01$, the spectrum (white bullets) lies entirely inside that region and the integration is therefore numerically stable.

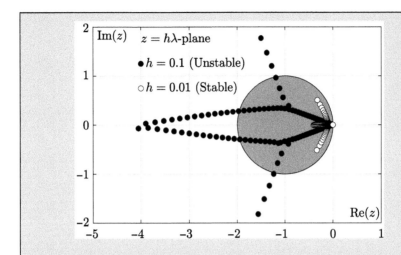

(b) For $h > 0$, the spectrum z_j lies within the stable region of the implicit Euler method. In particular, for $h = 0.1$ the implicit Euler method performs perfectly, as shown in the figure below:

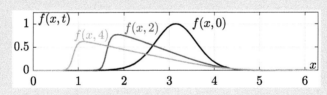

Glossary of Mathematical Symbols

- $\mathbb{N}$, $\mathbb{Z}$, $\mathbb{R}$, and $\mathbb{C}$: natural, integer, real, and complex numbers, respectively.
- $\mathbb{R}^n$: n-dimensional real vector space $\mathbb{R} \times \overset{(n)}{\cdots} \times \mathbb{R}$.
- $\ell_j(x)$: jth Lagrange cardinal polynomial.
- $\ell(x)$: nodal polynomial.
- λ_j: jth barycentric weight.
- $\lambda_n(x)$ and Λ_n: Lebesgue function and Lebesgue constant, respectively.
- $\displaystyle\sum_{j=0}^{n}{}' a_j \equiv \frac{a_0}{2} + a_1 + a_2 + \cdots + a_{n-2} + a_{n-1} + \frac{a_n}{2}$.
- $\mathbb{M}_n(\mathbb{R})$ and $\mathbb{M}_{nm}(\mathbb{R})$: $n \times n$ and $n \times m$ real matrices, respectively.
- $\mathbf{I}_n$: identity matrix in $\mathbb{M}_n(\mathbb{R})$.
- $\mathbb{R}_n[x]$: polynomials $p(x)$ with $\deg(p) \leqq n$.
- $\Pi_n(x)$: nth degree interpolant.
- $\Pi_n f(x)$: nth degree interpolant of the function $f(x)$.
- w_k: kth quadrature weight.
- $I_n(f)$: nth degree quadrature.
- $I_{n,m}(f)$: composite nth degree quadrature with m partitions.
- $\mathbf{u}^{\mathrm{T}}$: transpose of vector $\mathbf{u}$.
- $\mathbf{A}^{\mathrm{T}}$, $\mathbf{A}^{-1}$: transpose and inverse of matrix $\mathbf{A}$, respectively.
- $\mathbf{e}_k$: kth canonical vector.
- $\mathbf{M}_k$: Gaussian transformation matrix.
- $\mathbf{E}_{k_1 k_2}$: exchange matrix.
- $\langle \mathbf{x}, \mathbf{y} \rangle$: scalar product between vectors $\mathbf{x}$ and $\mathbf{y}$.
- $\|\mathbf{x}\| \equiv \sqrt{\langle \mathbf{x}, \mathbf{x} \rangle}$: 2-norm of vector $\mathbf{x}$.
- $\|\mathbf{A}\|$: 2-norm of matrix $\mathbf{A}$.
- $\mathbf{K}_m$, $\mathbb{K}_m$: Krylov matrix, Krylov space.
- $L^2(a,b)$: space of square integrable functions in (a,b).
- (f,g): scalar product between L^2-functions f and g.
- $\|f\|_{L^2(a,b)} \equiv \sqrt{(f,f)}$: norm of f in $L^2(a,b)$.
- $(f,g)_N$: discrete scalar product between functions f and g.
- $\varphi_k(x) \equiv e^{\mathrm{i}kx}$: kth element of Fourier basis.
- $\widetilde{\varphi}_k(x) \equiv e^{\mathrm{i}(k-\frac{N}{2})x}$: kth element of shifted Fourier basis.

Fundamentals of Numerical Mathematics for Physicists and Engineers, First Edition. Alvaro Meseguer.
© 2020 John Wiley & Sons, Inc. Published 2020 by John Wiley & Sons, Inc.

Bibliography

Acton, F. S. *Numerical Methods that Work*. The Mathematical Association of America, Washington, DC, 1990.

Allgower, E. L. and Georg, K. *Numerical Continuation Methods. An Introduction*. Springer-Verlag, Berlin, 1990.

Ascher, U. M. and Petzold, L. R. *Computer Methods for Ordinary Differential Equations and Differential-Algebraic Equations*. SIAM (Society for Industrial and Applied Mathematics), Philadelphia, PA, 1998.

Ascher, U. M., Mattheij, R. M. M., and Russell, R. D. *Numerical Solution of Boundary Value Problems for Ordinary Differential Equations*. SIAM (Society for Industrial and Applied Mathematics), Philadelphia, PA, 1995.

Bender, C. M. and Orszag, S. A. *Advanced Mathematical Methods for Scientists and Engineers*. McGraw-Hill, New York, 1978.

Boyce, W. E., DiPrima, R. C., and Meade, D. B. *Elementary Differential Equations and Boundary Value Problems*, 11e. Wiley, Hoboken, NJ, 2017.

Boyd, J. P. *Chebyshev and Fourier Spectral Methods*, 2e. Dover Publications, Mineola, NY, 2001.

Brass, H. and Petras, K. *Quadrature Theory, The Theory of Numerical Integration on a Compact Interval*. American Mathematical Society, Providence, RI, 2011.

Briggs, W. L. and Henson, V. E. *The DFT. An Owner's Manual for the Discrete Fourier Transform*. SIAM (Society for Industrial and Applied Mathematics), Philadelphia, PA, 1995.

Burden, R. L. and Faires, J. D. *Numerical Analysis*, 8e. Thomson-Brooks/Cole, Belmont, CA, 2005.

Butcher, J. C. *Ordinary Differential Equations*, 3e. Wiley, Chichester, 2016.

Dahlquist, G. and Björk, Å. *Numerical Methods for Scientific Computing*, vol. I. SIAM (Society for Industrial and Applied Mathematics), Philadelphia, PA, 2008.

Davis, P. J. *Interpolation & Approximation*. Dover Publications, New York, 1975.

Fundamentals of Numerical Mathematics for Physicists and Engineers, First Edition. Alvaro Meseguer.
© 2020 John Wiley & Sons, Inc. Published 2020 by John Wiley & Sons, Inc.

Davis, P. J. and Rabinowitz, P. *Methods of Numerical Integration*, 2e. Academic Press, London, 1984.

Dennis, J. E. and Schnabel, R. B. *Numerical Methods for Unconstrained Optimization and Nonlinear Equations*. SIAM (Society for Industrial and Applied Mathematics), Philadelphia, PA, 1996.

Eisberg, R. and Resnick, R. *Quantum Physics*, 2e. Wiley, 1985.

Epperson, J. F. *An Introduction to Numerical Methods and Analysis*, 2e. Wiley, Hoboken, NJ, 2013.

Folland, G. B. *Fourier Analysis and Its Applications*. Brooks/Cole Publishing Company, Belmont, CA, 1992.

Fornberg, B. *A Practical Guide to Pseudospectral Methods*. Cambridge University Press, Cambridge, 1996.

Golub, G. H. and Van Loan, C. F. *Matrix Computations*, 4e. The Johns Hopkins University Press, Baltimore, MD, 2013.

Gottlieb, D. and Orszag, S. A. *Numerical Analysis of Spectral Methods: Theory and Applications*. SIAM (Society for Industrial and Applied Mathematics), Philadelphia, PA, 1977.

Govaerts, W. J. F. *Numerical Methods for Bifurcations of Dynamical Equilibria*. SIAM (Society for Industrial and Applied Mathematics), Philadelphia, PA, 2000.

Greenbaum, A. *Iterative Methods for Solving Linear Systems*. SIAM (Society for Industrial and Applied Mathematics), Philadelphia, PA, 1997.

Griffiths, D. F. and Higham, D. J. *Numerical Methods for Ordinary Differential Equations. Initial Value Problems*. Springer-Verlag, London, 2010.

Hairer, E. and Wanner, G. *Solving Ordinary Differential Equations II. Stiff and Differential-Algebraic Problems*, 2e. Springer-Verlag, Berlin Heidelberg, 2002.

Hairer, E., Lubich, C., and Wanner, G. *Geometric Numerical Integration: Structure-Preserving Algorithms for Ordinary Differential Equations*. Springer-Verlag, Berlin Heidelberg, 2006.

Hairer, E., Nørsett, S. P., and Wanner, G. *Solving Ordinary Differential Equations I. Nonstiff Problems*, 2e. Springer-Verlag, Berlin Heidelberg, 2008.

Henrici, P. *Discrete Variable Methods in Ordinary Differential Equations*. Wiley, New York, 1962.

Isaacson, E. and Keller, H. B. *Analysis of Numerical Methods*. Wiley, New York, 1966.

Keller, H. B. *Numerical Solution of Two Point Boundary Value Problems*. SIAM (Society for Industrial and Applied Mathematics), Philadelphia, PA, 1976.

Kelley, C. T. *Solving Nonlinear Equations with Newton's Method*. SIAM (Society for Industrial and Applied Mathematics), Philadelphia, PA, 2003.

Kincaid, D. R. and Cheney, E. W. *Numerical Analysis: Mathematics of Scientific Computing, Series in Advanced Mathematics*, 3e. Brooks/Cole = Thomson Learning, Pacific Grove, CA, 2002.

Krommer, A. R. and Ueberhuber, C. W. *Computational Integration*. SIAM (Society for Industrial and Applied Mathematics), Philadelphia, PA, 1998.

Kythe, P. K. and Schäferkotter, M. R. *Handbook of Computational Methods for Integration*. Chapman & Hall/CRC Press, Boca Raton, FL, 2005.

Lambert, J. D. *Numerical Methods for Ordinary Differential Systems: The Initial Value Problem*. Wiley, Chichester, 1991.

Meyer, C. *Matrix Analysis and Applied Linear Algebra*. SIAM (Society for Industrial and Applied Mathematics), Philadelphia, PA, 2000.

Quarteroni, A., Sacco, R., and Saleri, F. *Numerical Mathematics*, 2e. Springer=Verlag, Berlin Heidelberg, 2007.

Rutishauser, H. *Lectures on Numerical Mathematics*. Birkhäuser, Boston, MA, 1990.

Saad, Y. *Iterative Methods for Sparse Linear Systems*, 2e. SIAM (Society for Industrial and Applied Mathematics), Philadelphia, PA, 2003.

Sanz=Serna, J. and Calvo, M. P. *Numerical Hamiltonian Problems*. Dover Publications, Mineola, NY, 2018.

Stoer, J. and Bulirsch, R. *Introduction to Numerical Analysis*, 2e. Springer=Verlag, Berlin, Heidelberg, 1993.

Süli, E. and Mayers, D. F. *An Introduction to Numerical Analysis*. Cambridge University Press, Cambridge, 2003.

Trefethen, L. N. Finite Difference and Spectral Methods for Ordinary and Partial Differential Equations. Unpublished text. http://people.maths.ox.ac.uk/trefethen/pdetext.html (accessed January 2020), Cornell, 1996.

Trefethen, L. N. *Spectral Methods in Matlab*. SIAM (Society for Industrial and Applied Mathematics), Philadelphia, PA, 2000.

Trefethen, L. N. *Approximation Theory and Approximation Practice*. SIAM (Society for Industrial and Applied Mathematics), Philadelphia, PA, 2012.

Trefethen, L. N. and Bau, D. *Numerical Linear Algebra*. SIAM (Society for Industrial and Applied Mathematics), Philadelphia, PA, 1997.

van der Vorst, H. *Iterative Krylov Methods for Large Linear Systems*. Cambridge University Press, Cambridge, 2003.

Watkins, D. S. *Fundamentals of Matrix Computations*, 3e. Wiley, Hoboken, NJ, 2010.

Wiggins, S. *Introduction to Applied Nonlinear Dynamical Systems and Chaos*, 2e. Springer=Verlag, New York, 2003.

Index

Fundamentals of Numerical Mathematics for Physicists and Engineers, First Edition. Alvaro Meseguer.

© 2020 John Wiley & Sons, Inc. Published 2020 by John Wiley & Sons, Inc.